Clean Technologies and Sustainable Development in Civil Engineering

Raluca- Andreea Felseghi
"Ştefan cel Mare" University of Suceava, Romania

Nicoleta Cobîrzan
Technical University of Cluj-Napoca, Romania

Maria Simona Raboaca
National Research and Development Institute for Cryogenic and Isotopic Technologies, Râmnicu Vâlcea, Romania

A volume in the Advances in Civil and Industrial Engineering (ACIE) Book Series

IGI Global
PUBLISHER of TIMELY KNOWLEDGE

Published in the United States of America by
IGI Global
Engineering Science Reference (an imprint of IGI Global)
701 E. Chocolate Avenue
Hershey PA, USA 17033
Tel: 717-533-8845
Fax: 717-533-8661
E-mail: cust@igi-global.com
Web site: http://www.igi-global.com

Library of Congress Cataloging-in-Publication Data

Names: Felseghi, Raluca-Andreea, 1978- editor. | Cobirzan, Nicoleta, 1977- editor. | Raboaca, Maria Simona, 1981- editor.
Title: Clean technologies and sustainable development in civil engineering / Raluca-Andreea Felseghi, Nicoleta Cobirzan, and Maria Simona Raboaca, editors.
Description: Hershey, PA : Engineering Science Reference, an imprint of IGI Global, [2022] | Includes bibliographical references and index. | Summary: "This book discusses how clean technologies can be considered in any process, product or service that reduces the negative impact on the environment through significant improvements in energy efficiency, sustainable use of resources or circular economy activities and offers knowledge on the implementation of clean technologies pertaining to Civil Engineering"-- Provided by publisher.
Identifiers: LCCN 2022970040 (print) | LCCN 2022970041 (ebook) | ISBN 9781799898108 (h/c) | ISBN 9781799898115 (s/c) | ISBN 9781799898122 (ebook)
Subjects: LCSH: Sustainable engineering. | Civil engineering. | Sustainable development.
Classification: LCC TA163 .C56 2022 (print) | LCC TA163 (ebook) | DDC 628--dc23/eng/20220314
LC record available at https://lccn.loc.gov/2022970040
LC ebook record available at https://lccn.loc.gov/2022970041

This book is published in the IGI Global book series Advances in Civil and Industrial Engineering (ACIE) (ISSN: 2326-6139; eISSN: 2326-6155)

British Cataloguing in Publication Data
A Cataloguing in Publication record for this book is available from the British Library.

All work contributed to this book is new, previously-unpublished material.
The views expressed in this book are those of the authors, but not necessarily of the publisher.

For electronic access to this publication, please contact: eresources@igi-global.com.

Advances in Civil and Industrial Engineering (ACIE) Book Series

Ioan Constantin Dima
University Valahia of Târgovişte,
Romania

ISSN:2326-6139
EISSN:2326-6155

MISSION

Private and public sector infrastructures begin to age, or require change in the face of developing technologies, the fields of civil and industrial engineering have become increasingly important as a method to mitigate and manage these changes. As governments and the public at large begin to grapple with climate change and growing populations, civil engineering has become more interdisciplinary and the need for publications that discuss the rapid changes and advancements in the field have become more in-demand. Additionally, private corporations and companies are facing similar changes and challenges, with the pressure for new and innovative methods being placed on those involved in industrial engineering.

The **Advances in Civil and Industrial Engineering (ACIE) Book Series** aims to present research and methodology that will provide solutions and discussions to meet such needs. The latest methodologies, applications, tools, and analysis will be published through the books included in **ACIE** in order to keep the available research in civil and industrial engineering as current and timely as possible.

COVERAGE

- Operations Research
- Engineering Economics
- Transportation Engineering
- Construction Engineering
- Materials Management
- Production Planning and Control
- Coastal Engineering
- Structural Engineering
- Urban Engineering
- Quality Engineering

IGI Global is currently accepting manuscripts for publication within this series. To submit a proposal for a volume in this series, please contact our Acquisition Editors at Acquisitions@igi-global.com or visit: http://www.igi-global.com/publish/.

The Advances in Civil and Industrial Engineering (ACIE) Book Series (ISSN 2326-6139) is published by IGI Global, 701 E. Chocolate Avenue, Hershey, PA 17033-1240, USA, www.igi-global.com. This series is composed of titles available for purchase individually; each title is edited to be contextually exclusive from any other title within the series. For pricing and ordering information please visit http://www.igi-global.com/book-series/advances-civil-industrial-engineering/73673. Postmaster: Send all address changes to above address.

Titles in this Series

For a list of additional titles in this series, please visit: http://www.igi-global.com/book-series/

Health and Well-Being Considerations in the Design of Indoor Environments
Roberto A. González-Lezcano (Universidad CEU San Pablo, pain)
Engineering Science Reference • © 2021 • 286pp • H/C (ISBN: 9781799872795) • US $215.00

Visibilities and Invisibilities in Smart Cities Emerging Research and Opportunities
H. Patricia McKenna (AmbientEase, Cnada)
Engineering Science Reference • © 2021 • 289pp • H/C (ISBN: 9781799838500) • US $195.00

Transforming Urban Nightlife and the Development of Smart Public Spaces
Hisham Abusaada (Housing and Building National Research Center, Egypt) Abeer Elshater (Ain Shams University, Egypt) and Dennis Rodwell (Independent Researcher, UK)
Information Science Reference • © 2021 • 290pp • H/C (ISBN: 9781799870043) • US $195.00

Handbook of Research on Driving Transformational Change in the Digital Built Environment
Jason Underwood (University of Salford, UK) and Mark Shelbourn (Birmingham City University, UK)
Engineering Science Reference • © 2021 • 569pp • H/C (ISBN: 9781799866008) • US $295.00

Structural Dynamics and Static Nonlinear Analysis From Theory to Application
Mourad Belgasmia (Setif 1 University, Algeria)
Engineering Science Reference • © 2021 • 347pp • H/C (ISBN: 9781799843993) • US $195.00

Handbook of Research on Developing Smart Cities Based on Digital Twins
Matteo Del Giudice (Politecnico di Torino, Italy) and Anna Osello (Politecnico di Torino, Italy)
Engineering Science Reference • © 2021 • 674pp • H/C (ISBN: 9781799870913) • US $295.00

IGI Global
PUBLISHER of TIMELY KNOWLEDGE

701 East Chocolate Avenue, Hershey, PA 17033, USA
Tel: 717-533-8845 x100 • Fax: 717-533-8661
E-Mail: cust@igi-global.com • www.igi-global.com

Editorial Advisory Board

Table of Contents

Detailed Table of Contents

The precast concrete industry has a relatively short history but a rapid development due to the increased demand for construction since the 1950s. Large or small-sized precast elements were used for construction of buildings making a continuous contribution to the development of the construction industry and the built stock. The development of new materials with the addition of natural and artificial fibers, binders, or aggregates as secondary raw materials, and high thermal performance materials, innovative joints for an easily disassemble or photovoltaic integrated in their structure are measures considered in the current context of climate neutrality as being necessary to increase the precast circularity or buildings energy performance. Precasts were mainly used due to their lower production and labor cost or reduced execution time, representing a good option for the future construction sector. The chapter presents the background, technical and economic shortcomings of the precast, their trends and development prospects in materials circularity, and recovery rate increase.

Developing countries like India are adopting new construction techniques for waste minimisation and practices due to the stringent rules and regulations from the past few years. However, some construction industries continuously fail to adopt and understand the legislation to minimise the limited use of natural resources. Recycling the construction waste during the construction phases is the new method currently adopted by the construction industries to reduce the waste. The current review attempts to identify the list of the techniques available for the minimization of waste in construction. Further, the authors have listed the various sources of waste generation in the construction industries and how the countries deal with the current waste.

The process of concrete obtaining involves high energy consumption, significant amounts of depletable natural raw materials, and greenhouse gas emissions. A greening variant of concrete is the substitution of the mineral aggregates from its composition with vegetal aggregates made by sunflower stalks. These vegetal aggregates have advantages from health, environment protection, comfort, energy saving, thermal, and/or sound insulation points of view. The aim of this chapter is to present the results of a research on concrete with sunflower stalk aggregates by applying a method of pretreatment of plant aggregates, and then a method of modifying the cement matrix, one of accelerating the hydration reaction of cement, and then a combination of cement matrix modification in parallel with the cement hydration reaction acceleration.

The authors would like to give their emphasis on advanced and economical green technologies due to COVID-19 pandemic situation. Due to coronavirus disease, not only the financial market but industrial and agricultural stock markets were affected. In all these references, the author created a positive side to go ahead with energetic stamina with a new name of "COVID: Create Opportunity Victory In Disaster." This chapter describes the most advanced sustainable and green building materials in construction (i.e., advanced green technology adopted in civil engineering). This is the time when we can do something new in the civil engineering area during this pandemic situation by using clean technologies for sustainable development in civil engineering construction and also can provide shelter for all who really feel helpless to construct their own homes in an economic way by using low-cost housing materials as a smart material. This chapter reveals advanced, economic, and sustainable green technologies that are being invented during the pandemic situation of COVID-19.

The development of populated centers has as a consequence the corresponding increase of the consumption of resources, first of all of the consumption of energy necessary for the transport of the additional flows of water. Their often-limited character materializes the desire to redistribute existing resources in time and space, in accordance with the requirements of the uses served. The widespread introduction, due to competitive prices, of composite pipes, but for which there is insufficient information on their behavior over time, has led to the need to research the sustainability and reliability of new systems in development for water transport. This chapter aims to provide an efficient tool for those interested in managing water supply systems with direct reference to the reliability and sustainability of water transport systems: supply and distribution networks.

The proposed process is a novel, inexpensive, and eco-friendly method to convert the sunburnt bricks into second-class bricks by increasing their compressive strength and other mechanical properties through the use of solar thermal energy. This is achieved by applying black lead-free paint on the whole surface of the brick and exposing it to solar radiation for about 28-30 days. As the black body absorbs radiation, there is an increased flux of radiation energy into the pores of the clay brick, and thus it increases the compression strength of brick and other its mechanical properties. Experimental samples were prepared and tested as per the prevailing industry standards. The results obtained proved that the tested samples manufactured by this novel process qualified as second-class bricks with a compressive strength of 65-70 kg/cm2. The proposed novel process is simple, cost-effective, and environmentally friendly as there is no use of fossil fuels in the conversion process. The current invention is filed for Indian patent with application number 201941036244 as in IPO publications.

Energy generation or green energy generation from waste materials is a necessity for the sustainable development of our society. In the last two decades, there is substantial scientific and technological development in waste to energy generation research. This chapter discussed the various aspects of the waste to energy generations technologies such as the hydrogen/energy generation from the waste materials. The chapter enlisted the advanced in the technology's development of waste to energy generation. It emphasized the most sustainable way to treat waste and their current state of the art in the waste to energy generation. This chapter also includes the biological, biochemical, and various fuel cells technologies for waste to energy generation. This chapter summarized the various state of the art of hydrogen production technologies from the waste materials or wastewater and emphasized the applications of nanotechnology in the cathode development for the bio-electrochemical system.

In order to achieve the goals of peak global greenhouse gas (GHG) emissions by 2030 and carbon neutralization by 2050, it is necessary to promote the transformation of China's energy sector dominated by fossil energy. Hydrogen energy is considered as the "ultimate energy" in the 21st century. It is redefining, restructuring, and transforming the China's energy sector. Therefore, Chinese enterprises need to reshape their value proposition through business model innovation. This chapter selects 10 leading enterprises in China's hydrogen energy industry. Through the business model canvas analysis of their business models, along with policies and development trends of China's hydrogen energy industry, this chapter provides the basic knowledge of popular business models in China's hydrogen industry and puts forward suggestions and research directions for the long-term development of its enterprises.

Difficulty in financing has always been a major problem faced by small companies or start-ups around the world, and China is no exception. This chapter aims to examine how crowdfunding can help startups in China to solve the problem of financing difficulties. By analyzing the Chinese startup marketplace as well as the sources of funding available to small businesses or startups, this chapter elaborates that it is reasonable to solve these funding problems and help startups survive through crowdfunding and draw forth regulations concerning it. Additionally, use cases are presented to help readers understand the development status of crowdfunding in China.

Preface

The construction sector is well known as the largest energy consumer in the share of total energy consumed worldwide (40%) and greenhouse gas (GHG) emissions (36%). At the same time, 75% of the building stock is energetically inefficient and only 1% of the existing buildings are renovated each year, therefore excellence in research in this field is of paramount importance.

European Green Deal priorities detailed by Renewable Energy Directive, Energy Performance of Buildings Directive and Energy Efficiency Directive are introducing new elements, with strong political signal on building renovation, promoting policies that will help to achieve highly energy efficient and decarbonised building stock by 2050.

Currently, the built environment operates in the linear economy mode in which large amounts of non-renewable (re)sources are used. In the current context, in which energy, ecological, economic and social concerns are becoming increasingly important, due to climate change, but also threats to energy security, depletion of traditional resources and human health, integration of green (re)sources and adoption of circular economy mode is of strategic importance for Civil Engineering, particularly for constructions.

New sustainable approaches in the built environment for Civil Engineering based on circular principles, closing the loop on building construction and demolition materials, the integration of green (re)sources and edificing circular constructions provide new opportunities, but its are also special challenges for actors involved in Civil Engineering researches. To preserve natural resources and promote responsible engineering, it is essential to be aware of the clean technologies emerging in the field of Civil Engineering.

Clean technologies can be considered any process, product or service that reduces the negative impact on the environment through significant improvements in energy efficiency, sustainable use of resources or circular economy activities. Clean technologies include a wide range of technologies related to the decarbonization process of construction industry, materials production, recovery and recycling of waste materials, green energy, information technology and more.

This book addresses clean technologies for the sustainable future of practical applications from Civil Engineering, the importance of the thematic area being topical from a scientific, technological point of view, but also from a socio-economic or environmental point of view.

The book entitled *Clean Technologies and Sustainable Development in Civil Engineering* was developed based on editors' motivation to provide an overview of the topic in the context of global interest increasing for climate neutrality. As the problem is complex, it requires an interdisciplinary approach, in which *material–system-building–clean technologies* are viewed in the same circumstances where the economic and social indicators as constituents of this process can no longer be neglected. The current challenges require thinking of global solutions for this "*common goal*". Natural resources are limited, and ecosystem are affected by our unsustainable lifestyle. The need for change is growing now and involves the setting of strategic ambitious plans to increase the energy efficiency and circularity beyond the modest values recorded so far. The measures considered must be feasible, in agreement with the stakeholders' requirements and easy to adopt. In this regard, the production of new waste mix materials using traditional or environmentally friendly technologies, the potential for recovery of precast elements and associated management strategies are among these objectives, presented and discussed in this book.

TARGET AUDIENCE

This book targets advanced-level students, researchers, specialists and practitioners in areas of interest: civil engineering, architecture and urbanism, mechanics, physics, chemistry, electronics, constructions and building services engineering, alternative energy resources, electrical engineering, environmental engineering, materials engineering and anyone interested in new clean technologies. Other stakeholders, who are interested in green technologies specific to the civil engineering domain will also find this book very useful in their research.

OBJECTIVE OF THE BOOK

This book's key objective is to inventory and disseminate the relevant knowledge regarding implementation of clean technologies dedicated to civil engineering, through overviews/reviews on the main components of the circular economy concept in the construction environment, through the presentation of the advanced research in the construction materials field that incorporates unconventional components, through theoretically and empirically based studies or through research approaches that highlight the practical applicability of clean technologies and sustainable development in Civil Engineering. The significance of the book resides in its capacity to enlarge our knowledge by emphasizing the complexity of a built environment based on the adoption of clean sustainable technologies by presenting in detail and analyzing the main issues related to: approach of circular economy in the field of buildings; advanced, sustainable and economical construction under green technologies spot; reliability and sustainability of water transport systems; clean technologies and unconventional green energies for the sustainable environment; hydrogen integration into the energy system as an energy vector, or ecological synthetic fuel; not least, business models of leading enterprises specific to the domain. Consequently, the book's main goal is to provide an expansive and nuanced understanding that will deepen the ontology of the subject of 'Clean Technologies and Sustainable Development in Civil Engineering'.

ORGANIZATION OF THE BOOK

Welcome to *Clean Technologies and Sustainable Development in Civil Engineering*. There are nine chapters in the book, each focusing on addressing current issues relevant to the main objective of the proposed topic, namely the clean technologies and sustainable development in Civil Engineering. The chapters are organized in such a way as to outline a logical structure and provide a clear perspective on the new sustainable approaches in the Civil Engineering environment, each chapter being approached to stand alone, advancing a broad perspective on a particular topic. A brief description of each chapter is as follows:

Chapter 1 presents the background, technical and economic limitations of the reuse of precast elements and their trends, or development prospects to sustain materials circularity and the extension of the rate of use/reuse in the field of construction. The precast concrete industry has a relatively short history, but a rapid development due to the increased demand for constructions since the 1950's. Large or small precast elements were used for construction of residential, socio-cultural, administrative buildings (educational units, hospitals, hotels etc.), industrial and agro-zootechnical constructions, terrestrial communication roads (bridges, viaducts, tunnels) having a continuous contribution to the development of the construction industry and built fund. Present and future trends of precast elements follow the same direction. The development of new innovative materials with the addition of natural and artificial fibers, binders and/or aggregates as secondary raw materials or photovoltaic panels integrated in their structure are measures considered necessary to increase the precast circularity or energy performance of buildings in the current context of climate neutrality. Precast elements were mainly used due to the reduced execution time, production and labor cost compared to cast on site elements.

Chapter 2 addresses the same direction of circularity through listing the different sources of waste generation in the construction industry and how countries treat current waste. Developing countries are adopting new construction techniques for waste minimization and practices due to the stringent rules and regulations from the past few years. However, some construction industries are continuously failing to adopt and understand the legislation and minimize the limited use of natural resources. Recycling the construction waste during the construction phases is the new method currently adopted by the construction industries to reduce the waste. The current review attempts to identify the list of the techniques available for the minimization of waste in construction.

Chapter 3 is dedicated to a methodology for integrating unconventional aggregates in the composition of building materials, namely, sustainability of vegetal concrete and means of improving its weaknesses. The process of concrete obtaining involves high energy consumption, significant amounts of depletable natural raw materials and greenhouse gas emissions. A greening variant of concrete is the substitution of the mineral aggregates from its composition with vegetal aggregates made by sunflower stalks. These vegetal aggregates have advantages from health, environment protection, comfort, energy-saving, thermal and/or sound insulation points of view. The aim of this chapter is to present the results of a research on concrete with sunflower stalk aggregates by applying a method of pretreatment of plant aggregates, and then a method of modifying the cement matrix, one of accelerating the hydration reaction of cement, and then a combination of cement matrix modification in parallel with the cement hydration reaction acceleration.

In Chapter 4, the author wanted to give their emphasis on advanced and economical green technologies due to COVID-19 pandemic situation. Due to coronavirus disease, not only the financial market but industrial and agricultural stock markets were affected. In all these references author created a positive side to go ahead with energetic stamina with a new name of COVID: "Create Opportunity Victory In Disaster". This chapter describes the most advanced sustainable and green building materials in construction i.e. advanced green technology adopted in Civil Engineering. This is the time when we can do something new in the Civil Engineering area during this pandemic situation by using clean technologies for sustainable development in Civil Engineering Construction and also can provide shelter for all who are really feel helpless to construct their own homes in an economic way by using low-cost housing materials as a smart material. This chapter reveals advanced, economical, and sustainable green technologies which are being developed during the pandemic situation.

Chapter 5 also addresses a crisis situation by realistically addressing the issue of reliability and sustainability of water transport systems. The development of populated centers has as a consequence the corresponding increase of the consumption of resources, first of all of the consumption of energy necessary for the transport of the additional flows of water. Their often limited character materializes the desire to redistribute existing resources in time and space, in accordance with the requirements of the uses served. The widespread introduction, due to competitive prices, of composite pipes, but for which there is insufficient information on their behavior over time, has led to the need to research the sustainability and reliability of new systems in development for water transport. This chapter aims to provide an efficient tool for those interested in managing water supply systems with direct reference to the reliability and sustainability of water transport systems: supply and distribution networks.

Chapter 6 presents the novel process of using solar thermal energy for improving the mechanical properties of bricks, research developed by the authors. The proposed process is a novel, inexpensive, and eco-friendly method to convert the sunburnt bricks into second-class bricks by increasing their compressive strength and other mechanical properties through the use of solar thermal energy. This is achieved by applying black lead-free paint on the whole surface of the brick and exposing it to solar radiation for about 28-30 days. As the black body absorbs radiation, there is an increased flux of radiation energy into the pores of the clay brick, and thus it increases the compression strength of brick and other its mechanical properties. Experimental samples were prepared and tested as per the prevailing industry standards. The results obtained proved that the tested samples manufactured by this novel process qualified as second-class bricks with a compressive strength of 65-70 kg/cm^2. The proposed novel process is simple, cost-effective and environmentally friendly as there is no use of fossil fuels in the conversion process. The current invention is filed for Indian patent with application number 201941036244 as in IPO publications.

Chapter 7 addresses to technologies for the clean and renewable energy production for the sustainable environment. Energy generation or green energy generation from waste materials is a necessity for the sustainable development of our society. In the last two decades, there is substantial scientific and technological development in waste to energy generation research. In this chapter, are discussed the various aspects of the waste to energy generations technologies, such as the hydrogen energy generation from the waste materials. Enlist the advanced in the technology's development of waste to energy generation. Emphasized on the most sustainable way to treat waste and their current state of the art in the waste to energy generation. This chapter also includes the biological, biochemical and various fuel cells technologies for waste to energy generation. This chapter summarizes the various state of the art of hydrogen production technologies from the waste materials or wastewater with emphasized on the applications of nanotechnology in the cathode development for the bio-electrochemical system.

Chapter 8 proposes a business approach to the issue and purpose of the book. In order to achieve the goals of peak global greenhouse gas emissions by 2030 and carbon neutralization by 2050, it is necessary to promote the transformation of energy sector dominated by fossil energy. Hydrogen energy is considered as the "ultimate energy" in the 21st century. It is redefining, restructuring and transforming the energy sector. Therefore, worldwide enterprises need to reshape their value proposition through business model innovation. This chapter selects ten leading enterprises in China's hydrogen energy industry. Through the Business Model Canvas analysis of their business models, along with policies and development trends of China's hydrogen energy industry, this chapter provides the basic knowledge of popular business models in China's hydrogen industry and puts forward suggestions and research directions for the long-term development of its enterprises.

Chapter 9 closes the present book in a financial manner. Difficulty in financing has always been a major problem faced by small companies or start-ups around the world. This chapter aims to examine how crowdfunding can help startups in China to solve the problem of financing difficulties. By analyzing the Chinese startup marketplace as well as the sources of funding available to small businesses or startups, this paper elaborates that it is reasonable to solve these funding problems and help startups survive through crowdfunding and draw forth regulations concerning it. Additionally, use cases are presented to help readers understand the development status of crowdfunding in China.

The technical information, discussions, studies, analyzes and research results presented during this book, but also the problems identified, the technical limitations encountered allow the identification and establishment of new future approaches in the field of *Clean Technologies and Sustainable Development in Civil Engineering*, each chapter representing a starting point in the elaboration of elaborated, exhaustive studies specific to the addressed subject that to further explore ways to reduce the gaps in the studies and the specialized technical literature.

Raluca - Andreea Felseghi
"Ştefan cel Mare" University of Suceava, Romania

Nicoleta Cobîrzan
Technical University of Cluj-Napoca, Romania

Maria Simona Raboaca
National Research and Development Institute for Cryogenic and Isotopic Technologies, Râmnicu Vâlcea, Romania

Chapter 1
Approach of the Circular Economy in the Field of Precast Elements and Buildings:
Limits, Trends, and Perspective

Nicoleta Cobîrzan
Technical University of Cluj-Napoca, Romania

Radu Muntean
https://orcid.org/0000-0001-9404-4060
Transilvania University of Brașov, Romania

Gyorgy Thalmaier
https://orcid.org/0000-0003-2762-6110
Technical University of Cluj-Napoca, Romania

Raluca-Andreea Felseghi
"Ștefan cel Mare" University of Suceava, Romania

ABSTRACT

The precast concrete industry has a relatively short history but a rapid development due to the increased demand for construction since the 1950s. Large or small-sized precast elements were used for construction of buildings making a continuous contribution to the development of the construction industry and the built stock. The development of new materials with the addition of natural and artificial fibers, binders, or aggregates as secondary raw materials, and high thermal performance

DOI: 10.4018/978-1-7998-9810-8.ch001

materials, innovative joints for an easily disassemble or photovoltaic integrated in their structure are measures considered in the current context of climate neutrality as being necessary to increase the precast circularity or buildings energy performance. Precasts were mainly used due to their lower production and labor cost or reduced execution time, representing a good option for the future construction sector. The chapter presents the background, technical and economic shortcomings of the precast, their trends and development prospects in materials circularity, and recovery rate increase.

INTRODUCTION

The field of construction due to materials production and buildings, annually consumes a large amount of natural resources and generates significant greenhouse gases emissions (*IEA, 2019*) contributing to resource depletion, environmental pollution, global warming and ecological disasters. In response to the current issues of human progress, a set of key policy initiatives have been taken, outlining strategies and measures for energy efficiency and decarbonisation in line with the sustainable development targets *(UN, 2015; European Commission, 2019*). Extending the annual renovation rate *(European Parliament, 2018*), using clean energy and adopting the circular economy model in all priority areas have now become necessary tools to achieve the climate neutrality *(European Commission, 2018; European Commission, 2019; European Commission, 2020*), and to sustain the growth of economy and peoples well-being improvement. It becomes an imperative to rethink the concept on which the building design was performed so far. The implementation of ecological and social principles in addition to the conventional ones considered usually in the design of buildings, are imperative elements that can no longer be neglected. "*Outdated*" thinking where the cost of construction was one of the most relevant issues in detriment of the environment quality, accessibility, affordability, equity, human needs, safety and local security or urban planning aspects, must be gradually changed having in mind the global context and current priorities. Our responsibility become more complex as humanity evolves and economy grows. Each person has the responsibility to get involved but also the resource to act.

The building industry must make a constant effort to adapt the technology to the new challenges in which sustainability and competitiveness are maintained *(UN, 2015; European Commission, 2018; European Commission, 2019; European Commission, 2020*), and where the consumer is part of this process. Materials, structural systems and construction elements need to be optimized for an easily disassembling to encourage reuse or recycle in the same or in other areas thus preserving its economic value *(Ellen MacArthur Foundation, 2013; European Commission, 2014; European Commission, 2015; European Commission, 2019; European Commission, 2020*), limiting the disposal of waste and avoiding the depletion of natural resources.

AN OVERVIEW OF PREFABRICATED ELEMENTS AND BUILDINGS

A Short Background

Concrete is one of the building materials widely used for century in construction of buildings and engineering structures. It is manufactured as an admixture of binder (cement, lime or natural pozzolans), aggregates (sand, gravel, etc.) water and chemical additives with the role of improving their technical properties (workability, settings, hardening, lowering of freezing temperature, strength, etc.).

The history of *cementitious material* date back to ancient civilizations. During the Roman Empire, a material called "*opus caementicum*" or "*roman concrete*" had superior mechanical properties (*Kyropoulou et al., 2022*) which led to the developments and evolution of materials and structural systems. The arch, vault, and dome built with this material, have reached larger span and diameter not exceeded before (Colosseum, Pont du Gard, Pantheon, etc). Another breakthrough came in 1824, when Joseph Aspdin invented Portland cement *(Ryan, 1929;Wankhade et al., 2016)*, a binder predominantly used for concrete production, due to its contribution in improving mechanical properties, required for construction of multistorey buildings or for structures with larger span.

The reinforced concrete appeared in 1849 and was patented in 1867 by Parisian gardener named Joseph Monier who first used the materials for pots and large tubes. In addition to garden pots, Joseph Monier sustained the utilization of reinforced concrete for application in structures or construction elements with different geometries and role (slabs, arches or bridges, etc.) *(Wankhade et al., 2016)*. Another Frenchman, Eugene Freyssinet, studied the pre-stressing of concrete elements as an innovative solution to remove the shortcomings of the concrete due to tensile deficiency *(Shushkewich, 2012)* and to improve the mechanical properties of materials. In the period that followed, significant advancement was performed in the developments of "*cementitious materials*" and type of structures.

A variant of the classic reinforced concrete, with steel reinforcement, is the dispersed reinforced concrete with different types of fibers (natural or artificial) such as carbon, mineral wool, steel, glass, polypropylene, etc. Fiber-reinforced concrete, although increasing tensile strength and reducing micro-cracks, cannot replace reinforced concrete in all types of applications as structural elements. Economic and technical advantages of the fiber-reinforced concrete ensure the required quality for being used in applications such as slender facade elements, formwork for beams, columns and lintels, floors, pipes, etc.

Precast concrete has been a popular material since the early of 1900's. Before the outbreak of World War I, British engineer John Alexander Brodie patented the technology of prefabrication, considered at that time as a revolutionary and innovative solution for the construction sector *(Moore, 1969)*. Since then, the precast concrete industry has grown steadily and continue to do so to meet the increased investors demands. Development of prefabrication technology offers unsuspected possibilities for production of a wider type of building elements with different architectural aspect, texture, and properties. Therefore, concrete remain an "*ideal material*" for builders, due to its good mechanical properties, high durability relatively low cost, or the possibility of being exposed to various environmental conditions being feasible for use in high-rise buildings and large span constructions (bridge, aqueduct, viaduct, etc.).

Large, Prefabricated Wall Panel

In Romania, as in other European countries large or small sized prefabricated panels were widely used in the past for buildings construction *(Hrabovszky-Horváth and Szalay, 2014; Csoknyai et al., 2016; Ungureanu and Muntean, 2019)*. During the years 1960-1989 the migration of the population in urban areas, favored the increase of the apartments demand *(Botici, et al., 2013*) and the rapid development of districts and neighborhoods. To meet this requirements and to facilitate the construction of a high number of apartments in a relatively short period of time, the buildings were made with a variable height up to 9 level *(Negoita et. al, 1976; P101-78, 1979)* determined from technical and economic reasons, based on standardized projects *(Pescari et al., 2015)* in accordance with seismic and geotechnical parameters specific to the Romanian territory. The structural wall panels arranged inside or outside the buildings were made in monolayer (14-16cm), two and three-layers (22-28cm) variants (*Negoita et al., 1976*), with the same size as the room. The external prefabricated panels had a low thermal performance, U=0.65…0.85W/m^{2}K *(Botici, et al., 2013*) established in accordance with regulations in force at the time of buildings design. Low comfort and energy inefficiency of buildings have made the renovations works indispensable over the last two decades and continue to be an important "tool" in joint efforts for achievement of climate neutrality commitments *(European Commission, 2019)*. Design and construction of large, prefabricated buildings has been gradually reduced since the 1990s, being replaced by other types of structural systems that were more attractive and energy efficient.

Buildings from large, prefabricated panels are found in other European countries such as Czech Republic, Estonia, Bulgaria, Hungary, France, Germany, Finland having an important share in the existing buildings stock. The external walls, have a thermal insulation layer made of different materials and variable thickness, the thermal performance of the panels being diversified depending on the period in which were used, and the climatic parameters *(Csoknyai et al., 2016)*. The energy efficiency of the buildings was influenced by the presence of linear thermal bridges created in the areas of joints or ribs (C*soknyai et al., 2016; Ungureanu and Muntean, 2019*), one of the major shortcomings of buildings made of large prefabricated panels.

Monotony of the facades, the small gross area of the apartments and the energy inefficiency have decreased the attractiveness and interest in designing new buildings made of precast elements. To counteract these drawbacks but also to widespread their utilization in the residential sector, many efforts has been made in the recent years which needs to be intensified, having as main priorities the improvement of energy efficiency and preservation of natural resources by acting for reducing the consumption of conventional materials.

High performance concrete, reinforced fibers or connectors made of composite materials (fiber reinforced polymer- FRP, carbon and glass fiber reinforced polymers (CFRP, GFRP), etc.) with lower conductivity value compared to steel reinforcement were used to reduce the panels weight and/or to improve mechanical properties. *(Einea et al., 1991; Shams et al., 2014; Daniel Ronald Joseph et al., 2019; O'Hegarty & Kinnane, 2020; Arevalo and Tomlinson, 2020; Yu et al., 2021)*. Thermal performances were improved by using materials as vacuum insulated panels; phenolic foam, etc. *(O'Hegarty and Kinnane, 2020; O'Hegarty et al., 2021)* with potential in reducing the overall panels thickness. New technologies and materials for finishing were adopted to increase the architectural and aesthetical aspect, and different panels geometry has been considered to allow application for a wider range of buildings.

Connectors designed to take over the shear forces and stresses that take place at the interface of concrete with thermal insulation layer, were in the past made in the form of cast concrete ribs (partially or continuous) (*Negoiță et al., 1976; P101-78,1979*). Later connectors with different geometries and materials were proposed to reduce the effect of heat losses caused by the presence of linear thermal bridges and to provide safety in operation. Based on a three-dimensional heat transfer models the authors *(Kim & Allard, 2014)* aimed to identify the best geometric shape of W, Z, J steel connectors, emphasizing that the geometry (diameter, size, shape) and spacings has an important influence on the thermal efficiency of the panel.

Other energy efficiency measures were consisting in integration of photovoltaic in the structure of precast façade elements (*Li et al., 2019*), or the uses of phase change materials (*Guerrero Delgado et al., 2020*) in the mixture of plaster or concrete *(Niall, et al., 2017)* to prevent the overheating and, consequently, to reduce the energy demand for cooling.

The construction of modular buildings made of precast elements can offer certain benefits in terms of time, safety or less resources consumption (*Ferdous et al, 2019*), but their joints must effectively be implemented on site for a good behavior to vertical and lateral actions. In case of spatial precast elements, used for modular buildings, the weight and size may raise restrictions and limitations due to higher costs for handling and transports.

Other Prefabricated Elements

Prefabricated elements like columns, stairs, beams are still used in administrative or commercial buildings, but their share of the total built fund is low. Their uses in the residential sector may have an important contribution in saving embodied carbon of buildings and waste disposal.

Prefabricated slabs have been used in the case of prefabricated, masonry, or frame buildings due to their good technical performance and economic efficiency. The prefabricated hollow-core slabs were used in the past for the construction of residential or administrative buildings. Their thickness was between 14 – 22 cm (*Negoiță et al., 1976*), depending on their span and structural performance in taking over the loads. The presence of hollows improved the thermal and acoustic comfort of rooms, for a lower weight in comparison with that of cast on site slabs. Currently, they can be used without restrictions in buildings located in non-seismic areas, but for those located in seismic areas, there are restrictions provided in the design codes. The provision of an over-concreting layer of minimum 6cm thickness *(P100-1/2013,2013)*, will increase the rigidity in their horizontal plane extending their field of utilization. From structural point of view this is a measure which must be taken to ensure the spatial character of structure but may have as effect an increase of building weight and embodied energy.

Prefabricated ceramic or concrete lintels are construction elements that represent an efficient and alternative solution to cast on site lintels. Ceramic lintels have superior thermal characteristics to monolithic or prefabricated reinforced concrete lintels, can reduce the amount of waste generated on-site, the labor cost, being easy to handle and install.

PREFABRICATED ELEMENTS AND BUILDINGS IN THE CONTEXT OF CIRCULAR ECONOMY

The prefabricated industry and the construction sector can make an important contribution to the current approach of climate neutrality. The reduction of greenhouse gas emissions from *"cradle to cradle"* can be achieved by using clean technology or low embodied materials in the phases of extraction, prefabrication, execution, maintenance, modernization, retrofit, or demolition/dismantling *(ISO, 2006a; ISO, 2006b; CEN, 2010; CEN, 2011; CEN, 2012; CEN, 2013; Estokova et al., 2017)*. Theoretically, at all phases may be proposed measures which may be adopted if it proves to be effective and there are no economic, technical, social, or legislative restrictions that make their application difficult. Therefore, in the preliminary phase of the buildings design or retrofit, several scenarios can be outlined, so that the materials, structural system, or intervention to be established accordingly. Climatic and ground parameters or investment cost will be assessed as primary parameters in designing and selecting the feasible package of measures in terms of performance criteria. Then, following a strategic thinking approach, the circularity, social and environmental performance of prefabricated will be assessed in accordance with *(ISO, 2006a; ISO, 2006b; CEN, 2010; CEN, 2011; CEN, 2012; CEN, 2013)* as relevant tools in making decission process, showing the strengths and the weakness of each proposed measures.

The differences between prefabricated and other types of structures was highlight in the studies conducted by the authors *(Aye et al., 2012; Mao et al., 2013; Cao et al., 2015*), and may represent useful information for all stakeholders (designers, real estate developers, materials industry, beneficiaries, etc.) involved in the selection or in the innovation process of precast technology.

The environmental performance of precast and the feasibility of their widespread utilization in construction sector from China was highlighted by the authors *(Cao et al., 2015*). The results showed a reduction in energy consumption and materials (thermal insulation, mortar), water and waste generated in case of prefabricated buildings in comparison with traditional one. Other studies conducted by *(Mao et al., 2013*) emphasize also the savings of greenhouse gases emissions, and embodied energy in the materials obtained through semi-prefabrication technology. The advantages of prefabrication over traditional technology, are obviously significant, therefore, expanding their applicability throughout the world could have a beneficial effect in achieving the climate neutrality commitment.

Design service lifespan of the structure normally ranges between 10 and 100 years (*CR0-2012; 2012*), set in accordance with the building destination, the material used and environmental conditions. Technical, economic, social, or environmental assessments are required in highlighting the potential for efficiency and in making the decision to demolish/dismantle or rehabilitate the building. The building service lifespan is prolonged when the strengthening techniques are implemented, and the energy or environmental performance are achieved by deeply retrofitting interventions. However, the interest in retrofitting is not always high, especially in degraded, insanitary, and low comfort buildings, located in industrial areas or infamous neighborhoods. There are several concerns that must be globally analyzed before to take the final decision. It is well known that strengthening works are time consuming, costly and requires the rental of houses where tenants to live during the interventions. In contrast, the demolish of buildings has a major pollution impact and requires treatment for recovery-reuse- recycle or for waste landfilling. As a result, no decision is easy to make. Each of these scenarios may have strengths but also weakness which require a rigorous assessment. In developed countries, the demolish making-decision of buildings is probably easier to be accepted, while in developing countries or in high-density districts the rehabilitation of buildings remain a favorite scenario. In recent years many efforts have been done in both directions, renovation, and recover/reuse. The large amount of waste resulting from demolition process has led to an increase in interest in findings effective ways for recover and reuse. Circular economy model based on waste management and hierarchy, established at the European level, by the EU waste Framework directive *(European Comission, 2008)* encouraging to reduce materials through an efficient design and to recover/reuse or recycle the waste in "*open loop*" or "*closed loop*" systems *(European Commission, 2015)* so that the waste landfilling to be quantitatively reduced and natural resources preserved.

The feasibility of reusing the prefabricated panels resulting from the dismantling of old structural system are mainly determined in terms of their technical parameters and the type of connection joints (wet or dry) adopted for assembly. Wet joints usually made of loop/hook steel rebars and concrete with higher mark as wall panel (*P101-78, 1979*) require a careful dismantling by removing the concrete from joints so that the rebars to remain unaffected (*Huuhka et al., 2015*). The cutting of both materials (rebars and concrete) may require additional measures (for ex. post installed rebars) to restore the connections and to ensure on vertical and horizontal directions, the structural continuity of the walls, monolithic character of structure, and an efficient assembly between the walls panels or between the walls and slab panels. In that circumstance the cost of construction may be increased but even so the overall environmental impact will continue to be low.

The authors *Huuhka et al., 2015* analyzed the potential of reusing dismantled prefabricated wall and hollowed strips resulted from block of flats built in Finland, in the years 1960s-1970s. Was found that recovered elements may be an important resource in the construction of detached houses.

The Circular Economy model may be easily adopted in the case of new buildings design, while in the case of existing buildings the decision-making process become more complex and sometimes very complicated. Designing precast elements for long lasting and easily disassembling to allow multiple reuses are necessary approaches in completing circularity process. For a faster assembly of precast elements dry joints connection (bolt, anchor rebar, gusset, rod etc.) may represent a good option, their type and number are defining elements in seismic performance of structures *(Guo et al, 2019; Fallah et al., 2020; Cheng et al., 2021; Zhang et al, 2022)* and buildings durability. The high recovery potential offers a wider perspective for large-scale use of prefabricated elements and buildings.

Recycling of waste resulted from production, construction and demolish is another ambition plan (*European Commission, 2014; European Commission, 2020*) for increasing the rate of materials circularity. Waste as secondary raw materials may be used in different percentage as binder or aggregates substitution being a valuable resource in production of new precast elements with similar or lower technical characteristics.

The recycling potential of waste resulted from construction and demolition of precast elements or that resulted in other sectors for production of concrete and finishing materials have been investigated by several authors highlighting their performance based on experimental test and/or numerical analysis.

The authors *Bagarić et al., 2021* proposed a new prefabricated ventilated façade panel made of reinforced concrete based on recycled aggregates (bricks and concrete) up to 50% promoting a "*closed loop solution*", for a lower environmental impact.

Laboratory tests and numerical simulations performed by the authors *Brandes & Kurama, 2018* on prestressed concrete beams made of recycled aggregate, resulted from precast elements have proven to be effective in replacing the crushed limestone with up to 100% by volume.

Other feasibility studies on waste recycling resulted from dismantling of one precast hotel built in 1981 in Czech Republic, were carried out by (*Venkrbec & Klanšek, 2020)*. Dismantling and sorting processes were carefully performed, crushing and preparation of aggregates was then followed and resulted aggregates were used in the concrete mixture. To maintain the workability of the concrete with the addition of 40% recycled aggregate, the amount of water was increased. The compressive strength of concrete was lower with one class than reference samples, a good argument to sustain the recycling process.

The authors (*Azúa et al., 2019)* made a stochastic simulation based on data provided by companies from USA, Canada, Chile to assess the quantity of natural resources, CO_2 emissions and costs incurred in the prefabrication panels having in their composition recycled concrete aggregates. The results confirmed that coarse recycled concrete aggregates represent a viable solution in regions with a deficit in natural resources but increase the water demand which make less efficient in areas where water is considered a limited resource. The partial substitution of cement with recycled materials or natural pozzolans (*Balog et al, 2014; Cobirzan et al, 2015)* with lower embodied carbon may also contribute to save greenhouse gas emissions, a viable solution for concrete and cement industry, considered as one of the main polluters globally.

The results obtained by scientific researchers clearly shows that recycling of waste may represent a good option for natural resources replacements and may be consider accordingly. The quality and size of aggregates as secondary raw materials will have an impact in mechanical properties of new products *(Padmini, et al., 2009)*. The percent of waste in the mass of new materials must be established considering the required properties of new materials, in accordance with the prescriptions provided by standards and design codes. As can be seen, many research progress has been made in the recent years that needs to be sustained, having in mind the society evolution and climate change concerns.

Design for easily disassembly, optimal utilization of natural resources, the uses of raw materials with low carbon and clean technologies are among the actions considered in the cement and concrete industry (*GCCA, 2021*) emphasizing the relevance for common efforts in achieving the targets proposed in climate agreement. An effective collaboration of producer, suppliers and customers must be currently strengthened for delivering a low embodied and/or energy efficient product.

Some perspectives and limitation in the use of prefabricated products are summarized in Table 1.

Digital technology can play an important role in optimizing process of prefabrications, transport, on-site and operation processes of buildings *(Zhengdao Li et al., 2018; Liu et al, 2020; Cretu et al., 2020)*, to meet new construction chalenges among development trends.

Some opportunities in terms of circularity indicators are presented in the table 2. It can be observed their interdependence in terms of sustainability.

Building design needs to be done in a broader context based on a common goal. Thus, the buildings must provide structural safety, have a low cost, be attractive, comfortable with an adequately gross living area, and located in areas with nearby facilities (school, shops, parks, etc.).

CONCLUSION

Precast concrete elements are used both in the superstructure and infrastructure of buildings, as large and small sized elements. Natural resources (materials, water, and energy) savings during the prefabrication and construction phases compared to cast on site buildings or other type of structures offer a higher benefit in terms of environment, which are good arguments in promoting the widespread utilization of precast elements. But even so, some other important elements in circular economy model as "recovery and recycle" the product at the end of their service time are needed to be considered in good agreement with technical, social, and economic indicators. Recovery of dismantled elements are a feasible option for the construction of new buildings but may require rehabilitation works to restore their integrity, process which may be time consuming and costly. Recovered precast elements may be less attractive but may represent a valuable resource for constructions especially for that located in scarce resources areas.

Table 1. Prefabricated elements used in construction

Element	Strength and Opportunities	Limits and weakness
Walls	ü can be dismantled for reuse or recycling ü short execution time and low maintenance cost may result in lower environmental impact compared to other traditional materials ü in execution does not raise problems due to cold weather ü the waste resulted during construction phase is reduced compared to cast on site elements ü may be made with addition of recycled/ low carbon materials in different percentages, contributing to materials circularity ü during operations may be replaced with other elements if it is demanded ü precast industry may facilitate new jobs and development opportunities for local communities.	ü façade monotony ü may require additional materials to correct the thermal bridges effect ü increased energy consumption and additional cost to transport the panels from factory to the site, especially in the case of large panels ü may be less attractive compared to other materials ü disassembled precast elements may be affected by external factors (weather, wind, humidity) that can affect their structural performance or aesthetical appearance
Hollowed strips	ü have good thermal and acoustical parameters compared to cast on site slabs ü cand be dismantled and reused in the same or other sectors ü can be made with addition of recycled materials to reduce the embodied energy ü have a lower weight compared with cast on site reinforced concrete slabs	ü cannot be used in high-rise buildings located in areas with high seismic zones without an over concreting layer with the role of stiffness increasing ü are not very effective in taking over the concentrated or large vertical loads ü due to structural constrains their uses is limited to buildings placed in non-seismic areas if their stiffness is not increased.
Lintels	ü proper for energy efficient buildings ü their use in the structural rehabilitation of buildings to reduce the waste ü high possibilities for product innovation	ü can support vertical loads but have limitation in seismic areas ü the presence on the construction market of some materials at low cost
Other elements (Columns, beams, stairs)	ü are good alternative to other materials used for construction of industrial, commercial, administrative buildings ü low execution time compared with cast on site concrete ü low maintenance cost ü can be recover/reused at the end of their service life ü can be made of concrete with high mechanical properties or concrete made with recycled/low carbon materials in different percentage	ü still limited utilization in residential buildings

Table 2. Circular indicators in terms of sustainability

	Reduce	Reuse	Recycle
Precast elements	ü the conventional energy consumption by replacing with clean energy in supply and production chains ü the waste disposal during raw materials extraction and production or construction phase of buildings, creating a healthy and safety environment ü the greenhouse gas emissions ü the embodied energy by partially/totally substituting natural resources with local or/and recycled materials ü the harmful nature of the product assuring workers, and consumers safety	ü disassembled precast elements for construction of new buildings or in other sectors ü can offer economic benefits, if are sold at the end of their service lifespan ü to preserve the natural resources in prefabrication ü to design buildings of lower cost especially in areas with low income and scarce resources ü as a solution which are ethically accepted from social communities	ü the secondary raw materials to limit the waste disposal (landfill) and to increase materials circularity beyond the current value ü to increase the human wellbeing and security in areas with landfills ü to develop local communities by increasing the attractive and peoples well being ü to stimulate local employments in areas with waste processing ü to support by fundings the concrete or cement industry which use recycled materials

At the level of precast industry, circularity require a detailed review of all feasible options and to rethink the supply chain to face the new challenges in which a competitive business model is demanded. Some of the technical solutions reviewed in this study shows their feasibility but must be adapted to the industry's business model, the type, and the role that the precast must play in the building structure, environmental conditions, buildings typology, the architectural style, and culture.

REFERENCES

P100-1/2013. (2013). Seismic design code - Part I - Design provisions for buildings (in romanian), 1–607, UTCB, Monitorul Oficial, Part I., No 558bis (2013)

P101-78. (1979). Instructiuni tehnice pentru proiectarea cladirilor de locuit cu structura de rezistenta din panouri mari [Technical instructions for the design of residential buildings with a structure made of large panels]. In *B.C.* (Vol. 5).

Arevalo, S., & Tomlinson, D. (2020). Experimental thermal bowing response of precast concrete insulated wall panels with stiff shear connectors and simple supports. *Journal of Building Engineering*, *30*, 101319. doi:10.1016/j.jobe.2020.101319

Aye, L., Ngo, T., Crawford, R. H., Gammampila, R., & Mendis, P. (2012). Life cycle greenhouse gas emissions and energy analysis of prefabricated reusable building modules. *Energy and Building*, *47*, 159–168. doi:10.1016/j.enbuild.2011.11.049

Azúa, G., González, M., Arroyo, P., & Kurama, Y. (2019). Recycled coarse aggregates from precast plant and building demolitions: Environmental and economic modeling through stochastic simulations. *Journal of Cleaner Production, 210*, 1425–1434. doi:10.1016/j.jclepro.2018.11.049

Bagarić, M., Banjad Pečur, I., & Milovanović, B. (2021). Application of developed facade panel from recycled CDW: A case study. *Journal of Physics: Conference Series, 2069*(1), 012187. doi:10.1088/1742-6596/2069/1/012187

Balog, A. A., Cobirzan, N., Aciu, C., & Ilutiu-Varvara, D. A. (2014). Valorification of volcanic tuff in constructions and materials manufacturing industry. *Procedia Technology, 12*, 323–328. doi:10.1016/j.protcy.2013.12.493

Botici, A. A., Ungureanu, V., Ciutina, A., Botici, A., & Dubina, D. (2013). Sustainable retrofitting of large panel prefabricated concrete residential buildings. *Proceedings of Central Europe Towards Sustainable Building 2013 Conference - Sustainable Building and Refurbishment for Next Generations*, 39-42. http://odolnesidliska.sk/wp-content/uploads/2018/10/CESB13_SUSTAINABLE-RETROFITTING-OF-LARGE-PANEL.pdf

Brandes, M. R., & Kurama, Y. C. (2018). Behavior of shear-critical prestressed concrete beams with recycled concrete aggregates under ultimate loads. *Engineering Structures, 165*, 237–246. doi:10.1016/j.engstruct.2018.03.029

CR0-2012. (2012). Design code. Basics of structure design. M. Of. 555bis/2013, in *B.C.*, 10/2012.

Cao, X., Li, X., Zhu, Y., & Zhang, Z. (2015). A comparative study of environmental performance between prefabricated and traditional residential buildings in China. *Journal of Cleaner Production, 109*, 131–143. doi:10.1016/j.jclepro.2015.04.120

CEN. (2010). *Sustainability of Construction Works - Sustainability Assessment of Buildings — Part 1: General Framework; EN 15643–1:2010–12*. European Committee for Standardization.

CEN. (2011). *EN 15978:2011–11 Sustainability of Construction Works - Assessment of Environmental Performance of Buildings — Calculation Method*. European Committee for Standardization.

CEN. (2012). *EN 15643–4:2012–04 Sustainability of Construction Works — Assessment of Buildings - Part 4: Framework for the Assessment of Economic Performance*. European Committee for Standardization.

CEN. (2013). EN 15804:2012+A1:2013 Sustainability of construction works—Environmental product declarations—Core rules for the product category of construction products. European Committee for Standardization.

Cheng, B., Cai, Y., & Looi, D. T. W. (2021). Experiment and numerical study of a new bolted steel plate horizontal joints for precast concrete shear wall structures. *Structures, 32,* 760-777. doi:10.1016/j.istruc.2021.03.043

Cobîrzan, N., Balog, A.-A., & Moşonyi, E. (2015). Investigation of the natural pozzolans for usage in cement industry. *Procedia Technology, 19,* 506-511. doi:10.1016/j.protcy.2015.02.072

Crețu, M., Czumbil, L., Bargauan, B., Ceclan, A., Berciu, A., Polycarpou, A., Rizzo, R., & Micu, D. D. (2020). Modelling and Evaluation of the Baseline Energy Consumption and the KPIs in TUCN Buildings within a Demand Response Programme – A Case Study. *IET Renewable Power Generation Journal, 14*(15), 2864–2875. doi:10.1049/iet-rpg.2020.0096

Csoknyai, T., Hrabovszky-Horváth, S., Georgiev, Z., Jovanovic-Popovic, M., Stankovic, B., Villatoro, O., & Szendrő, G. (2016). Building stock characteristics and energy performance of residential buildings in Eastern-European countries. *Energy and Building, 132*, 39–52. doi:10.1016/j.enbuild.2016.06.062

Daniel Ronald Joseph, J., Prabakar, J., & Alagusundaramoorthy, P. (2019). Experimental studies on through-thickness shear behavior of EPS based precast concrete sandwich panels with truss shear connectors. *Composites. Part B, Engineering, 166*, 446–456. doi:10.1016/j.compositesb.2019.02.030

Einea, A., Salmon, D. C., Fogarasi, G. J., Culp, T. D., & Tadros, M. K. (1991). State-of-the-art of Precast Concrete Sandwich Panels. *PCI Journal, 36*(6), 78–92. doi:10.15554/pcij.11011991.78.98

Ellen MacArthur Foundation. (2013). Towards the circular economy, economic and business rationale for an accelerated transition. Ellen MacArthur Foundation.

Estokova, A., Vilcekova, S., & Porhincak, M. (2017). Analyzing Embodied Energy, Global Warming and Acidification Potentials of Materials in Residential Buildings. *Procedia Engineering, 180*, 1675–1683. doi:10.1016/j.proeng.2017.04.330

European Comission. (2014). Communication from the commission to the european parliament, the council, the european economic and social committee and the committee of the regions Towards a circular economy: A zero waste programme for Europe /COM/2014/398 final. In *Com* (Vol. 398, pp. 1–8). https://eur-lex.europa.eu/legal-content/EN/TXT/?uri=celex%3A52014DC0398

European Comission. (2015). *Communication from the commission to the european parliament, the council, the european economic and social committee and the committee of the regions Closing the loop - An EU action plan for the Circular Economy*. COM/2015/0614 final.

European Commission. (2018). A Clean Planet for all. A European long-term strategic vision for a prosperous, modern, competitive and climate neutral economy. In *Com(2018) 773* (p. 25). https://eur-lex.europa.eu/legal-content/EN/TXT/PDF/?uri=CELEX:52018DC0773&from=EN

European Commission. (2019). *The European Green Deal*. COM/2019/640 final. https://eur-lex.europa.eu/legal-content/EN/TXT/?uri=CELEX:52019DC0640%0Ahttps://ec.europa.eu/info/strategy/priorities-2019-2024/european-green-deal_en

European Commission. (2020). Communication from the commission to the european parliament, the council, the European economic and social committee and the committee of the regions - A new Circular Economy Action plan for a cleaner and more competitive Europe. In COM(2020) 98 final (pp. 233–249).

European Parliament. (2018). Directive (EU) 2018/844 of the European Parliament and of the Council of 30 May 2018 Amending Directive 2010/31/EU on the Energy Performance of Buildings and Directive 2012/27/EU on Energy Efficiency (Text with EEA Relevance). In *Official Journal of the European Union* (Vol. 156, Issue 19.6.2018, pp. 75–91). https://data.europa.eu/eli/dir/2018/844/oj/eng

European Parliament and Council. (2008). Directive 2008/98/EC of the European Parliament and of the Council of 19 November 2008 on waste and repealing certain directives (Waste framework). In LexUriServ. do (pp. 3–30). .32008L0098 doi:2008/98/EC

Fallah, A., Vafaei, M., & Alih, S. C. (2020). A Review of Seismic Response of Precast Structures. *International Journal of Engineering Research & Technology (Ahmedabad)*, *13*(10), 2694–2705. doi:10.37624/IJERT/13.10.2020.2694-2705

Ferdous, W., Bai, Yu., Ngo, T. D., Manalo, A., & Mendis, P. (2019). New advancements, challenges and opportunities of multi-storey modular buildings – A state-of-the-art review. *Engineering Structures, 183,* 883-893. doi:10.1016/j.engstruct.2019.01.061

GCCA. (2021). *Concrete Future-The GCCA 2050 Cement and Concrete Industry Roadmap for Net Zero Concrete*. https://gccassociation.org

Guerrero Delgado, M., Sánchez Ramos, J., Álvarez Domínguez, S., Tenorio Ríos, J. A., & Cabeza, L. F. (2020). Building thermal storage technology: Compensating renewable energy fluctuations. *Journal of Energy Storage*, *27*, 101147. doi:10.1016/j.est.2019.101147

Guo, W., Zhai, Z., Cui, Y., Yu, Z., & Wu, X. (2019). Seismic performance assessment of low-rise precast wall panel structure with bolt connections. *Engineering Structures*, *181*, 562-578. doi:10.1016/j.engstruct.2018.12.060

Hrabovszky-Horváth, S., & Szalay, Z. (2014). Environmental assessment of a precast concrete building stock in a time perspective Environ. *Engineering Management Journal*, *13*(11), 2997–2804. doi:10.30638/eemj.2014.312

Huuhka, S., Kaasalainen, T., Hakanen, J. H., & Lahdensivu, J. (2015). Reusing concrete panels from buildings for building: Potential in Finnish 1970s mass housing. *Resources, Conservation and Recycling*, *101*, 105–121. doi:10.1016/j.resconrec.2015.05.017

IEA. (2019). *Global Status Report for Buildings and Construction 2019*. UN Environment Programme. https://www.iea.org/reports/global-status-report-for-buildings-and-construction-2019

ISO. (2006a). *ISO 14040 Environmental management — Life cycle assessment — Principles and framework*. ISO.

ISO. (2006b). *14044 (2006) Environmental management—life cycle assessment—requirements and guidelines*. ISO.

Kim, Y. J., & Allard, A. (2014). Thermal response of precast concrete sandwich walls with various steel connectors for architectural buildings in cold regions. *Energy and Building*, *80*, 137–148. doi:10.1016/j.enbuild.2014.05.022

Kyropoulou, D., Heliades, E., Karalis, P., Diamantopoulos, G., Gougoura, S., & Dotsika, E. (2022). Technological Evolution of Historic Mortars: From Lime-Based Mortars to Roman Opus Caementicium. In I. Vayas & F. M. Mazzolani (Eds.), *Protection of Historical Constructions. PROHITECH 2021. Lecture Notes in Civil Engineering*. Springer., doi:10.1007/978-3-030-90788-4_8

Li, M., Ma, T., Liu, J., Li, H., Xu, Y., Gu, W., & Shen, L. (2019). Numerical and experimental investigation of precast concrete facade integrated with solar photovoltaic panels. *Applied Energy*, *253*, 113509. doi:10.1016/j.apenergy.2019.113509

Li, Z. C., Xue, F., Li, X., & Hong, J. (2018). Geoffrey Qiping Shen, An Internet of Things-enabled BIM platform for on-site assembly services in prefabricated construction. *Automation in Construction, 89,* 146-161. doi:10.1016/j.autcon.2018.01.001

Liu, D., Li, X., Chen, J., Jin, R. (2020). Real-Time Optimization of Precast Concrete Component Transportation and Storage. *Advances in Civil Engineering*. doi:10.1155/2020/5714910

Mao, C., Shen, Q., Shen, L., & Tang, L. (2013). Comparative study of greenhouse gas emissions between off-site prefabrication and conventional construction methods: Two case studies of residential projects. *Energy and Building*, *66*, 165–176. doi:10.1016/j.enbuild.2013.07.033

Moore, R. (1969). An Early System of Large - Panel Construction. *Journal of the Royal Institute of British Architects*, 383.

Negoiță, A., Focșa, V., Radu, A., Pop, I., Tutu, L., Dumitras, M., & Negoita, I. (1976). *Constructii Civile*. Editura Didactica si Pedagogica.

Niall, D., Kinnane, O., West, R. P., & McCormack, S. (2017). Mechanical and thermal evaluation of different types of PCM–concrete composite panels. *Journal of Structural Integrity and Maintenance*, *2*(2), 100–108. doi:10.1080/24705314.2017.1318039

O'Hegarty, R., & Kinnane, O. (2020). Review of precast concrete sandwich panels and their innovations. *Construction & Building Materials*, *233*, 117145. Advance online publication. doi:10.1016/j.conbuildmat.2019.117145

O'Hegarty, R., Kinnane, O., Grimes, M., Newell, J., Clifford, M., & West, R. (2021). Development of thin precast concrete sandwich panels: Challenges and outcomes. *Construction & Building Materials*, *267*, 120981. Advance online publication. doi:10.1016/j.conbuildmat.2020.120981

Padmini, A. K., Ramamurthy, K., & Mathews, M. S. (2009). Influence of parent concrete on the properties of recycled aggregate concrete, *Construction and Building Materials, 23*(2), 829-836. doi:10.1016/j.conbuildmat.2008.03.006

Pescari, S., Stoian, V., Tudor, D., & Măduța, C. (2015). Energy demands of the existing collective buildings with bearing structure of large precast concrete panels from Timisoara. *Journal of Applied Engineering Sciences*, *5*(18), 75–81. doi:10.1515/jaes-2015-0010

Ryan, J. F. (1929). The story of portland cement. Part I. *Journal of Chemical Education*, *6*(11), 1855–1868. doi:10.1021/ed006p1854

Shams, A., Horstmann, M., & Hegger, J. (2014). Experimental investigations on textile-reinforced concrete (TRC) sandwich sections. *Composite Structures*, *118*(1), 643–653. doi:10.1016/j.compstruct.2014.07.056

Shushkewich, K. W. (2012). Eugène Freyssinet—Invention of Prestressed concrete and precast segmental construction. *Structural Engineering International*, *22*(3), 415–420. https://doi.org/10.2749/101686612X13363869853338

UN. (2015). *Transforming Our World: The 2030 Agenda for Sustainable Development*. A/RES/70/1. https://sdgs.un.org/sites/default/files/publications/21252030%20Agenda%20for%20Sustainable%20Development%20web.pdf

Ungureanu, V., & Muntean, D. M. (2019). Life Cycle Assessment of Retrofitting Large Prefabricated Panels Low-Rise Collective Dwellings. *IOP Conf. Ser. Materials Science and Engineering*, *471*, 112070. doi:10.1088/1757-899X/471/11/112070

Venkrbec, V., & Klanšek, U. (2020). Suitability of recycled concrete aggregates from precast panel buildings deconstructed at expired lifespan for structural use. *Journal of Cleaner Production*, *247*, 119593. https://doi.org/10.1016/j.jclepro.2019.119593

Wankhade, R. L., Landage, A. B., & Konnur, B. A. (2016). Experimental Investigation on Combined Effect of SBR and Steel Fiber on Properties of Concrete. *International Journal of Engine Research*, *5*(1), 257–261.

Yu, S., Liu, Y., Wang, D., Bahaj, A. B. S., Wu, Y., & Liu, J. (2021). Review of thermal and environmental performance of prefabricated buildings: Implications to emission reductions in China. *Renewable and Sustainable Energy Reviews, 137*. doi:10.1016/j.rser.2020.110472

Zhang, R., Zhang, Y., Li, A., & Yang, T. Y. (2022). Experimental study on a new type of precast beam-column joint. *Journal of Building Engineering, 51*, 104252. doi:10.1016/j.jobe.2022.104252

Chapter 2
Sustainable Current Waste Management Practices and Approaches in the Construction Industry of India

Madhavi Konni
https://orcid.org/0000-0002-2022-5528
Malla Reddy Engineering College (Autonomous), Secunderabad, India

Madhuri Pulavarthi
Department of Basic Science and Humanities, DADI Institute of Engineering and Technology, India

Manoj Kumar Karnena
https://orcid.org/0000-0001-9241-2990
Department of Environmental Science, School of Science, GITAM University (Deemed), India

ABSTRACT

Developing countries like India are adopting new construction techniques for waste minimisation and practices due to the stringent rules and regulations from the past few years. However, some construction industries continuously fail to adopt and understand the legislation to minimise the limited use of natural resources. Recycling the construction waste during the construction phases is the new method currently adopted by the construction industries to reduce the waste. The current review attempts to identify the list of the techniques available for the minimization of waste in construction. Further, the authors have listed the various sources of waste generation in the construction industries and how the countries deal with the current waste.

DOI: 10.4018/978-1-7998-9810-8.ch002

INTRODUCTION

The construction of buildings utilizes more resources than necessary and generates waste in more significant amounts (Osmani and Villoria-Sáez, 2019). Lu et al., 2021 stated that rapid industrialization in several countries showed a four-fold enhancement of annual material usage than the generation of the natural resources by the environment in one year. The wastage of material resources created environmental concerns and failed to consider the broader application of economic development with sustainability (Spišáková et al., 2021; Osmani and Villoria-Sáez, 2019).

The current situation triggered an attempt for developing national and regional strategic policies for the wise usage of the resources and prevention and minimization of demolition waste. Even though the worldwide construction industries acknowledged and accepted the reduction strategies, they failed to implement them (Na et al., 2021; Osmani and Villoria-Sáez, 2019). Conventionally, waste generated during construction activities is viewed as an inevitable by-product considered by investors.

Thus, management of generated waste has frequently should be addressed within the legislative and safety context. Accordingly, individuals' perception that the waste generated during the construction is unavoidable disallows trash management at the project level. The anecdotes shared a piece of evidence that over-ordering culture across the construction industries is prevalent and results in productivity issues, which might be due to the enhanced costs of the material than the wages of the labour (Osmani and Villoria-Sáez, 2019).

In contrast, many argued that frequent changes in the construction designs increase the material usage and generation of waste. The essential changes to the existing structure, management of waste and procurement of materials are required for the construction instruction industry to make them more sustainable. Current thinking on sustainability is redefined as the concepts of waste, i.e., by-products for improving performance, which missed the associated opportunity to reduce costs. Lauri, 1992 stated that waste generated adds costs to the industries and doesn't add value. Formoso et al., 2002, categorized waste as "unavoidable" when the expenses are lesser than the capital produced and "avoidable" when the higher investment is necessary to manage or reduce it. Thus, the conception of waste supposed to be looked at the activities relatively enhances the costs without adding value to the project.

The commonly accepted definition of waste is "any material discarded or intended to discard" (Directive, 2008; Osmani and Villoria-Sáez, 2019). This definition has been used for three decades and applied to different wastes irrespective of recovery and disposal operations. The revised concepts of the European Union related to waste management, including waste recycling and recovery; further in the revised version introduced the concepts of end of waste and by-products.

Waste Framework Directive in Article 5 recognised the conditions wherein the constituents might fall outside of the definition. These changes are aimed to identify the by-products that can be reused. This article describes the waste that ceases when it acts as a raw material for secondary products, distinguishes the by-products and waste, and emphasises waste prevention.

Figure 1 represents the waste hierarchy given by the European Union. The construction industry is the highest waste produced compared to others worldwide, reporting for thirty-five per cent of the total waste generation. This waste is equivalent to 4 folds of the household waste generated in India (Shrivastava and Chini, 2009; Ram and Kalidindi, 2017; Osmani and Villoria-Sáez, 2019).

Figure 1. Waste sources and their treatment
Redrawn with permission from Osmani and Villoria-Sáez, 2019

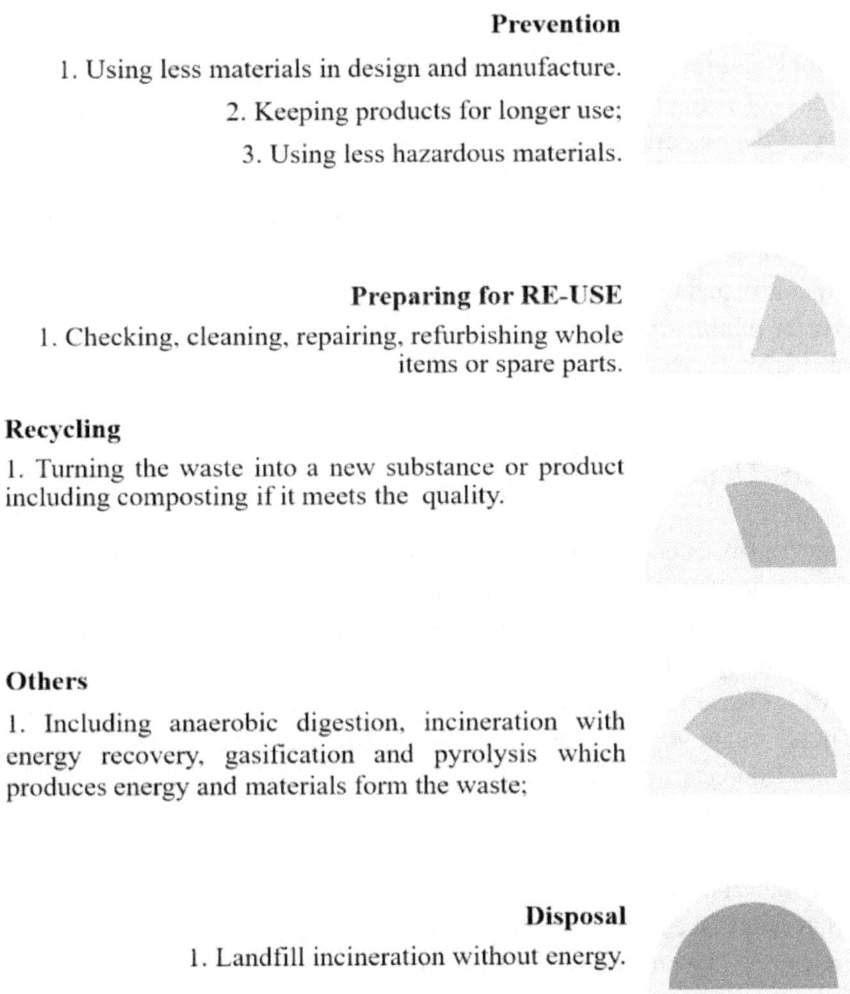

It was estimated that India's CD (Construction and Demolition) Waste generated is 431 million tonnes (Jain et al., 2021). The CD waste generated is an indicator for managing waste around the world. Nevertheless, many factors like GDP (Gross Domestic Product) and local regulatory actions influenced waste generation at lower levels. Zang et al., 2021 reported that waste generated per GDP is higher in developing nations like china compared to USA and EU. In general, waste generated is due to mineral wastage, i.e., excavating the soils and unnecessary dredging.

Further, they identified that mineral waste includes gypsum, concrete, wood, asphalt, etc. The difference in the trash generated in different countries might be due to the lack of proper regulations at a lower level for dealing with CD waste, low taxes on waste, no adequate recycling units, lower costs of primary materials, and lack of awareness in recycling the materials. Considering these factors, the European Construction Commission encouraged the developing countries to uptake the circular economy for increasing the CD waste recovery.

The European Union implemented an action plan for the circular economy for establishing and implementing the cycle from the design to waste generation (Cal et al., 2021). The main aim of this plan is to consider are for ensuring the complete recovery of the resources that are valuable from CD waste, proper management of resources in the construction industry, developing the guidelines for the demolition to encourage the recycling process and make the revisions in the legislative proposal on the CD waste. The current chapter aims to examine the concepts on construction waste, quantify and evaluate the CD waste, and highlight the current and emerging methods for minimising waste in the construction industry (Osmani and Villoria-Sáez, 2019).

COMPOSITION

Predicting the exact composition of the construction waste is difficult, nevertheless from the previous literature, it is evident that more than 30% of the waste of the total weight of the building materials are generated during the transportation, damage or over the ordering of the building materials (Mercader-Moyano et al., 2013; Mália et al., 2013; Osmani and Villoria-Sáez, 2019). The composition and waste streams of the waste might vary depending on the construction techniques.

Figure 2 depicts the construction waste quantification. Construction waste is categorized into three types

a. The materials that can't be recycled like chemicals, platers etc.
b. Materials that can't be recycled at the construction site can be recycled at other industries like paper, plastic, glass etc.

c. The valuable and recycled materials like bricks, tiles etc.

According to the waste streams, bricks and concrete have the more significant potential for recycling in the construction sector. This statement has been supported by the literature (Fořt and Černý, 2020; Gálvez-Martos et al., 2018) findings that were conducted across the world that was compared with volumes and streams of the waste generated during the construction.

Figure 2. Country-wise waste stream generation
Redrawn with permission from Osmani and Villoria-Sáez, 2019

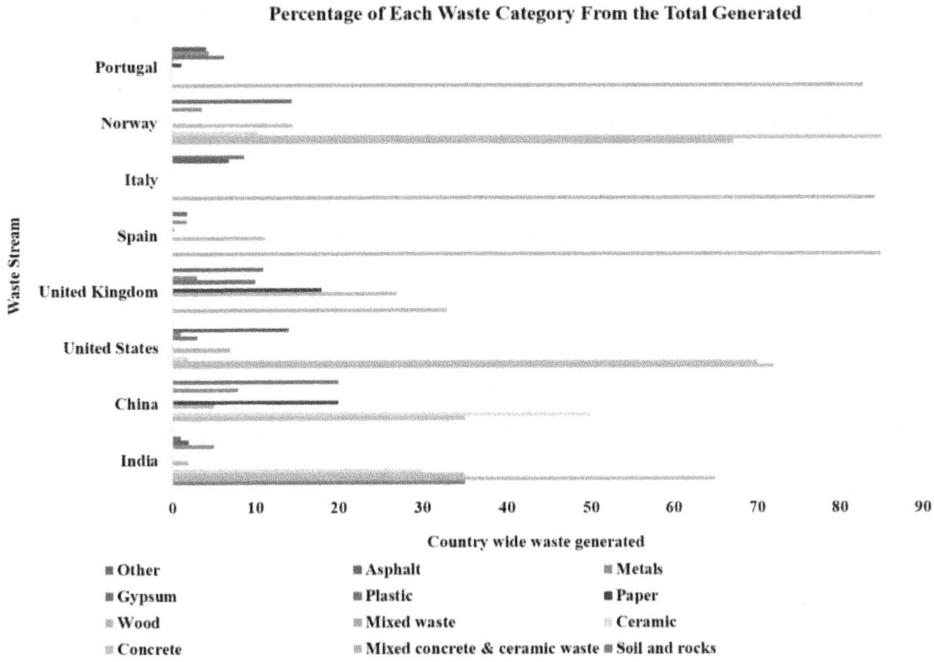

WASTE SOURCES

Various approaches are there to evaluate the origin of the waste, source and causes for waste generation. According to the available literature, the sources of waste generation are divided into eleven clusters. Table 1 shows how the waste is produced during construction and preconstruction activities. Omajuwa and Ngwu, 2021 stated that more than 33% of the waste produced during construction is due to architects' designs failures. Nevertheless, waste minimization via architect designs is complicated as buildings embody many materials.

Further, Omajuwa and Ngwu, 2021 stated that waste acceptance is inevitable; not defining the responsibilities and lack of training is a significant obstacle for the design to reduce the waste during the project properly. These processes become more complicated when stakeholders create other trash like suppliers, clients, and subcontractors. However, there is a general agreement that changing designs during operational activities is considered a critical origin of the waste during construction. The main factors for the variation in the structure during the construction activities are the inadequate understanding by the workers, shareholders changing requirements, complexity in the architect's plans, communication gap between the engineers and works and prolonged duration of the project (Doloi, 2013; Jaffar et al., 2013; Reyes-Veras et al., 2021).

The determining factors for the variations in the construction designs are as follows:

- Sudden changes in the construction are owing to stakeholder requirements.
- Insufficiency of designers in evaluating the methods and sequences of construction.
- Increasing the complexity of the construction designs.
- Unanticipated conditions of the construction sites.
- Prolonged duration.

However, the wastes generated by the procurement of the essentials required for the constructions fall under the Client's involvement during the project initiation, inefficiency in coordinating the project, uncertainties in allocating the workers' roles and responsibilities, and improper maintenance of purchase documents (Kurniawan et al., 2021).

MANAGEMENT

Despite the efforts by the government and stakeholders for continuous development for waste reduction, the uptake of these strategies around the globe is minimal. The increase in global research is required to reduce waste generated—the contemporary research for managing and minimizing construction waste into different categories in table 2. The research on construction waste is divided based on the waste hierarchy by R^3 concept, i.e., "reduce, recycle and reuse"'. The previous literature (Reike et al., 2018; Morseletto, 2020) shows that earlier studies promote awareness about the construction industry for minimizing waste to reduce the associated costs. The new onsite management resulted in conducting several studies on the construction industry. In addition, many studies have been undertaken to identify the behaviour and incentive-based approaches for improving the management of construction

Table 1. Sources of waste generation in the construction industry

S.no	Origin of Waste	Cause for waste generation
1.	Legal bindings 	➢ Documentation errors in the contract agreement. ➢ Incomplete contract documentation.
2.	Acquisition 	➢ Insufficient documentation during procurements. ➢ Absence of stakeholder's involvement during early purchase.
3.	Design 	➢ Complexity of design. ➢ Frequent changes in design. ➢ Poor coordination between designer and stakeholders.
4.	Planning and management 	➢ Improper management of waste and its planning. ➢ Improper supervision and material control in the construction site. ➢ Delay in information transfer regarding the components and material sizes.
5.	Concerned Services 	➢ Incident/accidents due to improper safety measures. ➢ Malfunction in the construction equipment. ➢ Work pressure due to lack of time.
6.	Transportation 	➢ Material damage during transit. ➢ Unloading methods of materials. ➢ Poor management in protecting the construction of the materials during loading and unloading.
7.	Ordered Materials 	➢ Errors in the ordering of the materials. ➢ Errors by the suppliers. ➢ Difficulties in ordering the materials in smaller quantities.
8.	Storage of Materials 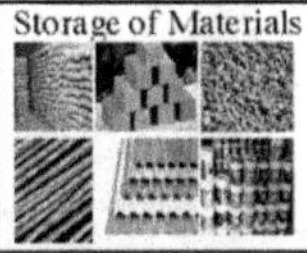	➢ Improper storage space for the storing of the material. ➢ Far away storage space. ➢ Practising improper storage methods.
9.	Handling of construction materials 	➢ Poor material handling. ➢ Suspending construction material in the loose form.
10.	Residuals 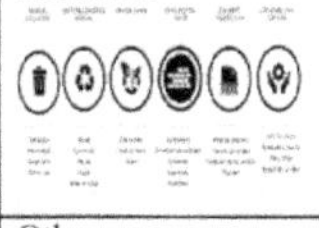	➢ Excess preparation of mortar than required. ➢ Not cutting the metals for perfect length.
11.	Others	➢ Theft. ➢ Climatic conditions. ➢ Hooliganism. ➢ Repairs. ➢ Demolition.

Reproduced with permission from Osmani and Villoria-Sáez, 2019 waste. Various tools and techniques have been developed to better control the waste onsite by segregation and quantifying the waste.

Moreover, multiple methods for evaluating the waste sources were developed for identifying the construction waste sources. Furthermore, many studies were conducted to assess the impacts of modern construction methods for waste reduction. New techniques like global positioning systems, geographical information systems and barcoding systems were introduced into the construction and demolition waste research. The various waste recycling, marketing methodologies and laboratory technologies have been introduced into the construction industry. These methods may be attempted to avoid the transfer the entire waste generated to the landfill. These methods facilitate assessing and auditing the waste already generated onsite as of date. There is no structured approach available to address the waste at the source and design. Thus, many argued that future construction waste management should be focused on design (Davis et al., 2021; Soharu et al., 2022). From the past few years, it is evident that limited design relating to waste research has increased (Soharu et al., 2022).

WASTE MINIMIZATION

The construction industry has utilized several tools and techniques to analyze the waste, manage the waste produced during the pre and post-construction and recycle the waste into valuable products.

Before Construction

There are not enough designs available in the literature to integrate waste produced from the construction industries. Nevertheless, the studies relating design to waste have been increasing from the past few years. The awareness among the people about construction waste is growing along with the legislation and standards. Specifically, the management plan for the waste reduction enhanced the stakeholders by providing a detailed plan with waste output and its costs during the construction activities. Langdon et al., 2009 stated that WRAP set guidelines for the architects to minimize waste in the construction industries.

These guidelines have five principles that are as follows:

i. Proper design for the recovery and reuse of waste in the preconstruction phase.
ii. Suitable design for the offsite construction.
iii. Optimization of materials.
iv. The appropriate design for the procurement of materials to avoid waste.

Table 2. Current tools and techniques for the management of construction waste

S.no	Stage	Waste management plan	Techniques
	All-purpose 	Legislation	Review of literature, case studies and questionnaire surveys
		Building Information Modelling	Interview methods, questionnaire surveys
	Scheming 	Benefits of minimization	Review of literature, case studies, questionnaire surveys and system dynamics.
		Prediction tools	Case studies and previous literature
		Waste evaluation	Interview method and previous literature
		Quantifying Waste	Statistical methods, mass balance principle and previous literature
		Waste minimization during the purchase of materials	Interview method and previous literature
		Waste minimization by design	Interview method and previous literature
	Structural 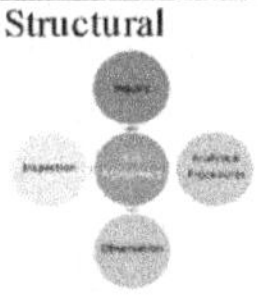	Modern methods	Workshops, observations and previous literature
		Onsite conservation	Interview method and previous literature
		Flow modelling	Life cycle analysis, Interview method and previous literature
		Auditing and assessment	Previous literature and computer tools
		Standards and guidelines	Workshops and previous literature
		Management studies	Previous literature and computer tools
	Life cycle 	Recycling	Case studies and previous literature
		Recovery	Case studies and previous literature
		Hazardous waste management	Case studies and previous literature

Reproduced with permission from Osmani and Villoria-Sáez, 2019

v. Designs should be flexible during deconstruction.

Even though the architects accepted WRAP guidelines, the WRAP has failed to consider all the parameters like environmental consequences, stakeholders' opinions and communications. Further, these guidelines did not consider the waste treatment at all the construction stages as these guidelines are critical in implementing and designing the waste strategies. The waste generation prediction tools help estimates the amount of construction waste generated during construction. The waste prediction tools rely on the waste indices acquired by diving the volume of waste generated during the construction per the area of the project (Letcher and Vallero,2019). Thus, estimation of the construction waste is required to plan an effective method for managing the waste in the construction. Methods introduced for estimating the waste generated might vary based on the geographical location and type of the project. The waste at a regional level can be estimated by multiplying the construction waste generated within the area by the rate of waste generated at onsite of the construction activity. Several studies aimed to study the amount of waste generated at the regional level by collecting data from the government (Zheng et al., 2017). Thus, the data play a vital role in this phase as they are related to the accuracy in estimating the waste generated. In addition, instances are generally considered to attain knowledge about the composition and generation of construction waste (Formoso et al., 2002). Many researchers have collected data using various tools and techniques for analysing waste. The generation of construction waste is analysed by considering factors like the topology of the buildings and the size of the dwelling; in a few cases, the accumulation rate throughout the construction activity is considered (Zhang et al., 2019; Faruqi et al., 2020; Zhang et al., 2021).

For instance, performance indicators for waste identification can be obtained from the smart waste plan projects that have been already completed. These indicators are used for the construction output to estimate the waste produced by new building materials and refurbish the waste generated onsite. Likewise, WRAP introduced a net waste tool to identify the onsite operations and reduce and recover the waste. Nevertheless, few studies used regression analysis and statistical tools to identify waste generation sources to minimise it (Aste et al., 2017). The construction, waste recycling, and floor plan compactness are crucial in the design variables. Later these design variables are restricted only to the high rising buildings. These models don't assess the overall impacts of the building design process and management of the waste at the construction site. The available literature understood that the BIM model (Building Information Modeling) could minimise the waste generated during construction (Eleftheriadis et al., 2017; Smith and Tardif, 2009). Few argued that the building planners might use these methods to minimise the waste during the projects. However, there is a need to research developing tools and techniques to minimise waste.

During Construction

The primary developers consider environmental issues and implement sustainable practices to minimise construction waste. An increase in construction industries created corporate social responsibility and awareness of the impact on natural resources (Akadiri et al., 2012). This awareness enhances the proper site management plan, workers' waste management training, and innovative methods for minimising and recycling the waste generated (Udawatta et al., 2015). Nevertheless, unlike large industries, which always thrives on reducing and managing construction waste at a larger scale, the small and medium construction industries do not follow the local legislation relating to managing the construction waste and causing environmental damage. The construction industries use toolbox talks, construction waste management plans, and barcoding systems during this phase (Fatimah et al., 2020). Waste management tools were developed to handle and quantify the waste generated, analyse the waste and reuse the waste generated onsite. In general, the waste quantification tool helps calculate the amount of waste generated from the construction project. The audit tools record the type, sources of the waste generated. Moreover, the studies on waste management have been increased for accessing the impacts of the offsite methods on reducing waste. State of the art technologies like a geographical information system, global positioning system etc., has been introduced into the construction industries to manage the waste onsite during construction (Aslam et al., 2020; Osmani and Villoria-Sáez, 2019).

Reduce, Recycle and Recovery (R^3)

Location-based waste management tools have gained much prominence in the past few years. They are enabled with the geographical information system that provides the location of the construction waste and helps manage the Waste (Labib, 2017). The waste management at the construction sites is influenced by the routine carried out in the construction industry. For example, one of the hurdles with waste recovery is the collection of mixed waste from the construction industry and the inefficiency of sorting of mixed Waste (Gálvez-Martos et al., 2018).

Thus, the training of the workers in the construction industry is essential for the proper management and segregation of waste from the mixed waste for the complete recovery and reuse of the waste onsite. The recycling recovery and reuse technologies are interrelated from the sources to end treatment (Ponnada and Kameswari, 2015; Osmani and Villoria-Sáez, 2019).

Challenges in the Construction Industry of India

Even though literature provides evidence for supporting the business by the benefits of waste management, the construction industries are still lagging in the implementation. The previous studies identified several factors that obstruct the adoption of sustainable waste management practices. Most of the elements are related to the conventional practice system in the industries. The few significant challenges of the construction industries facing are as follows:

- Improper organization at local levels.
- No rules and regulations.
- Failure to adopt new construction practices.
- Inadequacy of workers experiences in management of waste at the site.

- Perception of construction industries thinking that waste management methods are ineffective in cost reduction.
- Disinclination to reuse the waste owing to its low value.
- Added costs for the implementation and management of new initiatives.
- Improper defining of responsibilities of workers in the construction.
- Lack of waste management guidance (Osmani and Villoria-Sáez, 2019).

However, these challenges are not limited, and there is a need to research in-depth to manage the challenges occurring in the construction industry.

CONCLUSION

Construction industries are identifying the primary sources of waste generated during construction to prevent waste generation. According to the literature available, many researchers have failed to identify the primary sources of waste generated during construction. More research must be focused on, and developments are required to manage the waste generated in developing countries like India. "The methods discussed in the chapter are not adopted by the many construction industries like other industries; proper utilization and implementation will reduce the waste by fifty per cent" as stated by Osmani and Villoria-Sáez. Waste generated during construction can be considered a threat to the environment as this process is continuous in development. Thus, there is a need for sustainable practices, robust legislation, and an incentive to adopt new technological waste minimization and management methods. Adapting cyclic and linear approaches to design the building requires a cleaner environment. The building planners must take responsibility for the management of the project and be a key player for the eco-efficiency concept for defining the waste and reducing the waste.

REFERENCES

Akadiri, P. O., Chinyio, E. A., & Olomolaiye, P. O. (2012). Design of a sustainable building: A conceptual framework for implementing sustainability in the building sector. *Buildings*, *2*(2), 126–152. doi:10.3390/buildings2020126

Aslam, M. S., Huang, B., & Cui, L. (2020). Review of construction and demolition waste management in China and USA. *Journal of Environmental Management*, *264*, 110445. doi:10.1016/j.jenvman.2020.110445 PMID:32217323

Aste, N., Manfren, M., & Marenzi, G. (2017). Building Automation and Control Systems and performance optimization: A framework for analysis. *Renewable & Sustainable Energy Reviews*, *75*, 313–330. doi:10.1016/j.rser.2016.10.072

Cal, F. D., Garrido-Marijuan, A., Eguiarte, O., Arregi, B., Romero-Amorrortu, A., Mezzasalma, G., ... Bernardi, A. (2021). Energy Performance Assessment of Innovative Building Solutions Coming from Construction and Demolition Waste Materials. *Materials (Basel)*, *14*(5), 1226. doi:10.3390/ma14051226 PMID:33807904

Davis, P., Aziz, F., Newaz, M. T., Sher, W., & Simon, L. (2021). The classification of construction waste material using a deep convolutional neural network. *Automation in Construction*, *122*, 103481. doi:10.1016/j.autcon.2020.103481

Directive, E. C. (2008). 98/EC of the European Parliament and of the Council, on waste and repealing certain Directives. Official Journal of the European Union, (312), 3-30.

Doloi, H. (2013). Cost overruns and failure in project management: Understanding the roles of key stakeholders in construction projects. *Journal of Construction Engineering and Management*, *139*(3), 267–279. doi:10.1061/(ASCE)CO.1943-7862.0000621

Eleftheriadis, S., Mumovic, D., & Greening, P. (2017). Life cycle energy efficiency in building structures: A review of current developments and future outlooks based on BIM capabilities. *Renewable & Sustainable Energy Reviews*, *67*, 811–825. doi:10.1016/j.rser.2016.09.028

Faruqi, M. H. Z., & Siddiqui, F. Z. (2020). A mini review of construction and demolition waste management in India. *Waste Management & Research*, *38*(7), 708–716. doi:10.1177/0734242X20916828 PMID:32347194

Fatimah, Y. A., Govindan, K., Murniningsih, R., & Setiawan, A. (2020). Industry 4.0 based sustainable circular economy approach for smart waste management system to achieve sustainable development goals: A case study of Indonesia. *Journal of Cleaner Production*, *269*, 122263. doi:10.1016/j.jclepro.2020.122263

Fishbein, B. K. (1998). *Building for the future: strategies to reduce construction and demolition waste in municipal projects*. INFORM, Incorporated.

Formoso, C. T., Soibelman, L., De Cesare, C., & Isatto, E. L. (2002). Material waste in building industry: Main causes and prevention. *Journal of Construction Engineering and Management*, *128*(4), 316–325. doi:10.1061/(ASCE)0733-9364(2002)128:4(316)

Fořt, J., & Černý, R. (2020). Transition to circular economy in the construction industry: Environmental aspects of waste brick recycling scenarios. *Waste Management (New York, N.Y.)*, *118*, 510–520. doi:10.1016/j.wasman.2020.09.004 PMID:32980730

Gálvez-Martos, J. L., Styles, D., Schoenberger, H., & Zeschmar-Lahl, B. (2018). Construction and demolition waste best management practice in Europe. *Resources, Conservation and Recycling*, *136*, 166–178. doi:10.1016/j.resconrec.2018.04.016

Jaffar, N., Tharim, A. A., & Shuib, M. N. (2011). Factors of conflict in construction industry: A literature review. *Procedia Engineering*, *20*, 193–202. doi:10.1016/j.proeng.2011.11.156

Jain, S., Singhal, S., & Jain, N. K. (2021). Construction and demolition waste (C&DW) in India: Generation rate and implications of C&DW recycling. *International Journal of Construction Management*, *21*(3), 261–270. doi:10.1080/15623599.2018.1523300

Kurniawan, W., Endarto, B., Prasetyo, I., Aliyyah, N., Helmi, R., Indriastuti, D. E., ... Kalbuana, N. (2021). The role of law in public private partnership: Indonesia. *Journal of Management Information and Decision Sciences*, *24*, 1–11.

Labib, S. M. (2017, November). Volunteer GIS (VGIS) based waste management: A conceptual design and use of web 2.0 for smart waste management in Dhaka City. In *2017 Third International Conference on Research in Computational Intelligence and Communication Networks (ICRCICN)* (pp. 137-141). IEEE. 10.1109/ICRCICN.2017.8234495

Langdon, D. (2009). *Designing out waste: a design team guide for buildings*. WRAP.

Lauri, K (1992). *Application of the New Production Theory to Construction* (Vol. 72). Technical Report.

Letcher, T. M., & Vallero, D. (Eds.). (2019). *Waste: A handbook for management*. Academic Press.

Lu, W., Yuan, L., & Xue, F. (2021). Investigating the bulk density of construction waste: A big data-driven approach. *Resources, Conservation and Recycling*, *169*, 105480. doi:10.1016/j.resconrec.2021.105480

Mália, M., de Brito, J., Pinheiro, M. D., & Bravo, M. (2013). Construction and demolition waste indicators. *Waste Management & Research*, *31*(3), 241–255. doi:10.1177/0734242X12471707 PMID:23315370

Mercader-Moyano, P., & Ramírez-de-Arellano-Agudo, A. (2013). Selective classification and quantification model of C&D Waste from material resources consumed in residential building construction. *Waste Management & Research, 31*(5), 458–474. doi:10.1177/0734242X13477719 PMID:23446631

Morseletto, P. (2020). Targets for a circular economy. *Resources, Conservation and Recycling, 153*, 104553. doi:10.1016/j.resconrec.2019.104553

Na, S., Heo, S. J., & Han, S. (2021). Construction Waste Reduction through Application of Different Structural Systems for the Slab in a Commercial Building: A South Korean Case. *Applied Sciences (Basel, Switzerland), 11*(13), 5870. doi:10.3390/app11135870

Omajuwa, C. E., & Ngwu, C. (2021). Assessment of Construction Managers' Competency Performance on Project Delivery in Delta State,(Case Study of Isoko Community). *International Journal of Progressive Research in Science and Engineering, 2*(11), 76–90.

Osmani, M., & Villoria-Sáez, P. (2019, January). Current and emerging construction waste management status, trends and approaches. In Waste (pp. 365-380). Academic Press. doi:10.1016/B978-0-12-815060-3.00019-0

Ponnada, M. R., & Kameswari, P. (2015). Construction and demolition waste management–A review. *Safety, 84*, 19-46.

Ram, V. G., & Kalidindi, S. N. (2017). Estimation of construction and demolition waste using waste generation rates in Chennai, India. *Waste Management & Research, 35*(6), 610–617. doi:10.1177/0734242X17693297 PMID:28566032

Reike, D., Vermeulen, W. J., & Witjes, S. (2018). The circular economy: New or refurbished as CE 3.0?—exploring controversies in the conceptualization of the circular economy through a focus on history and resource value retention options. *Resources, Conservation and Recycling, 135*, 246–264. doi:10.1016/j.resconrec.2017.08.027

Reyes-Veras, P. F., Renukappa, S., & Suresh, S. (2021). Challenges faced by the adoption of big data in the Dominican Republic construction industry: An empirical study. *Journal of Information Technology in Construction, 26*, 812–831. doi:10.36680/j.itcon.2021.044

Shrivastava, S., & Chini, A. (2009). Construction materials and C&D Waste in India. *Lifecycle design of buildings, systems and materials, 72*.

Smith, D. K., & Tardif, M. (2009). *Building information modeling: a strategic implementation guide for architects, engineers, constructors, and real estate asset managers*. John Wiley & Sons. doi:10.1002/9780470432846

Soharu, A., Naveen, B. P., & Sil, A. (2022). An Approach Towards Zero-Waste Building Construction. In *Advances in Construction Materials and Sustainable Environment* (pp. 239–257). Springer. doi:10.1007/978-981-16-6557-8_19

Spišáková, M., Mésároš, P., & Mandičák, T. (2021). Construction Waste Audit in the Framework of Sustainable Waste Management in Construction Projects—Case Study. *Buildings*, *11*(2), 61. doi:10.3390/buildings11020061

Udawatta, N., Zuo, J., Chiveralls, K., & Zillante, G. (2015). Improving waste management in construction projects: An Australian study. *Resources, Conservation and Recycling*, *101*, 73–83. doi:10.1016/j.resconrec.2015.05.003

Zhang, L. W., Sojobi, A. O., Kodur, V. K. R., & Liew, K. M. (2019). Effective utilization and recycling of mixed recycled aggregates for a greener environment. *Journal of Cleaner Production*, *236*, 117600. doi:10.1016/j.jclepro.2019.07.075

Zhang, N., Zhang, H., Schiller, G., Feng, H., Gao, X., Li, E., & Li, X. (2021). Unraveling the Global Warming Mitigation Potential from Recycling Subway-Related Excavated Soil and Rock in China Via Life Cycle Assessment. *Integrated Environmental Assessment and Management*, *17*(3), 639–650. doi:10.1002/ieam.4376 PMID:33241889

Zheng, L., Wu, H., Zhang, H., Duan, H., Wang, J., Jiang, W., Dong, B., Liu, G., Zuo, J., & Song, Q. (2017). Characterizing the generation and flows of construction and demolition waste in China. *Construction & Building Materials*, *136*, 405–413. doi:10.1016/j.conbuildmat.2017.01.055

APPENDIX

Abbreviations

WRAP: Waste & Resources Action Programme
CD Waste: Construction and Demolition Waste
GDP: Gross Domestic Product

Chapter 3
Sustainability of Vegetal Concrete and Means of Improving Its Weaknesses

Cătălina Mihaela Grădinaru
"Gheorghe Asachi" Technical University of Iasi, Romania

Adrian Alexandru Şerbănoiu
"Gheorghe Asachi" Technical University of Iasi, Romania

ABSTRACT

The process of concrete obtaining involves high energy consumption, significant amounts of depletable natural raw materials, and greenhouse gas emissions. A greening variant of concrete is the substitution of the mineral aggregates from its composition with vegetal aggregates made by sunflower stalks. These vegetal aggregates have advantages from health, environment protection, comfort, energy saving, thermal, and/or sound insulation points of view. The aim of this chapter is to present the results of a research on concrete with sunflower stalk aggregates by applying a method of pretreatment of plant aggregates, and then a method of modifying the cement matrix, one of accelerating the hydration reaction of cement, and then a combination of cement matrix modification in parallel with the cement hydration reaction acceleration.

DOI: 10.4018/978-1-7998-9810-8.ch003

INTRODUCTION

Population growth, pollution caused by human activity, production processes of building materials, all are directly related to increasing the level of greenhouse gases, which determines the approach by research specialists to find alternative methods to build more sustainable, more environmentally friendly. Concrete is a material widely used in constructions, and its demand is constantly growing. A major disadvantage of this classic construction material is its highly polluting character due to its manufacturing process but also due to its components. The process of cement obtaining involves high energy consumption (36% of global energy consumption) (Global ABC/IEA/UNEP, 2020), significant amounts of depletable natural raw materials (1.5 tons of raw materials to obtain 1 ton of finished product) (Ali et al., 2015), and greenhouse gas emissions (39% of total global CO_2 emissions (Global ABC/IEA/UNEP, 2020), about one ton of CO_2 for every tonne of cement produced (Błaszczyński & Król, 2015), thereby causing global warming and climate change, this industry being one of the most important industries responsible for major greenhouse gas emissions worldwide. In addition, the manufacture of concrete involves the use of natural mineral aggregates, another category of depletable natural resources.

The greening variants of concrete researched so far include the cement substitution with various pozzolanic materials such as metakaolin, silica fume, fly ash (Walczak et al., 2015; Glazer et al., 2011; Sijakova-Ivanova et al., 2011; Badur & Chaudhary, 2008; Ignjatović et al., 2016; Sua-iam & Makul, 2015), which decrease the environmental pollution by reducing the cement amount produced on the one hand, and the limitation of waste from landfills, on the other hand (Huntzinger & Eatmon, 2009). Another greening option is the substitution of mineral aggregates in the concrete composition. It is a widespread natural resource, but it is also depletable and in some places non-existent. (Langer et al., 2004). The amount of mineral aggregates extraction in Europe, for example, is about 7 tons per citizen of the European Union, given a global demand of about 22 billion tons per year (Danielsen & Kuznetsova, 2016). The aggregate extraction industry is one of the most important mining industries in the world, producing about 16.5 billion tons annually (Langer et al., 2004).

In the direction of replacing the mineral aggregates in the concrete composition, a variant studied lately is represented by the use of aggregates of vegetal origin. They have advantages from several points of view: health, ecological character, comfort, energy saving, thermal and/or sound insulation (Amziane & Sonebi, 2016). Plant aggregates are raw materials whose processing relatively simple contributes to reducing pollution and environmental protection (Farias et al., 2017). They are obtained, in general, from agricultural waste. Concrete with agricultural waste has good thermal properties, examples of such waste being palm bark, coconut shell, corn cobs, sunflower stalks, rice husk.

Plant aggregates, in addition to the advantages they bring to the concrete characteristics and environmental protection, however, have a low durability in the alkaline environment determined by the cement in the concrete composition (Nozahic et al., 2012). Plant aggregates show severe degradation and become fragile in the alkaline and mineral-rich environment of the cement matrix, following the appearance of two mechanisms of aging: alkaline hydrolysis, and mineralization of the cell wall. To combat this degradation, there are mainly two ways: pre-treatment of plant aggregates and modification of the cement matrix. Pretreatment involves the use of chemical, physical and physico-chemical methods such as coating with silane, sodium silicate or potassium silicate (Harbulakova, 2014). This pretreatment may involve more effort and additional costs and must take into account the effect on the aggregate-cement matrix interface properties. The modification of the cement matrix can be achieved by replacing the cement as much as possible with other pozzolanic materials such as metakaolin, granular blast furnace slag, fly ash, silica fume, diatomite, and others. This solution causes a significant increase in the durability of cement composites, and the alkaline hydrolysis of amorphous components in natural aggregates and the mineralization of fiber cell walls are reduced (Harbulakova, 2014). In addition, ensuring a high degree of hydration of the cement further improves the durability of plant aggregates in the concrete composition (Jianqiang, 2016).

By virtue of the above, the purpose of this chapter is to present the results of a research on concrete with sunflower stalk aggregates by applying a method of pretreatment of plant aggregates, and then a method of modifying the cement matrix, one of accelerating the hydration reaction of cement, and then a combination of cement matrix modification in parallel with the cement hydration reaction acceleration.

Concrete with sunflower stalks is a material that combines the waste content with low-emission materials, locally available. Sustainable development at the local level is interdependent with global concerns, such as global warming or mineral resources depletion (Amziane & Sonebi, 2016). Their harvesting and storage is done with the help of existing agricultural equipment (Farias et al., 2017).

Globally, sunflower is a crop largely spreaded on *26,03 x* 10^6 *ha in 2020, (USDA, 2021)*, especially in Southern South America, Southern Europe, European Russia and South Africa (Leff et al., 2004). Over time, the main directions of sunflower stalk use have been as agricultural waste that is usually introduced into the ground under the furrow.

SUNFLOWER STALK CHARACTERIZATION

Over time, the advantages of using aggregates obtained from sunflower or hemp stalks in concrete have been analyzed. From the chemical composition point of view, the values from the literature show interesting similarities between hemp and sunflower. The sunflower stalk includes an important light, compressible part (marrow), and a woody, peripheral part, surrounded by the epidermis (Chabannes et al., 2015). The woody parts of hemp and sunflower have a cellulose content of about 50% and 40% in mass units, respectively. This percentage reaches 47.4% in the case of the sunflower stalk marrow. The lignin content seems to be higher for hemp (28%) than for the woody part of the sunflower (18.3%), but this result changes a lot with the evolution of plants and age. Sunflower marrow cellulose is a weak lignin (3.5%). An interesting aspect is the equal pectin content for hemp, the woody part and the sunflower marrow (6%). This value is very low compared to that determined for hemp fibers (20%). These comparisons vary depending on the methods of determination, the age of harvest and the species of plants used (Nozahic et al., 2012).

Aggregates from sunflower stalks are ultra-light, with a bulk density of 105 ± 2 kg/m^3 and a water content of 9,4% measured in a room at 20 ± 2 °C and 35 ± 5% relative humidity. Similar density results were obtained for hemp (103 ± 2 kg/m^3) and sunflower stalk (105 ± 2 kg/m^3) (Nozahic et al., 2012). By comparison, the relative density of wood particles is 305.8 ± 66 kg/m^3 (Nozahic et al., 2012).

The thermal conductivity of sunflower aggregates is very close to that of hemp, even lower (0.050 W/mK, compared to 0.055 W/mK). This can be explained by the presence of cellulosic marrow, that has a higher and better closed porosity than the wood particles (Chabannes et al., 2015).

Unlike hemp, which needs an adapted mechanism to remove the fibrous part of the stem, the entire stalk of the sunflower can be crushed and used as aggregates to obtain ecological concrete (Chabannes et al., 2015).

SUNFLOWER STALK USES IN THE CONCRETE COMPOSITION

Many physical characteristics of concrete with vegetal aggregates, especially the thermal and acoustic ones, are greatly altered by the presence and amount of absorbed water. The increased porosity and internal structure of sunflower aggregates are responsible for the high water absorption and its retention capacity (Nozahic et al., 2012). High water absorption is not desirable for a raw material used in concrete composition, given the binder-aggregate interface problems of this type of material, but it can be discussed as a benefit as soon as the stalk marrow supports the internal water supply to avoid concrete contraction (rupture), a benefit that can be valued in areas with warm climates (Chabannes et al., 2015).

Chabannes et al. (2015) studied the properties of concrete with sunflower aggregates, respectively with hemp aggregates, in a binder/aggregate ratio of 2/1. Tests performed on concrete with sunflower aggregates demonstrated a thermal conductivity of 0.096 W/mK and an average compressive strength of 0.5 N/mm^2 at 60 days. These results were almost similar to those recorded by concrete with hemp aggregates. Also, Nozahic et al. (2012) demonstrated that aggregates from sunflower stalks and those from hemp stalks have a similar honeycomb structure and chemical composition. The authors developed a concrete with sunflower aggregates and binder based on limestone and pumice, with a binder/aggregate ratio equal to 18. The compressive strength at 28 days of this type of concrete was 2.52 MPa. The same concrete recipe, but which included hemp aggregates, recorded a compressive strength of 2.77 MPa. Mati-Baouche et al. (2014) developed a composite material of particles of sunflower and chitosan with a thermal conductivity of 0.056 W/mK and a compressive strength of 2.0 MPa, in the case of a recipe with 4.3% chitosan (a biopolymer made of chitin - a substance obtained from the waste of shrimp shells and other crustaceans) and sunflower stalk particles bigger than 3 mm.

***

In their research, this chapter authors developed a series of concrete compositions with vegetal aggregates from sunflower stalks and analyzed the density, compression and tensile strength, elasticity, resistance to freeze-thaw cycles and thermal conductivity. For obtaining suitable aggregates to be introduced into the concrete composition, the sunflower stalks were shredded and then treated in order to reduce their water absorption capacity.

METHOD OF OBTAINING AGGREGATES FROM SUNFLOWER STALKS AND THEIR TREATMENT

The procedure by which the aggregates were obtained from sunflower stalks was as follows:

1. Only the stalk, without leaves and hat, was harvested from the sunflower plants.
2. After harvesting, the sunflower stalks were left to dry in stacks in the environment (figure 1a).

3. The sunflower stalks were crushed using a mill used for animal feed crushing, in granules with a diameter smaller than 5-6 mm and fibers with lengths smaller than 25 mm (figure 1b).
4. After crushing, the granules and fibers of sunflower stalks were treated with sodium silicate solution (SS) (figure 1c).
5. After immersion in the SS solution, the sunflower granules and fibers were left to dry on a surface heated to around 50-55 °C, in a room with an ambient temperature of 25-27 °C, naturally ventilated.
6. After drying to a constant mass, the granules and fibers from sunflower stalks were used as vegetable aggregates in the concrete composition (figure 1d).

Figure 1. Sunflower stalks aggregates obtaining: a. drying of sunflower stalks; b. appearance of sunflower stalks after shredding; c. appearance of crushed sunflower stalks treated with sodium silicate solution, in the wet state; d. appearance of sunflower stalks after treatment with SS, in dry state

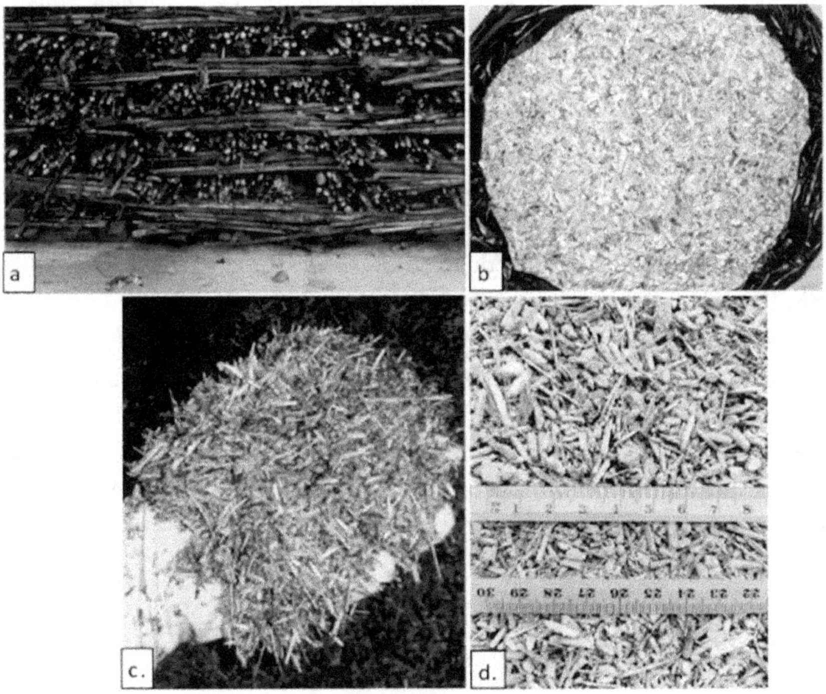

DETERMINING THE OPTIMAL VARIANT FOR TREATING THE SUNFLOWER STALK AGGREGATES

In order to study the effects of the treatment applied to the sunflower stalk aggregates (SSA), micro-concrete recipes were made with these vegetable aggregates treated with sodium silicate solution, in two concentration variants, 25% and 40%.

As a starting point, a conventional micro-concrete recipe of C30/37 class was used, with a maximum aggregate size of 8 mm (RC_1): natural sand up to 4 mm in diameter; river gravel with a diameter of 4-8 mm; cement type CEM II/A-LL 42.5R MPa, produced in Romania; superplasticizer additive based on polycarboxylate ether; rhodanide-based accelerator additive. The C30/37 class concrete was used as the reference concrete, as it was foreseeable that the concrete strength will decrease by replacing the mineral aggregates with the vegetable ones (because lignocellulosic concretes generally have a lower strength compared to conventional concrete with mineral aggregates). Using a superior class of concrete as a starting point, it was intended to obtain an acceptable strength of the vegetal concrete.

In the second stage, a concrete with vegetal aggregates was made, starting from the RC_1 recipe, replacing the mineral aggregates (sand and gravel) with SSA (figure 2), obtained according to the method described above. The SSA were treated with SS in two variants: with a solution with a concentration of 25%, and with a solution with a concentration of 40%. Treatment of SSA with 25% SS resulted in a reduction in water absorption from 402% (measured for untreated vegetal aggregates) to about 200% (a 50% reduction). When using the more concentrated SS of 40%, a water absorption capacity of 100% (a reduction of 75%) was obtained. After immersion in the SS, the vegetal aggregates were left to dry on a surface heated to about 50-55 °C, in a room with an ambient temperature of 23-25 °C. After drying to constant mass, CTSS 25/50 (concrete with 50% SSA treated with SS 25%) and CTSS 40/50 (concrete with 50% SSA treated with SS 40%). A concrete with untreated sunflower stalk aggregates, CUSS 50 (concrete with untreated sunflower stalks) was also made. Table 1 presents the values for the density and water absorption capacity of the SSA in the three variants.

Table 1. The bulk density and water absorption capacity of SSA

Treatment applied to SSA	SSA bulk density [kg/m^3]	SSA absorption capacity [%]
No treatment	207,00	402
SS 25%	173,00	192
SS 40%	328,10	100

The water/cement ratio for the reference concrete was set to 0.50, but using a super plasticizer additive was reduced to 0.43. In the compositions with SSA, an additional amount of water was added, its amount varying depending on the specific absorption capacity of the sunflower stalks.

Figure 2. Transversal section appearance of the concrete with SSA

Characteristics of Concrete with 50% of Treated and Untreated SSA

The test specimens were 150 mm side cubes for compressive strength testing, prisms with 100x100mm^2 transversal section and 550mm length for flexural tensile strength testing, and cylinders with 100mm in diameter and 200mm in length for splitting tensile strength testing, three specimens for each test. The tests were performed at the age of 28 days, in accordance with the standards in force (EN 12390-3:2019, EN 12390-5:2019, EN 12390-6:2010).

The densities of the concrete compositions determined at the age of 28 days are shown in fig. 6. All concrete compositions with SSA belong to the category of lightweight concrete, with a density smaller than 2000 kg/m^3 (according to C155:2013 standard). From this point of view, the lightest concrete is the one made with untreated SSA.

Regarding the compressive strength (figure 4a), the replacement of 50% of the mineral aggregates with SSA led to a significant decrease of this parameter compared to the reference concrete. On the other hand, the treatment of SSA with SS 40% solution (resulting in a reduction in absorption capacity by 75%) resulted in a significant improvement of the concrete compressive strength, compared to the one with untreated SSA.

Figure 3. Concrete density with 50% SSA in the three treatment variants, at 28 days [kg/m³]

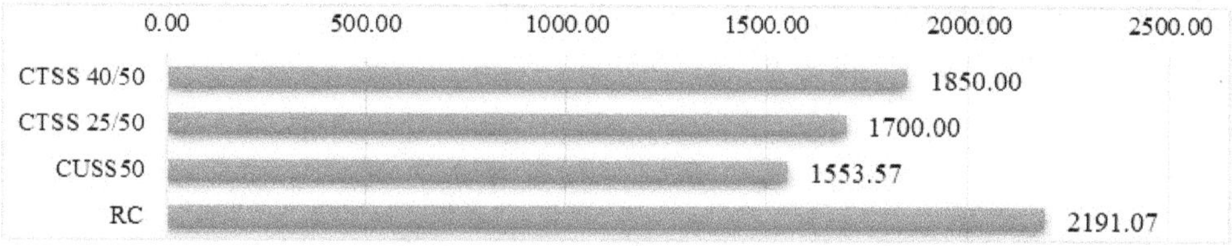

Figure 4. Mechanical properties of the concrete with 50% SSA in the three treatment variants [N/mm²]: a. compressive strength; b. flexural tensile strength; c. splitting tensile strength

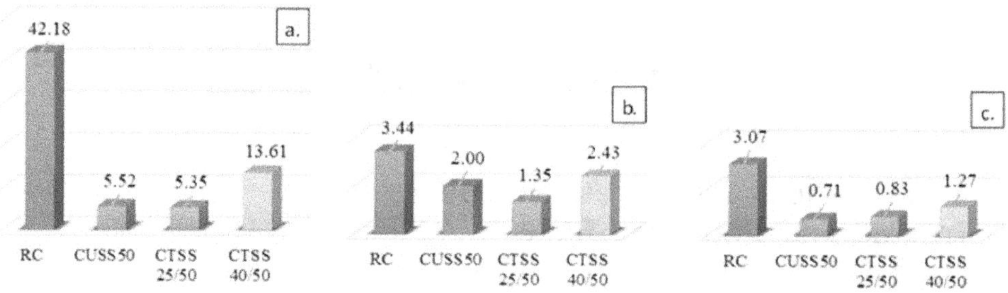

Treatment of SSA with more concentrated SS solution led to an increase of compressive strength by 146.5% compared to the compressive strength of concrete with untreated SSA (figure 4a).

Treatment of SSA with less concentrated SS solution resulted in an unexpected decrease of the flexural tensile strength of the concrete (figure 4b) by 32.5% (1.35 N/mm²) in the case of CTSS 25/50 compared to CUSS50, most likely due to the stiffening of the wood fibers from the SSA composition, which led to a decrease in their flexibility and the development of a breakable character. The SSA treatment with the more concentrated SS determined the increase of the flexural tensile strength of the concrete by up to 2.43 N/mm², this increase being attributed to the stiffening of the wood fibers in the SSA composition but with the development of an increased bending strength this time, compared to the first treatment option (the higher density of the solution of SS 40% compared to that of SS 25% caused an increase in the SSA mass, this additional mass opposing a higher resistance to the bending forces).

Regarding the splitting tensile strength (figure 4c), the concrete with untreated SSA registered a value of 0.71 N/mm^2. The application of the first treatment on the SSA led to a slight increase in the concrete strength, while the second treatment determined a value of about 79% higher than in the case of the one with untreated SSA.

***

The experimental results led to the conclusion that the replacement of mineral aggregates in a proportion of 50% with SSA led to a decrease in the mechanical properties of concrete. On the other hand, the treatment of SSA with the SS 40% resulted in a significant increase in the compressive and tensile strengths of concrete, compared to the values obtained by the concrete with untreated SSA.

Although the mechanical properties of the concrete with SSA were below those of the reference, the reduction of the absorption capacity of the SSA in parallel with the strength increase of the concrete made with them, and the lightweight concrete obtaining lead to the conclusion that sunflower stalks can be used as a partial substitute for the mineral aggregates in the concrete composition.

DETAILED ANALYSIS ON THE EFFECT OF MINERAL AGGREGATES SUBSTITUTION IN CONCRETE COMPOSITION WITH SSA TREATED WITH SS 40%

After determining the optimal treatment variant of SSA, a series of micro concrete recipes made with SSA in several percentage variants were analyzed: 20%, 35%, 50%, 65%, 80%, 100% of the total volume of aggregates in the composition. This study aimed to analyze the effect of substituting mineral aggregates in the concrete composition with SSA treated with SS 40%, making determinations of density, compressive and tensile strength, elastic modulus, durability to freeze-thaw and thermal conductivity.

The analysis of figure 5 shows the concrete **density** tendency to decrease as the weight of vegetal material increases.

Replacing 20% of mineral aggregates with SSA reduced the density of the concrete by around 8%, while their replacement in proportions of 35%, 50%, 65% and 80% resulted in decreases of 15.70%, 20.62%, 28.72% and 37.97%, respectively, compared to reference concrete. The density of the vegetal concrete decreased by 6-9% from one compositional variant to another, in the case of CSS20, CSS35, CSS50, CSS65 and CSS80. The 100% replacement of the mineral aggregates with SSA determined the decrease of the reference concrete density by approx. 62%, this compositional variant registering a decrease in density by around 25% compared to the previous variant that involved the replacement of 80% of the mineral aggregates.

According to figure 6, the same decreasing trend of **compressive strength** (determined on cylinders with 100 mm in diameter and 200 mm length) is observed as the share of SSA increased.

Figure 5. Apparent density of concrete with 20, 35, 50, 65, 80 and 100% SSA, measured 28 days after casting [kg/m³]

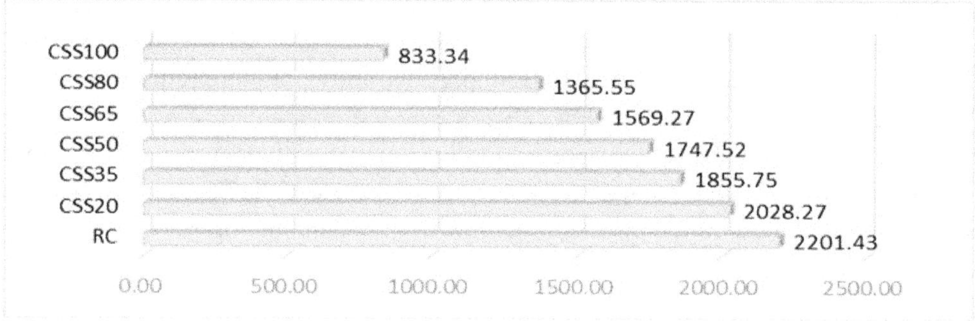

Figure 6. Compressive strength of concrete with 20, 35, 50, 65, 80 and 100% SSA, determined 28 days after casting [N/mm²]

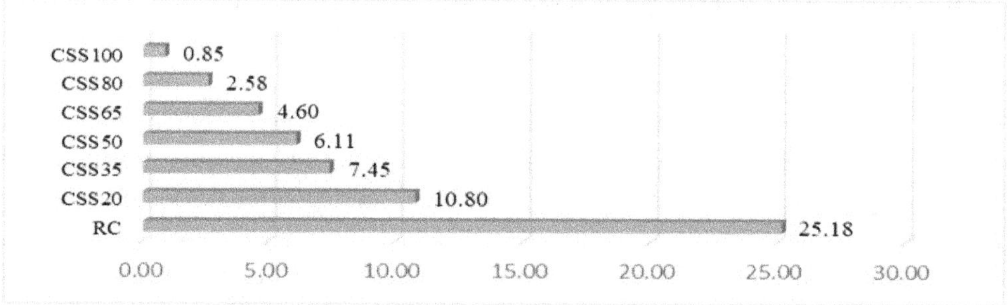

In case of 20% replacement of the mineral aggregates with SSA, it was obtained a compressive strength decrease of the concrete by around 57% compared to RC, and in case of 35% replacement rate, at a decrease of 30% compared to the previous variant. Therefore, a surplus of 15% SSA resulted in a double decrease in compressive strength (30%), but in a much smaller rate than the initial replacement of 20% of mineral aggregates with SSA (30% vs. 57%). The further increase of the share of SSA from 35% to 50% determined the decrease of the compressive strength by approx.

76% compared to RC, but with only 18% compared to the previous version. From the 50% level, the decrease rate of the compressive strength started to increase from one variant to another, as the weight of the vegetal material in the total volume of the concrete increased by 15%. Thus, in the case of CSS65, the compressive strength decreased by 25% compared to CSS50, and in the case of CSS80 and CSS100, by 44% and 67%, respectively, compared to their previous version. Compared to reference concrete, the replacement of mineral aggregates over 65% resulted in a compressive strength decrease by more than 80%, CSS100 registering a lower value of this parameter by 96.62% than RC. The value recorded for compressive strength decreased below 5 N/mm² in the case of using only SSA in concrete composition, CSS100 registering 0.85 N/mm².

In the case of **splitting tensile strength**, there was a decrease also as the weight of the vegetal material increased (figure 7). The SSA rate increase from 20% to 35%, from 50% to 65%, and from 80% to 100% determined the decrease of this parameter by 44.51%, 18.64% and 82.69%, respectively, compared to the previous variant. On the other hand, the increase of this rate from 35% to 50%, and from 65% to 80% determined the improvement of the splitting tensile strength by 3.34% and 7.78%, respectively, compared to the previous variant. The best value of this parameter was recorded by CSS20 (1.72 N/mm²), while CSS35, CSS50, CSS65, and CSS80 recorded values quite similar, with small variations between 0.80 - 0.99 N/mm² (figure 7).

Figure 7. Splitting tensile strength of concrete with 20, 35, 50, 65, 80 and 100% SSA, determined 28 days after casting [N/mm²]

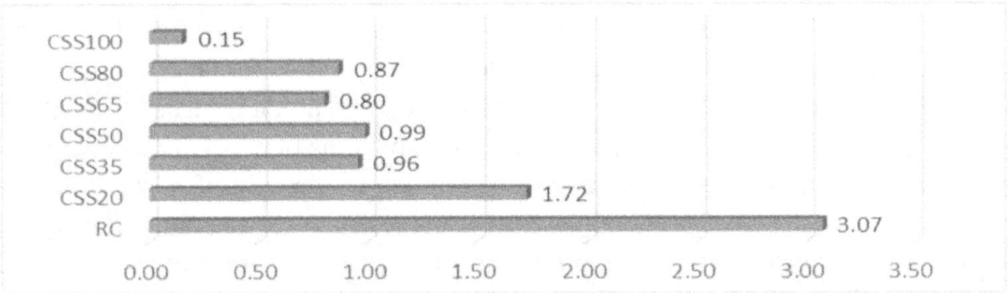

In order to determine the **modulus elasticity in compression** of the concrete developed in this research, the second method from the SR EN 13412:2007 standard was applied. The second method applies to products based on cement binders. The equipment used to perform these determinations was a 100 tf hydraulic press and a specific displacement measuring device as it is showed in figure 8.

Figure 8. Equipment used for deformation measuring in determining the modulus of elasticity of concrete

As can be seen in the figure 9, the modulus of elasticity of concrete compositions with SSA decreases as the share of vegetal material in their structure increases.

Figure 9. Modulus of elasticity of concrete with 20, 35, 50, 65, 80 and 100% SSA

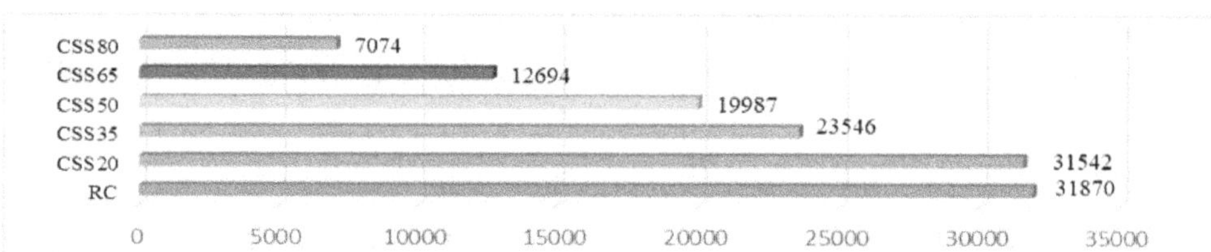

The **freeze-thaw resistance** test was performed according to SR EN 3518: 2009. The standard assumes the existence of a destructive method for determining these performances by testing the compressive strength, and a non-destructive method, which involves measuring the relative dynamic modulus of elasticity. As the non-destructive method is recommended to be applied in the case of normal density concrete, the destructive method was used in the research carried out in this study, as it is recommended in the case of both normal density and lightweight concrete. This method involved the realization of cube-type specimens with sides of 100 mm. The test was performed after the concrete reached the age of 28 days. The specimens were subjected to 50 freeze-thaw cycles. For this number of cycles, the specification from the standard in force of using 6 specimens from each concrete composition was taken into account, 3 of them serving as control specimens and the other 3 being effectively subjected to the freeze-thaw cycles. After completing 50 cycles of 4 hours of frost - 4 hours of thawing, the specimens are tested for compression (figure 10).

Compressive strength loss (η) was determined using equation 1:

$$\eta = \frac{R_m - R_t}{R_m} \bullet 100 \qquad (1)$$

where,

R*m* – average compressive strength of the control specimens [N/mm^2];
R*t* – average compressive strength of the specimens subjected to freeze-thaw cycles [N/mm^2].

Figure 10. Freeze-thaw resistance test: a. vegetal concrete specimen subjected to freeze-thaw cycles, before (a1) and after (a2) compressive strength testing; b. vegetal concrete control specimen, before (b1) and after (b2) compressive strength testing

The 50 freeze-thaw cycles decreased the compressive strength of the concrete. Thus, in the case of CSS20, the decrease of this parameter was around 21%, in the case of CSS50, 28%, and in the case of CSS65, around 34%. CSS35 registered a compressive strength decrease by 44%, and CSS80 by 51% (figure 11). The concrete with 100% SSA could not be tested as it did not withstand the 50 freeze-thaw cycles.

The method by which the **thermal conductivity** was determined was the one provided in SR EN 12667: 2009, namely the thermoflux meter method. From the figure 12 analysis, it is observed the decrease of the thermal conductivity of the vegetal concrete as the SSA rate increases. Thus, the replacement of 50% of the mineral aggregates with SSA led to thermal conductivity decrease of the reference concrete by 50.21%, the value registered being 0.4465 W/(mK), and the total replacement of these, at a reduction of thermal conductivity by approximately 83%, obtaining a value of 0.1537 W/(mK).

Figure 11. Loss of compressive strength of concrete with 20, 35, 50, 65, 80 and 100% SSA, following freeze-thaw testing [%]

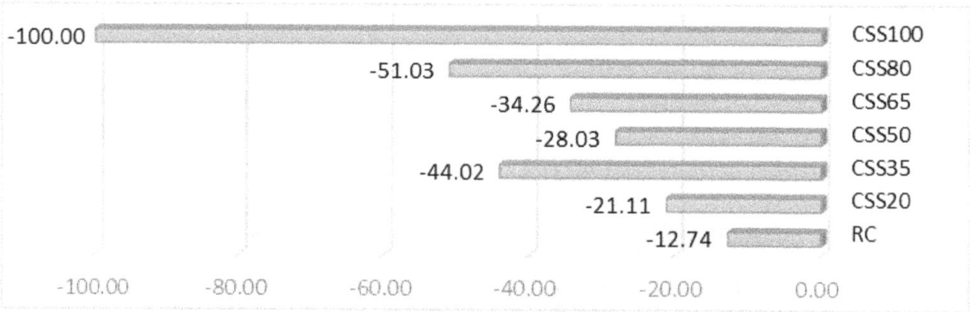

Figure 12. Thermal conductivity of concrete with 20, 35, 50, 65, 80 and 100% SSA [W/mK]

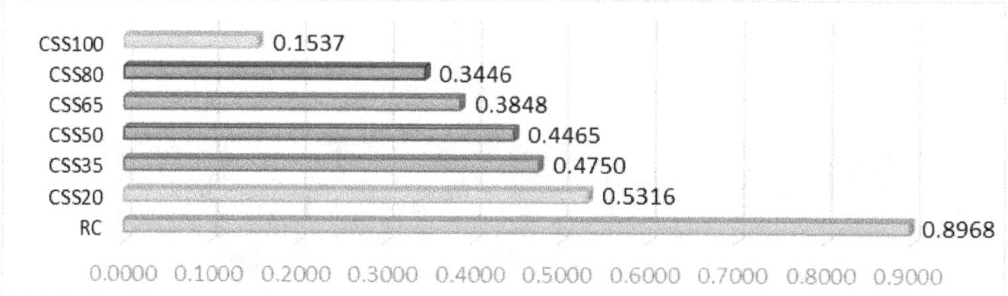

The increase by 15% (up to 80%) of the SSA share in the concrete composition determined decreases of its thermal conductivity by 6-14% from one compositional variant to another (figure 12). Thus, CSS35 and CSS80 registered a lower thermal conductivity by around 10.5% compared to CSS20 and CSS65, respectively. CSS50 recorded a decrease in thermal conductivity by 6% than CSS35, and CSS65 by about 14% compared to CSS50. In the case of CSS100, there was a decrease in thermal conductivity by 55.40% compared to CSS80.

VARIANTS FOR IMPROVING THE PERFORMANCE OF CONCRETE WITH SSA

In order to improve the properties of concrete with vegetal aggregates, with the aim to obtain a material with a low thermal conductivity and an adequate level of mechanical strength, the cement matrix modification, the acceleration of the hydration reaction of cement or both these methods simultaneously can be used. In order to closely analyze the effects that can be obtained by applying these methods, a study was performed that involved:

A. ***Modification of the cement matrix*** by using fly ash as a cement substitute in a proportion of up to 30% of its volume in the reference concrete recipe;
B. ***Acceleration of the cement hydration reaction*** by using 100% sodium silicate solution as an additional additive, in proportion of 5% of the binder volume, in addition to the use of a commercial rhodanide-based accelerating additive. By using the hardening accelerator additive and the SS of 100% concentration, the aim was to obtain an improvement of the hydration reaction of the cement. This measure was justified by the fact that the water added during the process of making the concrete with SSA is absorbed relatively quickly by the vegetal part of the concrete, thus calling into question the percentage of accomplishment of the hydration reaction of the cement. As an additive in fresh concrete mix, sodium silicate acts as a setting accelerator for Portland cement (Sellami et al., 2013). In the research from this stage, the SS of 100% concentration in a ratio of 5% of the cement volume was used.
C. ***The cement matrix modification*** by using ultrafine silica as a substitute for cement in proportion of 10% of its volume, ***simultaneous with the accelerating the hydration reaction of cement*** by using 100% sodium silicate solution as supplementary additive, in proportion of 5% of the binder volume, in addition to the use of a commercial accelerating additive based on rhodanide, and the formation of air gaps in the concrete structure by using an air entraining additive in the proportion of 0.4% and 0.8% of the total volume of binder. By using the substitution of a part of the cement with ultrafine silica, the aim was to improve the durability of the vegetal matter in the concrete structure and to obtain an improved interface of vegetal aggregates - matrix. During this stage, an air entraining additive was added. The benefits of incorporating this additive are related to the formation of fine air pores in the concrete structure, with a uniform distribution, which allows the expansion of water infiltrated into the concrete mass on frost, without destroying the concrete structure. Concretes made by incorporating this additive are characterized by an improvement in workability as a consequence of the bearing effect.

In all the concrete compositions made, a water/cement ratio of 0.43 was applied, regarding the mixing water. However, as the general objective has been to ensure that concrete has approximately the same level of workability that allows it to be easily handled during pouring, the total amount of water used in a concrete composition has varied depending on the absorption capacity of the SSA or on the effect of replacing cement with fly ash or ultrafine silica or the use of sodium silicate or air entrainment additive. Also, in all compositions, a super plasticizer additive was used, this being added to the concretes for a better maintenance of their workability, by better dispersion and hydration of the cement, decrease of water requirement and constant maintenance of water/cement ratio, decrease of friction forces between cement and aggregates, resulting in a more homogeneous concrete, with a longer working time even in the case of high temperatures, with a reduction in shrinkage and slow flow and an increase in resistance to carbonation.

PRESENTATION OF CONCRETE COMPOSITIONS WITH SSA

Following the application of the three variants of modification of concrete compositions, a series of recipes were developed, as follows:

A. Cement matrix modification

To carry out this study, concrete recipes with 10%, 20% and 30% fly ash as a substitute for cement were developed. These served as reference recipes for the subsequent realization of other concrete recipes, which involved maintaining the same matrix, and the mineral aggregates were replaced by volume with SSA in 20% and 50%. Thus, the following concrete compositions were developed:

- RC - reference concrete of 25/30 strength class;
- CFA10 - concrete with fly ash as a substitute for 10% of the cement volume in the RC recipe;
- CFA20 - concrete with fly ash as a substitute for 20% of the cement volume in the RC recipe;
- CFA30 - concrete with fly ash as a substitute for 30% of the cement volume in the RC recipe;
- CSSFA 20/10 - concrete with fly ash as a substitute for 10% of the volume of cement in the RC recipe, and with SSA as a substitute for a volume of 20% mineral aggregates;

- CSSFA 20/20 - concrete with fly ash as a substitute for 20% of the volume of cement in the RC recipe, and with SSA as a substitute for a volume of 20% mineral aggregates;
- CSSFA 20/30 - concrete with fly ash as a substitute for 30% of the volume of cement in the RC recipe, and with SSA as a substitute for a volume of 20% mineral aggregates;
- CSSFA 50/10 - concrete with fly ash as a substitute for 10% of the volume of cement in the RC recipe, and with SSA as a substitute for a volume of 50% mineral aggregates;
- CSSFA 50/20 - concrete with fly ash as a substitute for 20% of the volume of cement in the RC recipe, and with SSA as a substitute for a volume of 50% mineral aggregates;
- CSSFA 50/30 - concrete with fly ash as a substitute for 30% of the volume of cement in the RC recipe, and with SSA as a substitute for a volume of 50% mineral aggregates;
- CSS 20 - concrete with SSA as a substitute for 20% of the volume of mineral aggregates;
- CSS 50 - concrete with SSA as a substitute for 50% of the volume of mineral aggregates.
- The composition of reference concrete,, involved the use of the following components:
- Portland cement CEM II / A-LL42.5R;
- two sorts of river aggregates: sand with a diameter of up to 4 mm and gravel sort 4-8 mm;
- additives: a polycarboxylate ether superplasticizer and a rhodanide-based curing accelerator.

B. Acceleration of the Cement Hydration Reaction

To carry out this study, concrete recipes with 100% sodium silicate solution (SS) were developed as an additive to accelerate the hydration reaction of cement, in a proportion of 5% of the volume of cement. Mineral aggregates were replaced with 20%, 50% and 80% SSA. Thus, the following recipes were developed:

- RC - reference concrete of 25/30 strength class;
- CSS20 - concrete with SSA as a replacement for 20% of mineral aggregates volume;
- CSS20.SS - concrete with SSA as a substitute for 20% of mineral aggregates volume and with SS 100% additive, in proportion of 5% of the cement volume;

- CSS50 - concrete with SSA as a replacement for 50% of mineral aggregates volume;
- CSS50.SS - concrete with SSA as a substitute for 50% of mineral aggregates volume and with SS 100% additive, in proportion of 5% of the cement volume;
- CSS80 - concrete with SSA as a replacement for 80% of mineral aggregates volume;
- CSS80.SS - concrete with SSA as a substitute for 80% of mineral aggregates volume and with SS 100% additive, in proportion of 5% of the cement volume.

C. Cement Matrix Modification and Cement Hydration Reaction Acceleration

As the mechanical properties of CSS concrete have decreased significantly compared to RC, several options have been considered to find a solution to improve their mechanical properties by replacing 10% of the cement volume by fly ash (FA), and ultrafine silica (US), respectively, and by the SS addition of SS in 5% of the cement volume, and air entraining additive (AE) in 0.4% and 0.8% of the cement volume. The mixing rates of the variable components in the developed concrete compositions are shown in Table 2.

Therefore, the concrete compositions developed for this study were:

- RC - reference concrete of 25/30 strength class;
- CSS - concrete with SSA;
- CSS.SS - concrete with SSA and SS;
- CSS.FA - concrete with SSA and FA;
- CSS.SS.US.AE1 - concrete with SSA, SS, ultrafine silica and 0.4% AE;
- CSS.SS.US.AE2 - concrete with SSA, SS, ultrafine silica and 0.8% AE.

Table 2. Mixing ratios of variable components in the analyzed concrete compositions

Concrete composition	SSA [vol. % of total aggregates]	FA [vol. % of cement]	US [vol. % of cement]	SS [vol. % of cement]	AE [vol. % of cement]
RC	0	0	0	0	0
CSS	50	0	0	0	0
CSS.SS	50	0	0	5	0
CSS.FA	50	10	0	0	0
CSS.SS.US.AE1	50	0	10	5	0,4
CSS.SS.US.AE2	50	0	10	5	0,8

Analysis of the Properties of Modified Concrete Compositions with SSA

The analyzes performed on the concrete specimens aimed to determine the density of hardened concrete, the compressive strength, the splitting tensile strength, the modulus of elasticity, the resistance to repeated freeze-thaw cycles and its thermal conductivity. The methods by which these determinations were made are the same as those applied in the analyzes performed in the previously presented research.

Density

A. Cement Matrix Modification

From the figure 13a analysis, it is observed that the partial replacement of cement with FA (10%, 20%, 30%), in parallel with the 20% mineral aggregates replacement with SSA caused a decrease in density by around 3.0 - 4.5% compared to the value of concrete without fly ash, only with 20% SSA. Replacing 50% of the mineral aggregates with SSA reduced the density of reference concrete by around 20%. In the case of concrete with 50% SSA, the replacement of 10%, 20% or 30% of the cement volume with FA did not lead to significant density variations.

B. Acceleration of the Cement Hydration Reaction

The effect of the addition of SS was a slight decrease in the density of the related concrete in the case of CSS20.SS and CSS50.SS, and a slight increase in the case of CSS80.SS (figure 13b). At 28 days, the addition of SS had an insignificant effect on the density in all three variants analyzed.

C. Cement Matrix Modification and Cement Hydration Reaction Acceleration

All the developed concrete compositions were category of lightweight concrete category, with a density smaller than 1800 kg/m^3 (figure 13c). The lightest concrete was CSS.SS.US.AE2, with a density of 1628.47 kg/m^3. Compared to RC, CSS, CSS.SS, CSS.FA, and CSS.SS.US.AE1, they recorded 20-22% smaller densities. CSS.SS.US.AE2 had a smaller density by around 26%.

In conclusion, the density of concrete with sunflower stalk, measured 28 days after pouring, decreased as the share of vegetal material in the concrete composition increased. Lightweight concrete can be obtained with a percentage of more than 50% of vegetal matter.

According to the Regulation on the production of lightweight concrete C155-2013, concrete with lightweight aggregates can be classified in the following classes, according to their density: D0.8; D1.0; D1.2; D1.4; D1.6; D1.8 and D2.0. Table 3 presents these density classes, together with the distribution of concrete compositions developed and analyzed in this research on these classes.

Figure 13. Density of the modified concrete with SSA, at 28 days [kg/m³]: a. concrete with 20% and 50% SSA, and 0, 10, 20 and 30% FA; b. concrete with 20%, 50% and 80% SSA and SS additive; c. concrete with 50% SSA, with modified cement matrix and/or additives

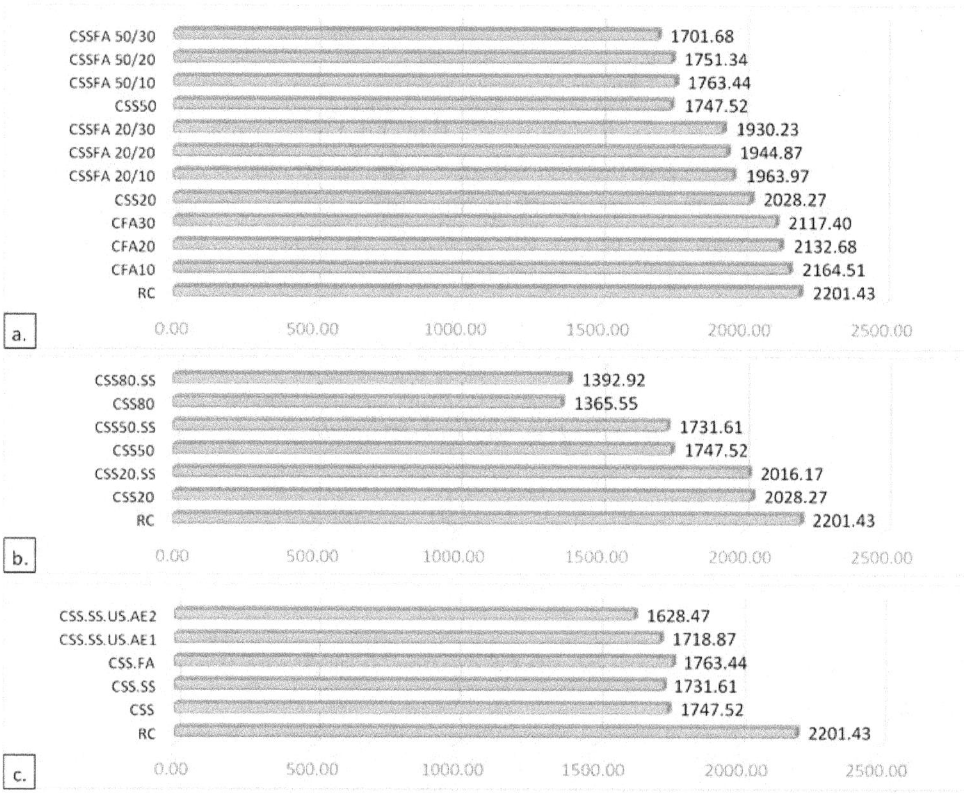

Compressive Strength

A. Cement Matrix Modification

As the FA amount increased, the compressive strength of the concrete decreased (figure 14a). For a replacement rate of 10% of the cement volume with FA, the obtained concrete had a smaller compressive strength with 8% than RC. For a 20% replacement rate, the decrease was 10%, while a 30% replacement rate led to a decrease of about 17%. Therefore, although the amount of replaced cement increased in equal rates, the effect on the compressive strength of the concrete was not in the same rates.

Table 3. Density classes for lightweight aggregate concretes and classification of concretes developed in this research

Density mass class	Density mass range [kg/m³]	Vegetal concrete compositions
D0.8	(400, 800]	-
D1.0	(800,1000]	CSS100
D1.2	(1000, 1200]	-
D1.4	(1200, 1400]	CSS80, CSS80.SS
D1.6	(1400, 1600]	CSS65
D1.8	(1600, 1800]	CSS50 CSSFA50/10, CSSFA50/20, CSSFA50/30 CSS50.SS CSS.SS.US.AE1, CSS.SS.US.AE2
D2.0	(1800, 2000]	CSS35 CSSFA20/10, CSSFA20/20, CSSFA20/30

B. Acceleration of the Cement Hydration Reaction

The SS addition in the concrete compositions with SSA determined the improvement of their compressive strength (figure 14b). Thus, in the case of CSS20, it led to the increase of this parameter by 7.41%, and in the case of CSS80, by 16.87%. In the case of CSS50, it caused the greatest improvement in compressive strength, by around 34%. The best compressive strength was 11.60 N/mm^2 in the case of CSS20. SS. It can be concluded that the optimal effect of the SS addition was obtained by replacing 50% of the mineral aggregates with SSA.

C. Cement Matrix Modification and Cement Hydration Reaction Acceleration

The SS use in the concrete composition had a positive effect on the compressive strength, the increase being of around 35% (figure 14c). When replacing 10% of the cement volume with FA, the compressive strength increased by about 38%. Replacing cement with ultrafine silica and adding 0.4% AE increased the compressive strength of concrete with SSA by 27.78%. An increase of AE rate led to a decrease in the compressive strength of concrete with SSA by 17.36%.

Figure 14. Compressive strength of the modified concrete with SSA, at 28 days [N/mm²]: a. concrete with 20% and 50% SSA, and 0, 10, 20 and 30% FA; b. concrete with 20%, 50% and 80% SSA and SS additive; c. concrete with 50% SSA, with modified cement matrix and/or additives

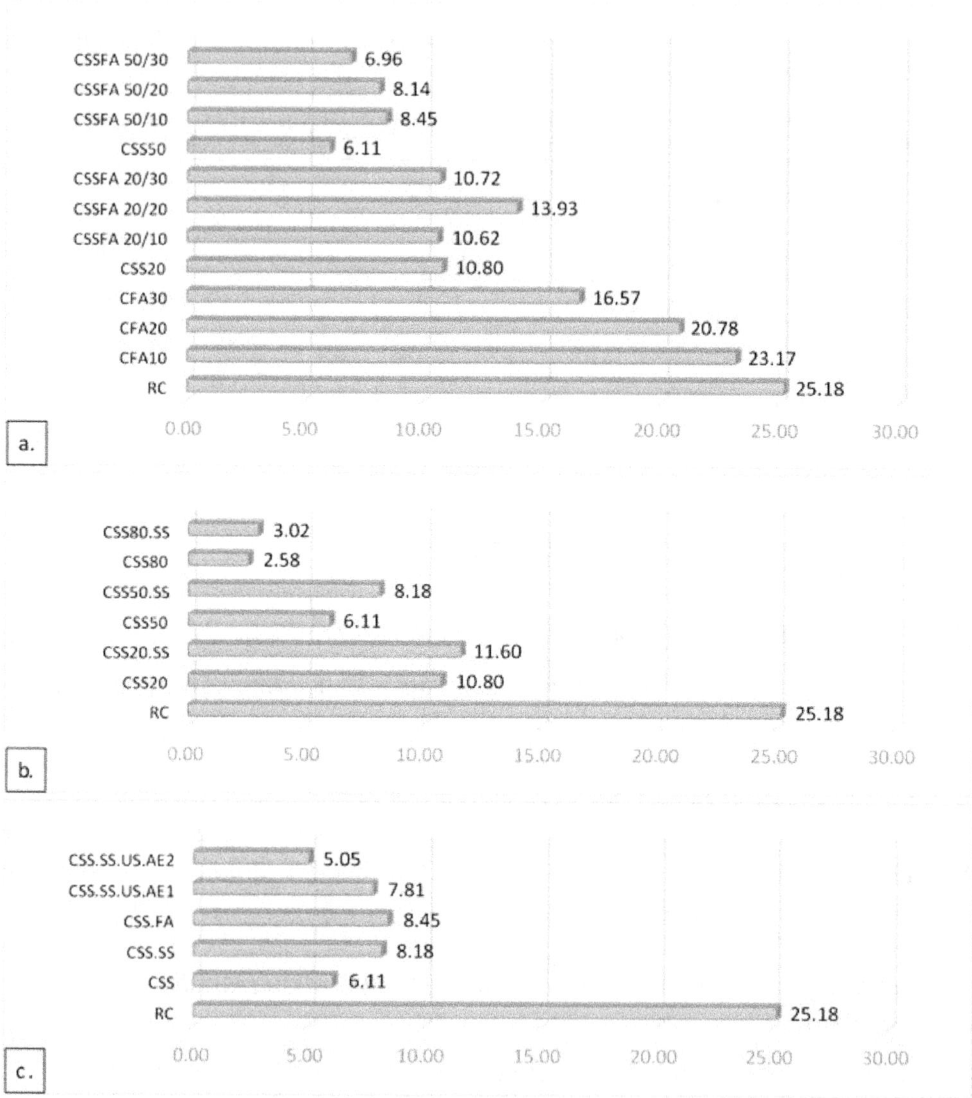

According to NE 012/1, the strength classes of lightweight concretes are different from those of normal mass concrete. The Table 4 presents the strength classes of lightweight concretes, highlighting at the same time the distribution of the compositions developed in this research. The notation of these classes consists of the code LC (lightweight concrete) followed by the minimum characteristic strength on cylinders/cubes, expressed in [N/mm^2].

Of all compositions developed in this research, only those listed in Table 4 can be classified according to the recorded compressive strength. Thus, CSSFA 20/20 falls into the strength class LC12/13, and in the lower strength class, LC8/9, the concretes marked with CSS20, CSSFA 20/10, CSSFA 20/30, CSSFA 50/10, CSSFA 50/20, CSS20.SS and CSS50.SS. The rest of the developed concrete compositions registered compressive strengths below the level of LC8/9 strength class.

Of the variants applied as cement substitutes or as additives, the best results, in terms of density-compressive strength, were obtained in case of FA using in the concrete composition with 50% SSA, but also the case of combination SS + ultrafine silica + 0.4% AE, improving the performance of concrete compositions SSA. SS used as an additive had a positive effect on concrete with SSA. A significant improvement effect was also obtained in the case of replacing 20% of the cement volume with FA in the concrete composition with 20% SSA.

Table 4. Strength classes of the developed lightweight concretes

Strength classes of lightweight concrete	Concrete compositions
LC8/9	CSS20 CSSFA 20/10, CSSFA 20/30, CSSFA 50/10, CSSFA 50/20 CSS20.SS, CSS50.SS
LC12/13	CSSFA 20/20

Splitting Tensile Strength

A. Cement Matrix Modification

In the case of 20% cement replacement with FA, an improvement by up to 16% compared to CFA10 was observed. A surplus of another 20% FA led to a decrease in splitting tensile strength by around 27% compared to RC. In the case of concrete with 20% SSA, the partial replacement of cement resulted in a decrease in splitting tensile strength, but it still improved as the FA percentage increased (figure 15a).

As regards the concrete with 50% SSA, FA improved the splitting tensile strength in the case of all three variants of cement substitution. Thus, the application of a substitution rate of 10% led to an improved strength of around 38%, applying a substitution rate of 20%, to an improvement of around 33%, and one of 30%, to an improvement of around 14% (figure 15a).

B. Acceleration of the Cement Hydration Reaction

The splitting tensile strength decreased with SSA rate increasing in the concrete composition. The use of SS as an additive in the concrete determined the decrease of this parameter by around 34% in the case of concrete with 20% SSA (figure 15b). In the case of concrete with 80% SSA, this additive caused a slight increase in the splitting tensile strength, by around 0.4%, and in the case of concrete with 50% SSA, an increase of 6.22%.

C. Cement Matrix Modification and Cement Hydration Reaction Acceleration

Regarding the splitting tensile strength by splitting, all the concrete variants studied in this subsection met the objective of improving this property compared to simple concrete with SSA (figure 15c).

The use of SS additive improved the splitting tensile strength (figure 15c) by about 6% of the concrete with SSA. Replacing cement with FA increased this strength by more than 20%. Ultrafine silica and AE also increased this parameter by 16%. In the case of concrete with 50% SSA, improvements of this parameter were obtained even in the case of using a 30% substitution rate of cement with FA. An effect of splitting tensile strength improving was also obtained in the case of concomitant use of SS, ultrafine silica and AE in the concrete composition with 50% SSA.

Modulus of Elasticity

A. Cement Matrix Modification

The association FA-SSA determined the same type of evolution of the modulus of elasticity as in the case of the reference series, CFA10, CFA20, CFA30 only in the case of using 20% SSA (figure 16a). In the case of concrete with 50% SSA, the replacement of 30% of the cement volume with FA contributed to the decrease of the modulus of elasticity compared to its replacement by 20%, CSSFA 50/30 registered the lowest value of this parameter in this series of concrete with SSA.

B. Acceleration of the Cement Hydration Reaction

The effect of SS on the modulus of elasticity of concrete with SSA was generally to decrease it. In the case of concrete with 80% SSA, the effect was the opposite, of growth, compared to the variant without additive (figure 16b).

Figure 15. Compressive strength of the modified concrete with SSA, at 28 days [N/mm²]: a. concrete with 20% and 50% SSA, and 0, 10, 20 and 30% FA; b. concrete with 20%, 50% and 80% SSA and SS additive; c. concrete with 50% SSA, with modified cement matrix and/or additives

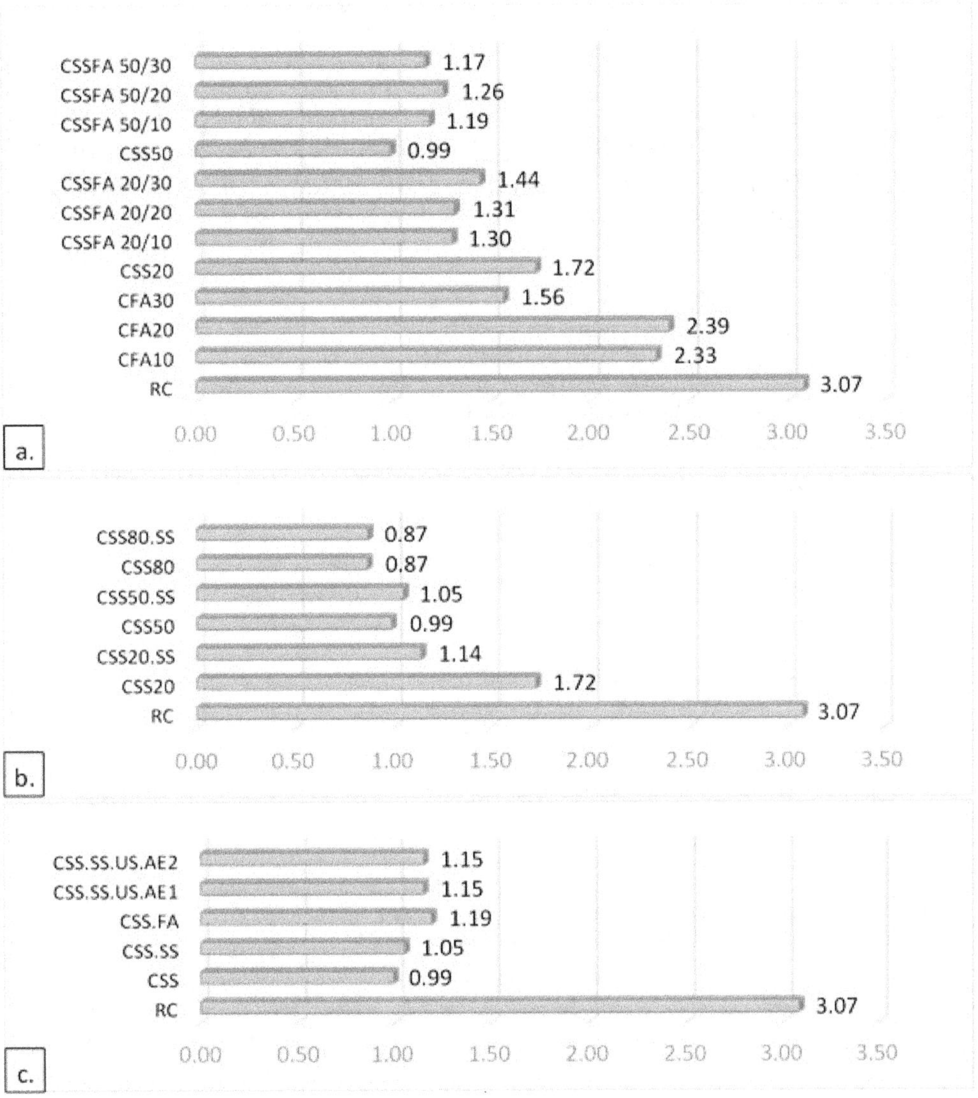

C. Cement Matrix Modification and Cement Hydration Reaction Acceleration

The use of SS as an additive in the concrete composition reduced the modulus of elasticity of CSS and the application of 10% replacement of the cement volume with FA also led to its decrease, but to a lesser extent than in the case of SS use (figure 16c). The replacement of 10% of the cement volume with ultrafine silica at the same time with the addition of 5% of SS and AE determined the increase of the elastic modulus, both in case of adding 0.4% AE and in the one of 0.8% AE, the difference between the two variants being relatively small.

Figure 16. Modulus of elasticity of the modified concrete with SSA, at 28 days: a. concrete with 20% and 50% SSA, and 0, 10, 20 and 30% FA; b. concrete with 20%, 50% and 80% SSA and SS additive; c. concrete with 50% SSA, with modified cement matrix and/or additives

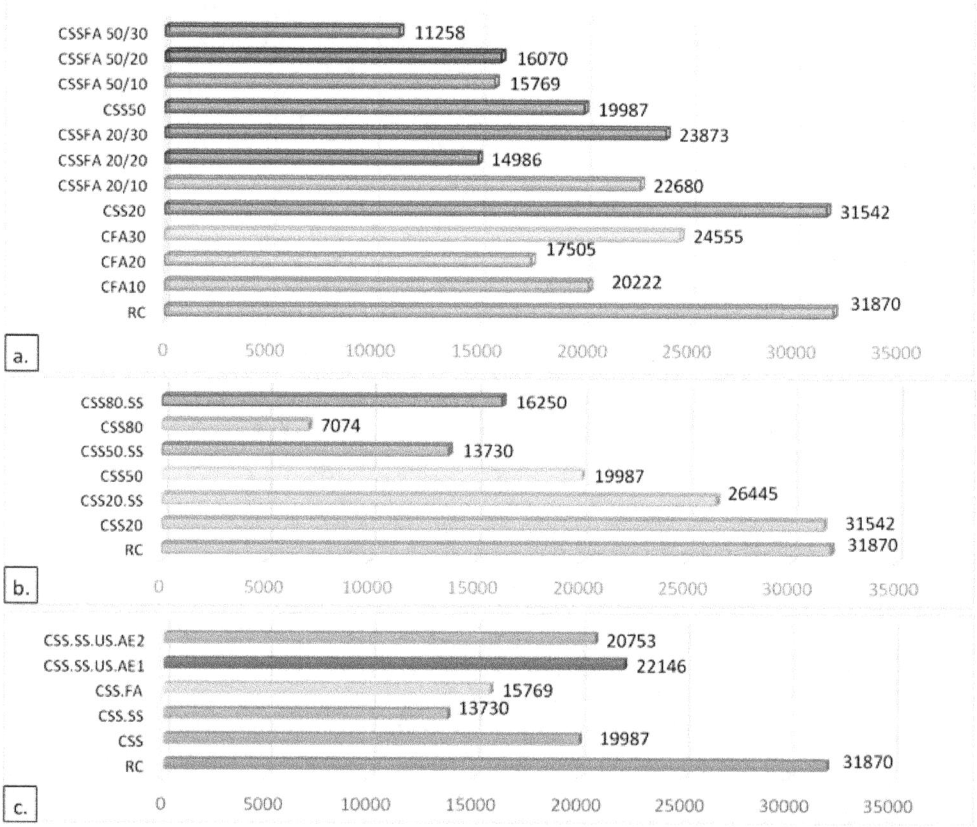

Freeze-Thaw Resistance

A. Cement Matrix Modification

Concrete with SSA registered a decrease in compressive strength, following the 50 freeze-thaw cycles, of around 21% (figure 17a). The combination of 20% replacement of mineral aggregates with SSA, and replacement of 10%, 20%, and 30% of the cement volume with FA, had positive effects on the loss of compressive strength caused by the freezing-thawing process. Thus, in the case of CSSFA 20/20, the loss was 16%, and in the case of CSSFA 20/10, 5.64%. In the case of concrete with 50% SSA, the compressive strength recorded a loss of 28%. The use of a cement substitution rate of 10% and 20% with FA, slightly reduced this loss to 25.12%, and 23.95%, respectively. The use of a cement replacement rate of 30% with FA, on the other hand, resulted in a substantial reduction in the compressive strength loss following the 50 freeze-thaw cycles, CSSFA 50/30 having a lower compressive strength with 1.19% than the control sample.

In conclusion, the use of a replacement rate of 30% of the cement volume with FA led to the most favorable results in terms of compressive strength after freeze-thaw testing of concrete with sunflower stalks.

B. Acceleration of the Cement Hydration Reaction

The addition of SS resulted in a greater loss of compressive strength, regardless of the percentage of plant aggregates involved (figure 17b). The lowest loss of compressive strength among the concretes analyzed at this stage was recorded by CSS20.

C. Cement Matrix Modification and Cement Hydration Reaction Acceleration

Compressive strength decreased by 28% following the freeze-thaw cycles (figure 17c). The SS addition resulted in a slightly higher loss of compressive strength, up to 31%, by 10.75% more than in the case of CSS. Replacing 10% of the cement volume with FA, on the other hand, resulted in a lower loss of compressive strength, by 10.40% compared to CSS. In the case of the simultaneous application of the substitution of 10% of the cement volume with ultrafine silica and the addition of 5% of SS, and of 0.4% AE determined a significant improvement in the resistance of this concrete to freeze-thaw, recording a loss of compressive strength of 11.43%, with about 59% less than in the case of CSS. The increase of the AE rate, from 0.4% to 0.8% of the cement volume, determined, instead, a drastic decrease of the compressive strength of the afferent concrete of 58.39%, with around 108% more than CSS. The lowest loss of compressive strength was recorded by CSS.SS.US.AE1, among the concretes analyzed at this stage. From the point of view of resistance to repeated freeze-thaw cycles, a positive influence on the performance of concrete with SSA had the replacement by up to 30% of the cement volume of cement with FA in concrete compositions with 20% and 50% SSA. The simultaneous use of SS, ultrafine silica and AE has led to an improvement in the resistance of freeze-thaw in the case of the concrete with 50% SSA, but only if AE is used in 0.4% rate in the composition.

Figure 17. Compressive strength loss of the modified concrete with SSA, [%]: a. concrete with 20% and 50% SSA, and 0, 10, 20 and 30% FA; b. concrete with 20%, 50% and 80% SSA and SS additive; c. concrete with 50% SSA, with modified cement matrix and/or additives

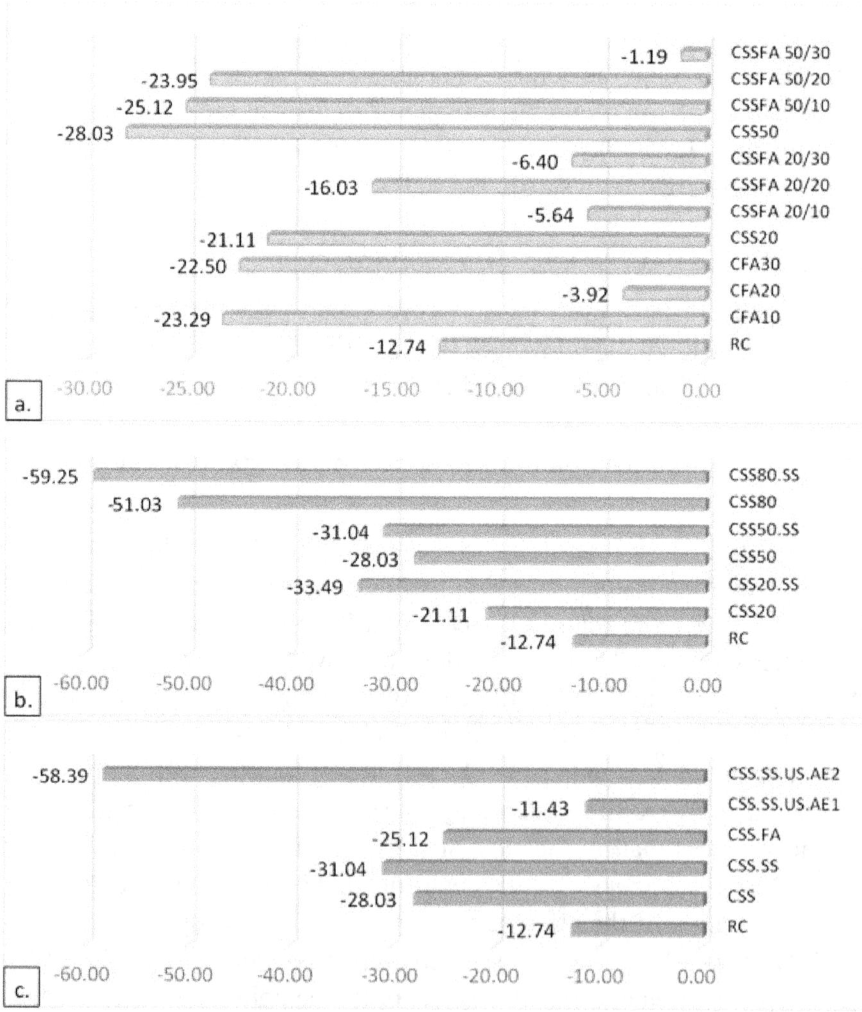

Thermal Conductivity

A. Cement Matrix Modification

Concrete with 20% SSA was characterized by a thermal conductivity value of 0.5316 W/(mK) (figure 18a). The association SSA-FA determined the decrease of the thermal conductivity of the afferent concrete by 2.20 - 4.20%. The use of a 50% replacement of mineral aggregates with SSA resulted in a value of the thermal conductivity of the concrete of 0.4465 W/(mK). The replacement of 10% of the cement volume with FA determined the decrease of the thermal conductivity of this concrete by 3.23%; the replacement of 20% of cement with fly ash resulted in a decrease of thermal conductivity of 8.32% and a replacement of 30% resulted in a decrease of 5.40%. In conclusion, the lowest value of thermal conductivity was recorded by CSSFA 50/20, 0.4094 W/(mK).

B. Acceleration of the Cement Hydration Reaction

The SS use as an additive had a positive effect in terms of decreasing the thermal conductivity of CSS20 and CSS80, by 0.14% and 3.28%, respectively, and a negative effect, increasing the thermal conductivity by approx. 2%, in the case of CSS50 (figure 18b). The lowest thermal conductivity was recorded by CSS80.SS, 0.3333 W/(mK).

C. Cement Matrix Modification and Cement Hydration Reaction Acceleration

The concrete with SSA registered a thermal conductivity of 0.4465 W/(mK) (figure 18c). The SS use as an additive in this concrete composition caused a slight increase in thermal conductivity by around 2%. The replacement of 10% of the cement volume with FA caused the decrease of this parameter by 3.23%. In the case of SS use as an additive in the composition, the replacement of 10% of the cement volume with ultrafine silica and the use of 0.4% AE led to a very small increase in the thermal conductivity of CSSS, by 0.74%.

Instead, the increase of AE rate from 0.4% to 0.8% determined the decrease of the thermal conductivity of the afferent concrete by around 11% compared to CSS. In conclusion, the lowest value of thermal conductivity, of 0.3968 W/(mK), was recorded by CSS.SS.US.AE2.

According to C155-2013, concretes with lightweight aggregates are classified, according to the value of their thermal conductivity, in the following classes: CT24, CT29, CT40, CT55, CT78 and CT105. Concretes that are part of the thermal conductivity CT24 class are insulation concretes, those from the CT29, CT40 and CT55 classes are insulation and strength concretes, and those from CT78 and CT105 classes are strength concretes. Table 5 divided the concrete compositions developed in this research by thermal conductivity classes. From Table 5 it can be seen that most of the developed concrete compositions belong to the category of Insulation and Strength Concrete. The compositions with 65% and 80% SSA and those with 50% SSA and AAA are part of the thermal conductivity class CT40, the compositions with 100% vegetal aggregates of CT24 class, and the rest of the vegetal concretes of CT55 class.

Figure 18. Thermal conductivity of the modified concrete with SSA, [W/mK]: a. concrete with 20% and 50% SSA, and 0, 10, 20 and 30% FA; b. concrete with 20%, 50% and 80% SSA and SS additive; c. concrete with 50% SSA, with modified cement matrix and/or additives

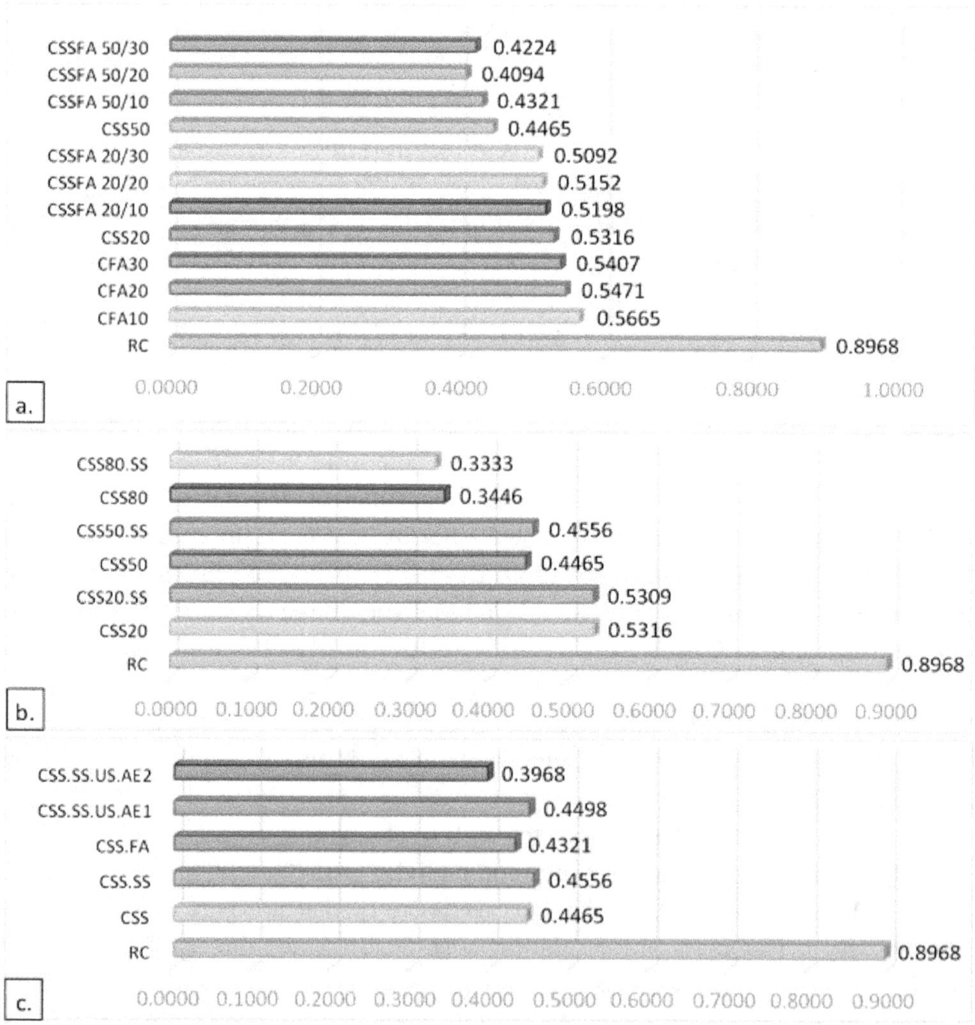

Table 5. Distribution of developed concrete compositions by thermal conductivity classes

Thermal conductivity class	Thermal conductivityvalue, λ [W/(mK)]	Concrete compositions	Concrete cathegory
CT 24	0,18 ÷ 0,24	CSS100	Insulation concrete
CT 29	0,24 ÷ 0,29	-	Insulation and strength concrete
CT 40	0.30 ÷ 0,40	CSS65, CSS80, CSS80.SS CSS.SS.SU.AE2	
CT 55	0,40 ÷ 0,55	CSS20, CSS35, CSS50, CSSFA 20/10, 20/20, 20/30 CSSFA 50/10, 50/20, 50/30 CSS20.SS, BFS50.SS CSS.SS.US.AE1	
CT 78	0,55 ÷ 0,78	-	Strength concrete
CT 105	0,78 ÷ 1,05	RC	

CONCLUSION

Most of the vegetal concrete compositions developed in this complex study fall into the category of lightweight concrete for insulation and strength, according to the Regulation on the production of lightweight concrete C155-2013. According to the same regulation, this type of concrete can be used in the construction of closing elements (reinforced or unreinforced), brick replacement elements, wall panels for buildings for civil, agro-zootechnical or industrial purposes, partition walls, parapets, filling of the floors from prefabricated beams or for the realization of the monolithic over-concreting. Concretes that are part of the Lightweight Insulation Concrete category can be used in making screeds or thermal insulation plasters. In this research, a lightly insulating concrete composition was also obtained, by totally replacing the mineral aggregates with aggregates from sunflower stalks.

REFERENCES

C155. (2013). *Normativ Privind Producerea Betoanelor Ușoare (Regulation on the Production of Lightweight Concrete); ASRO—Asociatia de standardizare din Romania*. Romanian Standardization Association.

Ali, N., Jaffar, A., Anwer, M., Alwi, S. K. K., Anjum, M. N., Ali, N., Raja, M. R., Hussain, A., & Ming, X. (2015). The Greenhouse Gas Emissions Produced by Cement Production and Its Impact on Environment: A Review of Global Cement Processing. *International Journal of Research*, *2*(2), 488. http://internationaljournalofresearch.org

Amziane, S., & Sonebi, M. (2016). Overview on Biobased Building Material made with plant aggregate. *RILEM Technical Letters*, *1*, 31–38. doi:10.21809/rilemtechlett.2016.9

Badur, S., & Chaudhary, R. (2008). Utilization of hazardous wastes and byproducts as a green concrete material through S/S process: A review. *Reviews on Advanced Materials Science*, *17*, 42–61.

Błaszczyński, T., & Król, M. (2015). Usage of green concrete technology in civil engineering. *Procedia Engineering*, *122*, 296–301. doi:10.1016/j.proeng.2015.10.039

Chabannes, M., Nozahic, V., & Amziane, S. (2015). Design and multi-physical properties of a new insulating concrete using sunflower stem aggregates and eco-friendly binders. *Mater. Struct. Constr.*, *48*(6), 1815–1829. doi:10.161711527-014-0276-9

Danielsen, S. W., & Kuznetsova, E. (2016). *Resource management and a Best Available Concept for aggregate sustainability*. doi:10.1144/SP416.7

EN 12390-3:2019 Testing Hardened Concrete Part. 3: Compressive Strength of Test. Specimens; *European Committee for Standardization*: Brussels, Belgium, 2019.

EN 12390-5:2019 Testing Hardened Concrete Part. 3: Flexural Tensile Strength of Test. Specimens; *European Committee for Standardization*: Brussels, Belgium, 2019.

EN 12390-6:2010 Testing Hardened Concrete; Part 6: Split Tensile Strength of Test Specimens; *European Committee for Standardization*: Brussels, Belgium, 2010.

EN 12390-7:2019 Testing Hardened Concrete Part. 7: Density of Hardened Concrete; European Committee for Standardization: Brussels, Belgium, 2019.

EN 13412:2006 Products and Systems for the Protection and Repair of Concrete Structures. Test Methods. Determination of Modulus of Elasticity in Compression; *European Committee for Standardization*: Brussels, Belgium, 2006.

EN 197-1:2011. Cement. Composition, Specifications and Conformity Criteria for Common Cements; *European Committee for Standardization*: Brussels, Belgium, 2011.

Farias, R. D., García, C. M., Cotes Palomino, T., Andreola, F., Lancellotti, I., & Barbieri, L. (2017). Valorization of agro-industrial wastes in lightweight aggregates for agronomic use: Preliminary study. *Environmental Engineering and Management Journal, 16*(8), 1691–1699. doi:10.30638/eemj.2017.184

Glazer, B., Graber, C., Roose, C., Syrett, P., & Youssef, C. (2011). *Fly ash in concrete*. The Perkins+Will Team Publisher.

GlobalABC/IEA/UNEP (Global Alliance for Buildings and Construction, International Energy Agency, the United Nations Environment Programme). (2020). *GlobalABC Roadmap for Buildings and Construction: Towards a zero-emission, efficient and resilient buildings and construction sector*. IEA. https://www.iea.org/reports/globalabc-roadmap-for-buildings-and-construction-2020-2050

Harbulakova, V. O., Estokova, A., Luptakova, A., & Korenova, S. (2014). Study of durability of fibrous cement based materials exposed to microorganisms. *The 9th International Conference "Environmental Engineering"*. http://enviro.vgtu.lt

Huntzinger, D. N., Eatmon, T. D. (2009). A life-cycle assessment of Portland cement manufacturing: Comparing the traditional process with alternative technologies. *Journal of Cleaner Production, 17*, 668–675. ro.2008.04.007 doi:10.1016/j.jclep

Ignjatović, I., Sas, Z., Dragaš, J., Somlai, J., & Kovács, T. (2016). Radiological and material characterization of high volume fly ash concrete. *Journal of Environmental Radioactivity, 168*, 38–45. doi:10.1016/j.jenvrad.2016.06.021 PMID:27400654

Jianqiang, W., Siwei, M., & D'Shawn, G. T. (2016). Correlation between hydration of cement and durability of natural fiber-reinforced cement composites. *Corrosion Science, 106*, 1–15. doi:10.1016/j.corsci.2016.01.020

Langer, W. H., Drew, L. J., & Sachs, J. S. (2004). Aggregate and the environment. American Geological Institute Environmental Awareness Series. American Geological Institute.

Mati-Baouche, N., De Baynast, H., Lebert, A., Sun, S., Lopez-Mingo, C. J. S., Leclaire, P., & Michaud, P. (2014). Mechanical, thermal and acoustical characterizations of an insulating bio-based composite made from sunflower stalks particles and chitosan. *Industrial Crops and Products, 58*, 244–250. doi:10.1016/j.indcrop.2014.04.022

NE 012/1-2007. Normativ Pentru Producerea Betonului Și Executarea Lucrarilor Din Beton, Beton Armat Și Beton Precomprimat Partea 1: Producerea Betonului (Regulations for the Production of Concrete and the Execution of Concrete, Reinforced Concrete and Prestressed Concrete); ASRO—Asociatia de standardizare din Romania. Romanian Standardization Association, 2007.

Nozahic, V., Amziane, S., Torrent, G., Saïdi, K., & De Baynast, H. (2012). Design of green concrete made of plant-derived aggregates and a pumice–lime binder. *Cement and Concrete Composites, 34*(2), 231–241. doi:10.1016/j.cemconcomp.2011.09.002

Sijakova-Ivanova, T., Panov, Z., Blazev, K., & Zajkova-Paneva, V. (2011). Investigation of fly ash heavy metals content and physico chemical properties from thermal power plant, Republic of Macedonia. *International Journal of Engineering Science and Technology, 3*(12).

SR 3518:2009. Încercări Pe Betoane. Determinarea Rezistenţei La Îngheţ-Dezgheţ Prin Masurarea Variaţiei Rezistenţei La Compresiune¸ Si/Sau Modulului de Elasticitate Dinamic Relativ (Tests Hardened Concrete. Determination of Freeze-Thaw Resistance by Measuring the Variation of the ComPressive Strength and/or the Relatively Dynamic Modulus of Elasticity); *ASRO - Asociatia de standardizare din Romania (Romanian Standardization Association)*: Bucharest, Romania, 2009.

Sua-iam, G., & Makul, N. (2015). Rheological and mechanical properties of cement–fly ash self-consolidating concrete incorporating high volumes of alumina-based material as fine aggregate. *Construction & Building Materials, 95*, 736–747. doi:10.1016/j.conbuildmat.2015.07.180

USDA (United States Department of Agriculture). (2021). *World Agricultural Production*. Circular Series WAP 10-21 2021, 38. Available at https://apps.fas.usda.gov/psdonline/circulars/production.pdf

Walczak, P., Szymański, P., & Różycka, A. (2015). Autoclaved Aerated Concrete based on fly ash in density 350 kg/m^3 as an environmentally friendly material for energy-efficient constructions. *Procedia Engineering, 122*, 39–46. doi:10.1016/j.proeng.2015.10.005

KEY TERMS AND DEFINITIONS

Agro-Waste: It consists in general of lignocellulosic biomass and it is produced as a consequence of different agricultural processes; it includes farm waste, harvest waste and others.

Ecological Concrete: Concrete that contains unconventional raw materials that are eco-friendly or its production involves sustainable fuels or other processes with a reduced energy consumption than the usual; in general, the production of ecological concrete assumes smaller greenhouse gas emissions than the normal concrete.

Green Building Materials: Building materials that imply in their production, placing or maintenance, processes with low environmental impact.

Lightweight Concrete: Concrete with a density between 800 and 2000 kg/m^3, made entirely or partially with light aggregates.

Mechanical Properties: Physical characteristics exhibited by a material when one or multiple forces are applied upon it.

Natural Fibers: Fibers that are the result of geological processes or they are parts of plants or animals.

Sunflower Stalk: That part of the sunflower plant grown above the soil till under the hat with the seeds; it does not include the leaves.

Thermal Conductivity: The degree by which a material let the heat pass through it.

Vegetal Aggregates: Raw material obtained by the processing of different plants that are included in the concrete as partial or total replacement of the sand and/or gravel from its composition.

Chapter 4
Advanced, Sustainable, and Economic Construction Under Green Technologies During COVID-19:
A Review of Construction Material During COVID-19

Shubha Khatri
Samrat Ashok Technological Institute, Vidisha, India

ABSTRACT

The authors would like to give their emphasis on advanced and economical green technologies due to COVID-19 pandemic situation. Due to coronavirus disease, not only the financial market but industrial and agricultural stock markets were affected. In all these references, the author created a positive side to go ahead with energetic stamina with a new name of "COVID: Create Opportunity Victory In Disaster." This chapter describes the most advanced sustainable and green building materials in construction (i.e., advanced green technology adopted in civil engineering). This is the time when we can do something new in the civil engineering area during this pandemic situation by using clean technologies for sustainable development in civil engineering construction and also can provide shelter for all who really feel helpless to construct their own homes in an economic way by using low-cost housing materials as a smart material. This chapter reveals advanced, economic, and sustainable green technologies that are being invented during the pandemic situation of COVID-19.

DOI: 10.4018/978-1-7998-9810-8.ch004

INTRODUCTION

All over the world COVID-19 is spreading its effect on human health as well as on sustainable growth of economies. Due to corona virus disease not only financial market but industrial and agricultural stock market are so much affected. Most essential material which play an important role for any kind of construction in civil engineering i.e. cement, sand, aggregate, reinforcing bar/steel, brick, stone, admixture etc. have been also affected due to low productivity and its slow popularization amongst the world. It's a well known fact that without these materials we cannot imagine a civil engineering structure. Due to pandemic corona virus disease civil engineering construction works has been slow down and in many places it is noticed that already stocked materials are just stopped to transfer at various parts of the world because of uncertainty of production of conventional materials in a vast manner. In this sense financial crises begun activated all over the world. Normal prices of a regular using material are so high up and even after constant COVID-19 situation we all are facing socio-economic problems in our day to day daily routine of life. In this book chapter author would like to give her emphasis on advanced and economical green technologies during COVID-19 pandemic situation to overcome financial problems for construction.

With reference to above situation if we can think a positive side to go ahead so we can give new name of COVID and it is "Create Opportunity Victory in Disaster". This new name is just an invention by Dr. Janardan Singh Chauhan (Ex. Director) Senior Professor of Department of Civil Engineering of SATI, Vidisha, M.P. to motivate itself and as per such kind of wonderful inspiration author decided to introduce green technologies as a partial or full replacement of conventional material by clean and green smart materials as an advanced and sustainable construction material in civil engineering construction industry. This is most important how we can use conventional materials in a socio – economic – eco friendly way without sacrifices in structural strength. In this regard various methodologies are being invented and used by Human Civilization which supports day to day activities which are more economical and environmental friendly for supporting construction work. However, new technologies are related to green technologies to make it more effective and environment friendly. Awareness about research and development increases when our requirements increase. In research areas of green management these technologies are characterized as green or clean technology. Green technologies involve: Unconventional energy efficient, recycling of waste materials, safety, health and socio-economic concerns, renewable energy resources, sustainable development with clean and green techniques and many more.

Transports being limited and even restricted among countries have slowed down global economic activities. Most importantly, consumers and firms have prevented their usual consumption patterns due to the creation of panic among them and created market abnormality, causing significant economic impact all over the globe affecting both advanced and emerging economies such as the United States, Italy and India. COVID-19 has collapsed the backbone of the financial market. To boost up the stock market proper policy measures must have to be adopted by the government. In India, it's more pandemic from economical point of view. So as per situation of world it's very clear that there is an only way and that is to be "self dependent". It's not a self centered system but it is about all happiness. To increase the demand of non-conventional materials from industrial areas, urban areas and rural areas, India's Prime Minister Mr. Narendra Modi started "Aatmnirbhar Bharat Abhiyan" during COVID-19 pandemic situation. Aaatmnirbhar Bharat is not an organization but it is a new policy to promote the local development by which national development as well as global development can take place. It's a kind of road map for next generation and it is directly related to the neat-clean-green technologies by using non-conventional/ wastage materials for development and management of smart infrastructures through recycling process. This is a new policy and focus would be on making agricultural sector a profitable business by construction of roads and infrastructure related work and utilization of each thing which is in the form agricultural waste and industrial waste material as well as locally available material in most economic way.

As per new name of COVID "Create Opportunity Victory in Disaster", it is considered that costly conventional construction materials for any type of structure can be partially replaced with agricultural waste or industrial waste materials or locally available materials for best infrastructure development in the field of civil engineering. For example Cement is very high demanding conventional construction materials and it is made up of several chemical components in which Silica (component) plays an important role with respect to imparting strength. We can partially replace cement with other materials like wheat husk or wheat husk ash/rice husk or rice husk ash/sugarcane baggase ash *(Nithyambigai, G. and Ash, F., 2015)*, Fly ash and Silica Fumes - industrial wastage materials *(Shariati, M., et.al. 2021)*. If we are facing global health emergency as well as global economic downturn so definitely we should go with green technologies and it is matter of create opportunity victory in disaster specially concentrated during COVID-19 pandemic situation.

Agricultural wastage materials from agricultural land actually dumped under the ground or in river so we can reutilize by recycling them for new construction. Wheat Husk or wheat husk ash/Rice Husk or rice husk ash/Sugarcane Baggase ash - agricultural wastage materials *(Munir, M. J., et.al., 2021)* are having high silica content which is equal to cement's silica content as per experimental research. These materials are green materials and eco friendly environment materials. Practically we should use environmentally friendly technology for building construction. We traditionally use earth bricks, concrete, and wood in construction. All mentioned above materials being used in every construction work. Similarly, the continue destruction of trees for timber and the mining of resources to produce cement for binding sand, gravel, and bricks, reduction in generation of cement has been point out somewhere in last 5-6 years. In this regard for a better world, there are new processes, and sustainable green building material alternatives that can be used in construction today. Eco-Friendly building material is a type of material that doesn't harm the environment, whether in its production, use or disposal and can easily be recycled *(Cobîrzan, N., et.al., 2022)*. Using Eco-Friendly materials is hugely beneficial in the long run. Building a green home reduces carbon emissions significantly and saves energy, which results in saving money on energy bills also. This chapter describes most advanced sustainable and green building materials in construction i.e. green advanced technology adopted in Civil Engineering Construction which are also called as clean green technologies for sustainable development in Civil Engineering.

Civil engineering construction is a big responsibility of a civil engineer and if neat-clean-green economical eco friendly environmental friendly technologies are available and they are adopted at construction site so it will be a golden side of a civil engineer. It's a great opportunity for being a good civil engineer if he is using low price renewable green energy efficient technologies to construct a building. India is an agricultural country and world's most responsible industry. A farmer grows crop but he didn't know that how to utilize its bi-product or wastage products in engineering way i.e. in the form of science and development. Now-a-days it's an essential activity for a civil engineer that how he has taken initiatives from old technology to new technology in civil engineering construction. He has to observe in broad sense all the parameters that can be promoted in construction technology in the form of neat and clean green technology. In the present scenario it has been observed that conventional materials like cement, sand and aggregates are partially or fully replaced by non-conventional materials like agricultural/industrial/local green materials for any type of construction work like road, dam, bridges, multistory buildings etc.

By keeping in view of new government policy of India i.e. Aatmnirbhar Bharat during COVID-19 pandemic situation and to reach the utmost economical way of civil engineering construction, this chapter describes various green materials which are useful in engineering science and practically adopted in sustainable construction and development in civil engineering in the area of research and development. Green tech–or green technology–is an umbrella term that describes the use of technology and science to reduce human impacts on the natural environment. Green technology encompasses a wide area of scientific research, including energy, atmospheric science, agriculture, material science, and hydrology.

Few of green materials for green technologies are:

1. Husk or husk ash is a wastage material such as wheat husk/wheat husk ash (WHA), rice husk/rice husk ash (RHA), sugarcane husk/sugarcane baggase ash (SBA). These all are from agricultural land and waste materials can be considered under green materials and environmental friendly materials. An experimental analysis has been carried out in Samrat Ashok Technological Institute's concrete technology lab to find out their physical and chemical compositions for comparison study with cement's physical and chemical composition. It is observed that few of agricultural waste materials like WHA & RHA are much useful because of similarities with cement's physical and chemical composition and wheat husk or wheat husk ash and rice hush or rice husk ash can be partially or fully replaced with cement. Rice husk ash (RHA) and wheat husk ash (WHA) are rice milling and wheat milling products. Their uses as a soil stabilizer are an alternative to the final disposition with environmental benefit. Because RHA and WHA are not self-cementitious, a hydraulic binder such a lime must be added to form cements to improve the soil strength. Researches on stabilization by applying RHA and WHA and lime combinations were conducted in sandy soils. Liquid limit tests were conducted on black cotton soils with different RHA and WHA with lime. Results show strength improvements for all RHA & WHA with lime contents.
2. M-Sand can be used in place of engineering sand or river sand or natural sand. Manufactured sand is an alternative for river sand. Due to fast growing construction industry, the demand for sand has increased tremendously, causing deficiency of suitable river sand in most part of the word. Due to the depletion of good quality river sand for the use of construction, the use of manufactured sand has been increased. Another reason for use of M-Sand is its availability and transportation cost. Since manufactured sand can be crushed from hard granite rocks, it can be readily available at the nearby place, reducing the cost of transportation from far-off river sand bed. Thus, the cost of construction can be controlled by the use of manufactured sand as an alternative material for construction. The other advantage of using M-Sand is, it can be dust free, the sizes of m-sand can be controlled easily so that it meets the required grading for the given construction.
3. Various industrial waste materials like quarry dust, glass powder, ceramic dust and coal dust are used as partial replacement of fine aggregate and assessed the strength parameters and compared the profit percentages after replacement with waste materials.
4. Waste plastic or crumb rubber can be used as a partial replacement of aggregates in concrete *(Bahoria, B., et.al., 2013)*.

With reference to the above mentioned points author has concentrated only on point no. 1 and point no. 2 i.e. agricultural waste materials like RHA & WHA and M-Sand respectively particularly in this chapter which deals with importance of utilization of economic, eco-friendly, sustainable green materials construction in civil engineering during COVID-19 period and it is further concentrated to use same materials even after COVID-19 situation also. In this chapter a literature survey has been carried out in specific areas which are related to point no. 1 and 2 i.e. agricultural waste materials like RHA & WHA and M-Sand/crushed sandy soils only. This chapter is just an overview of sustainable green construction materials such as waste agriculture materials and locally available crushes stone sand which is also called M-Sand to popularize the farming and to grow up about vocal for local i.e. to promote and to appreciate local available products.

Background

With the increasing demand of high quality infrastructure, the lack of conventional material in cities becomes the main problem. Due to this limitation in conventional material such as ordinary sand that is excavated from the river and harmful for the environment due to the reduction in ground water level, can be partially/fully replaced by M-Sand/Crushed stone rub down as an economic and ecological alternative for sustainable concrete structures. The use of M-Sand is a great effort towards sustainable enlargement in India. The life of a structure is affected by various parameters i.e. moisture content, free surface moisture (real time sample received in the lab), saturated surface dry (SSD), specific gravity, grain size analysis (to find out the sand zone), workability, water-cement ratio, impact value of aggregates etc. In this situation, while code provisions are also not available for M-Sand, a design parametric study was conceded out in the laboratory with respect to all necessary parameters with dissimilar grades of concrete and with different percentage of M-Sand which highly affect the design mix of concrete. In this article a study has been carried out for M-30 grade concrete including river or natural sand with fully and partially replacement of M-Sand/Crushed sand likewise 0.0%, 10%, 40%, 45%, 50%, 55% and 100%. An advanced design mix procedure has been initialized here by using British codes, ACI and IS. It is observed in the current project result analysis that most of the properties are similar of River/Natural Sand and M-Sand/Crushed Sand and it is also concluded that natural sand able to be partially or fully replaced by M-sand as a substitute sustainable construction material in Civil Engineering Construction.

Similarly, to assess the feasibility of utilizing the agricultural waste materials after harvesting of crops from the agricultural land like rice husk/ ash, wheat husk/ash, sugarcane baggase/ash, coconut/jute fibers for producing an economical concrete as well as bricks/pavor blocks by studying the properties like Compressive Strength, flexural strength, Water Absorption and Slump Retention, to develop a suitable technique for enhancement of quality control product in the form of agricultural waste material as a partial replacement of cements for concrete and bricks both after analyzing the chemical composition so that balance should be there in case of bearing capacity and strength are few factors creating big realization of the proposed chapter and it is really connected to the development of a practical system which will be more durable in all the sense of Civil Engineering Structures and will also be more economic. Utilization of waste materials for Ground Improvement Techniques specially for poor soil like Black Cotton Soils by improvement of index properties of soil as well as by improving OMC (optimum moisture content) and MDD (maximum dry density) to make transportation system more economical.

Due to COVID-19 Pandemic situation concentration was going on new advanced construction technology in Civil Engineering Construction in form of green technology to be adopted to provide good serviceability to the whole world. Few of recent green technologies are discussed below for sustainable development in Civil Engineering which is advanced and green technologies. Authors feel that before going to start a new work and before advertisement of product in the market a literature survey should be adopted for better understanding the surrounding things. It's very much true that surely this back ground will help us to go ahead. Authors are sharing few of activities with conventional materials like cement, sand because these are very costly materials now-a-days but important part of the structures. In this sense authors have been conducted research work in the laboratory to check that what type of materials can be used in place of cement and sand. It's also noted down that what impacts have been felt or observed during the research work. So here are few of these studies are of importance level.

Impact of Substitution of Crushed-M-Sandy Soil on Concreting under Green Technology in Civil Engineering Structural Development

Now-a-days in our country, the natural river sandy soil which hollow out from watercourse couch is used to make conservative concreting. Due to exhaustion of river soil source the ecological difficulty are induced and hence soil excavation is controlled by ministry governing body which created deficiency in good sand as well as significantly increases in its demanding prices also *(Boopathi, Y. and Doraikkannan, J., 2016; Jayaraman A., 2014)*. To utilize man-made sand has been taken as an attempt towards workable enlargement in our Country. It is helpful to get feasible explanation to the retreating availability of river sandy soil to make eco-balance *(Monika, N.R. and Sarankokila, S., 2015)*. Due to shortage of river sandy soil the only solution to this problem is the use of crushed sandy soil or crushed M-sandy soil *(Boopathi, Y. and Doraikkannan, J., 2016)*. The crushed M-sandy soil is developed by devastating the rocky structures *(Jayaraman, A., 2014)*.

The design mix criterion with crushed (M)-Sand is very complex as compare to natural sand. Till date, mix design in India is carried out with the help of different Indian codal provision and a Codal provision i.e. IS: 456-2000, IS: 10263-2009 and IS: 383-1970, in which at present still no provision of crushed-M-sand. M-Sand preserve as excellent fine aggregates in soaring strength concrete design mix with dissimilar grades of concrete. Research scholars in 2016 inspected that the impact of crushed M-Sandy soil in concreting by replacement river sandy soil and developed a highly performing concreting *(Meghashree, M., 2016)*. The replacement of 55% of fine aggregates by Crushed-M-Sandy soil included highly flexural stress, highly ultimate strength and high compress strength, *(Kankam, C.K., Meisuh, B.K., Sossou, G. and Buabin, T.K, 2017)*. Steady increase in strength when using crushed M-sandy soil up to 65%, so it might be used in concrete as feasible alternative to river sand, *(Meghashree M., 2016; Kankam, C.K., Meisuh, B.K., Sossou, G. and Buabin, T.K, 2017)*. Here Natural River sandy soil is replaced by man-made sandy soil with a range of 0% no replacement, 40%, 60% and 80% replacement and gained high strength at 65% *(Meghashree, M., 2016)*. It was observed that 2.7% and 4.5% increment in the concreting flexural stress by 55% and 100% M sand replacement compare to concrete with 0% substitution of M sand *(Jayaraman, A., 2014; Mani,K.U., Sathya,N. and Sakthivel,R., 2014)*. The properties of M sand is same as the River sand Therefore, man-made sand can be used as a replacement for natural sand which maintains the eco balance, *(Mundra, S., Sindhi, P.R., Chandwani, V., Nagar, R. and Agrawal, V., 2016)*.

The artifact of admixture (PE-GLYCOL-400) on ultimate strength, flexural stress and compression stress by changeable quantity of PE-GLYCOL-400 by volume of cement by 0.2%, 0.4%, 0.8% were considered for 40-50% substitution of crushed-M-Sandy soil on M-20, *(Meisuh, B.K., Kankam, C.K. and Buabin, T.K., 2018)*.

The motorized characteristics of concreting mix were experienced by complete replacement of natural sandy soil by crushed-M-sandy soil. Introduction of silica fumes at the progressive interval of 2.9%, the substitution of silica fume, recover the compression, ultimate and flexure strength *(Shanmugapriya T.R.N.U., 2012; Priyanka, D.K.K., Jadhav,A., 2013)*.

The concreting characteristics by utilization of crushed-M-sandy soil are roughly unique to the unadventurous concreting for dissimilar mix designing prepared for different concreting grading on the basis of British codes, IS and ACI *(Meghashree, M., 2016)*. For overall crushed-sandy soil substitution leveling, the strain-stress curves were similar. 8.8%. Lowest MoE concluded by crushed-M-sandy soil at 95%

natural river sand substitution level by crushed-M- sandy soil *(Subramani,T.and Ramesh,K.S., 2015; Jaishankar, P. and Saravana Raja Mohan,K., 2015).*

The compression strength increased by 50% replacement of natural sandy soil by Crushed-M-sandy soil with 5% to 10% replacement of wastage plastic materials, which results in decreasing the stress with additional increment to 10-40% replacement of wastage plastic materials and results in increment with the time *(Suresh,S. and Revathi,J., 2016).*

Impact of Rice Husk Ash-Wheat Husk Ash-Sugarcane Baggase Ash on Concrete and Soil Under Green Technology in Civil Engineering

Compressive strength and flexural strength Comparative studies with 15%, 20% and 50% replacement of cement by RH/WH/SBH OR RHA/WHA/SBHA (Sugarcane baggase husk ash) will be carried out for concrete cubes and Bricks *(Habeeb, G. A. and Mahmud, H. B., 2010 and Ghassan, H. and Hilmi, B.M., 2009).* Development of new structural properties with M-40, M-60, M-80 Concrete by using RH/WH/SB OR RHA/WHA/SBA as a partial replacement of cement and Development of new structural properties for high class Bricks by using above products by conducting experimental study of compressive and flexural strength (Structural Properties Of Rice Husk Ash Concrete with M-30 GRADE *(Godwin, A., Maurice, E.E., Akobo, I.Z.S. and Joseph, O.U., 2013).* Study on structural properties of admixture including RHA/WHA/SBHA in concrete mix design,Impact of Admixture and Rice Husk in Concrete Mix Design. Preliminary work has been done on "Enhancement of Index and Engineering Properties of Black Cotton Soil Using Rice Husk Ash and Wheat Husk Ash Including Lime" by Santosh Prajapati-M.Tech, Thesis of student under my guidance in Radharaman Engineering College, Bhopal (M.P.) in 2018 Experimental analysis on black cotton soils using admixtures in RH/WH/SBHA with 5% to 30% for improving index properties and engineering properties *(Santosh, P. and Shubha, K., 2018)*, Enhancement of Index and Engineering Properties of Black Cotton Soil Using Rice Husk Ash and Wheat Husk Ash Including Lime *(Prajapati,S. and Khatri,S., 2018).* Preliminary work has been done on "Impact of Admixture and Rice Husk in Concrete Mix Design in 2014 in Civil Engineering Department Lab, OPJIT College, Raigarh (C.G.) by students of final year for their major project work under my guidanceAn Experimental analysis was carried out on BCS by using Coconut coir fiber for stabilization of BCS under GIT,Improvement of Index and Engineering Properties of Black Cotton Soil By Using RHA and WHA Including Lime *(Khatri, S., 2014).*

FOCUSED AIMS OF THE CHAPTER

The aims of this chapter are as follows which highlighting Civil Engineering-Janpad Abhiyantriki in a broad form during COVID-19 in terms of green technology:

- Scrutinizing the characteristics of Crushed-M-sandy soil.
- Analyse of swot up the impact of artificial sandy soil on concreting characteristics.
- Development surroundings information on sandy soil aggregates and mix design.
- Analyse the alive concreting products which are formed by using man-made sandy soil.
- Define conclusions, giving future scope and indicate areas for further work.
- How to utilize the wastage material in field of civil construction era in terms of Scientific / Technological with respect to RHA/WHA/SBHA/COCONUT FIBER are partial replacement materials of cement and it's so useful i.e. it improves the compressive strength of concrete cubes as well as ultimate strength of cylinders.
- How It will provide more economic system in terms of Industrial / on regional economic growth especially for rural areas people because we know that now-a-days each and every construction material is so costly and construction demands are increasing day by day, so these are useful in regional economic growth of people especially in rural areas.

Justification of Chapter Objectives

- Innovation- Proposed research project will be a new emerging trend in the field of Civil Engineering construction era. There are so many innovative ideas are applied in the construction field now-a-days and construction demands are never ending and cost of construction materials is a big issue in this fast forward and luxurious life so agricultural wastage materials as a partial replacement of cement and mortar an economic, eco-friendly, socio-economic idea for civil engineers. This will give high strength to the concrete structures.
- Capacity building / development of research- This project can be used in capacity building for the growth and as a more accurate and durable cum economic system in rural and urban areas while it's a challenged for Civil Engineers to developed high strength concrete structures and high class grade bricks economically in India especially. This project may be a new beginning step in the direction of innovation in Civil Engineering field.

- Practicability of its applications- It can be used as advances in structural engineering field because by using RHA/WHA or any other wastage agricultural materials in cement and BCS. Compressive strength of concrete structures continuously increases. Its application is directly related to workability, durability, compatibility and early high strength of concrete structures and high class grade bricks with an economic and eco-friendly system.

PROPOSED METHODOLOGY UNDER GREEN TECHNOLOGY

It has been always noted down from the subject point of view that a good methodology should be adopted so that people who are really involved in research work are to be able to better understand the things. With reference to this above statement authors decided to explain each and every point which discussed in literature survey of this chapter to reach the objectives of this chapter and for further results point of view also. So here certain methodologies which have been adopted in form of laboratory/ field research works to satisfy the objectives of this chapter are described below.

The experimental analysis conducted out on M-Sandy Soil in concrete technology lab of Samrat Ashok Technological Institute, Vidisha (M.P.) India.

First of all authors itself conducted Grain Size Analysis shown in Figure 1 and Figure 2 for river sand and M-Sand, respectively. Mechanical properties of concrete with and without M-Sand are found suitable for M-30 grade concrete design mix.

As per IS-456-2000 codal provision regarding specifications for grades of concrete, selection of water content, cement content, percentage of volume of coarse or granular aggregates and fine aggregates *(Manikandhan, K. U., et.al. 2015)*, mix proportion for trial quantities mandatory for the mix design for each bag of cement, adjustment for moisture with respect to a) water b) authentic amount of very well aggregate necessary after allowing for mass of free wetness c) authentic amount of coarse aggregate requisite d) real quantities of dissimilar constituent required for one bag mix and finally actual mix proportions and per cubic meter quantities were calculated.

Therefore, compressive strength and ultimate tensile strength were calculated for 0% M-Sand and then partially replaced river/natural sand with Crushed-M-Sandy soil in diverse ratios i.e. 0.0%, 10.0%, 40.0%, 45.0%, 50.0%, 55.0% and 100.0% as fine aggregates in concrete design mix for M-30 grade of concrete.

Figure 1. Grain Size Distribution for River Sand

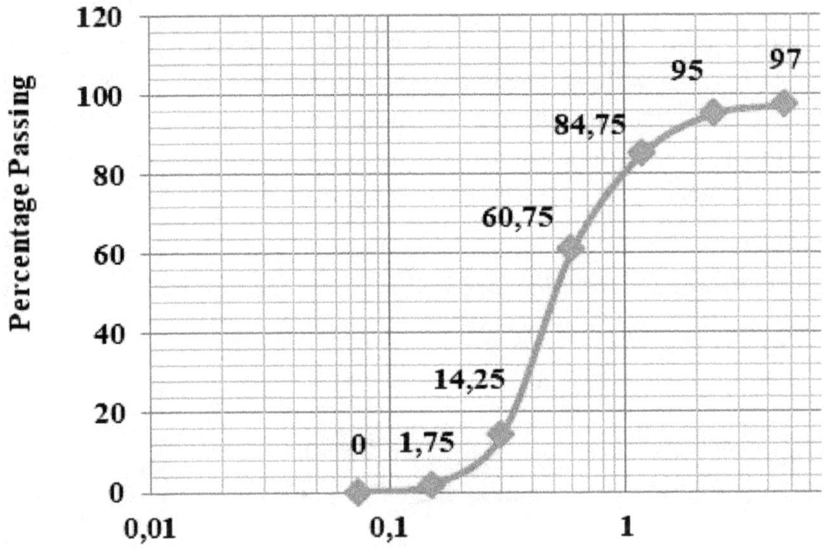

Figure 2. Grain Size Distribution for M-Sand

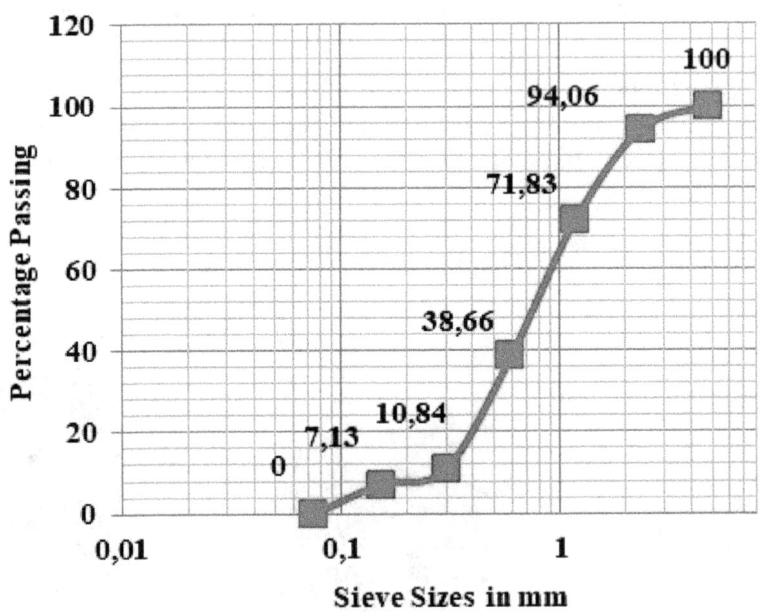

Test Method of Groundwork of Sampling

Batch mixes of ingredients- In this research work weigh batch mix process is done. Varieties of batch mix design technologies are available on construction ground. It's a truthful technique to measure the resources. The batch mix procedure was conducted as per the arranged conventional mix proportions. In Figure 3 we can see batching materials like cement, sand and gravel.

Figure 3. (a) Cement; (b) Fine aggregates river sand and M-Sand; (c) Coarse aggregates

Amalgamation of Supplies and Assemble of Samples

Assembling or compounding is such an imperative factors to improve the workability of concrete. Amalgamation equipment was used throughout this research work. From Figure 4 we can see combination of batching materials like cement, sand and gravel.

Figure 4. (a) Materials; (b) Slump Cone Test; (c) Moulds filled with concrete

Afterward the mix concreting was kept in a metallic plate and then workability test was done by Slump cone apparatus. Afterward the same concreting mix was poured inside the iron square shaped I.S. containers mold for ultimate tensile strength and compressive strength testing.

Afterward those containers or mold were kept on the vibrator machinery for obtaining the largest settlement of material in the mould in proper way. Figure 4 showing iron containers molds packed up with unmarked concreting mix and workability checker slump test. The specimens were reserved in clean water for 24 hours for the curing.

The experimental analysis conducted out on Agricultural wastage material like WHA, RHA including Lime in Concrete Technology Lab and Geotechnical Engineering Lab of Samrat Ashok Technological Institute, Vidisha (M.P.) India.

During COVID-19 it is came to know that low cost housing materials can be earned form agricultural farm houses. Authors have been studied the black cotton soil situation and it's noticed that pavement construction materials can be very economical and suitable by using ashes of wheat husk, rice husk etc. In India, one-fifth of our land area is covered by Black Cotton Soil which is also known as expansive soil. These soils are mostly found in arid and semi-arid regions. These soils are found to be highly problematic in constructional activities. It causes severe damages to the structure because of its alternate swelling and shrinkage nature. This happens due to alternate drying and wetting of soil. To avoid these circumstances, soil must be stabilized and strength is to be increased. In present investigation the type of solid wastage materials namely RICE HUSK ASH (RHA), WHEAT HUSK ASH (WHA) are selected to study the effects of the index and engineering characteristics of problematic soil LIKE BLACK COTTON SOILS. In order to utilize the rice husk ash and wheat husk ash for the improvement of problematic clay a detailed programmed studies have been formulated. In the longer term, LIME STABILIZATION provides performance benefits that reduce maintenance costs. In addition to stabilization of new materials, lime is an excellent choice for the reclamation of road bases. In this project work black cotton soil has been used for which soil stabilization has been done using Rice Husk Ash and Wheat Husk Ash with some percentage of lime content mixed in the soil to impart durability.

Various tests have been performed in laboratory on BCS to find and improve INDEX PROPERTIES OF SOIL MASS and tests which have been performed are: Liquid Limit, Plastic Limit, Shrinkage Limit, Grain Size Analysis, Specific Gravity, Moisture Content, Heavy compaction to find out optimum moisture content and maximum dry density.

Compressive strength and flexural strength Comparative studies were carried out with 15%, 20% and 50% replacement of cement by RHA/WHA for concrete cubes and Bricks related to "Study on Properties of Rice Husk Ash and Its Use as Cement Replacement Material". Development of new structural properties with M-40, M-60, M-80 Concrete by using RH/WH/SB OR RHA/WHA/SBA as a partial replacement of cement and Development of new structural properties for high class Bricks by using above products by conducting experimental study of compressive and flexural strength related to "Structural Properties of Rice Husk Ash Concrete with M-30 GRADE". Study on structural properties of admixture including RHA/WHA/SBHA in concrete mix design related to "Impact of Admixture and Rice Husk in Concrete Mix Design".

ANALYSIS AND RESULTS

Literature survey and methodology show the background study and step procedure respectively. After reaching on analysis level, authors divided analysis in two parts. There are two types of analysis which are correlated to research as described below.

Analysis with Sandy Soil Material

Compression Strength/Stress Test- Compression stress is the capability of a material to withstand against the compressive load which is dependent up on numerous factors i.e. ratio of water-cement, grading of concrete, quality of cement etc.

- Compression test is a standardized procedure which was carried out in iron molds of Indian standard sizes 150 mm x150 mm x150 mm. Then next step is curing of the cubes which were done at three, seven, fourteen and twenty eight days. Afterward that hard-dry square cubes were put up in to the motorized compression testing equipment and after that the loading condition was given until unless the failure of the samples. Figure 5 is showing Compression test on concreting mold cubes. Equation 1 is the formula for finding out compressive strength of concrete cubes.

$$Fc = Pu / Ac \tag{1}$$

where, Fc = Compressive stress [N/mm^2],
Pu = ultimate applied loading values to the cubes [kN],
Ac = concrete specimen cross sectional area [mm^2].

Figure 5. (a) Moulded cubes; (b) Testing on cubes; (c) Cracks developed in cube during compressive testing

(a) (b) (c)

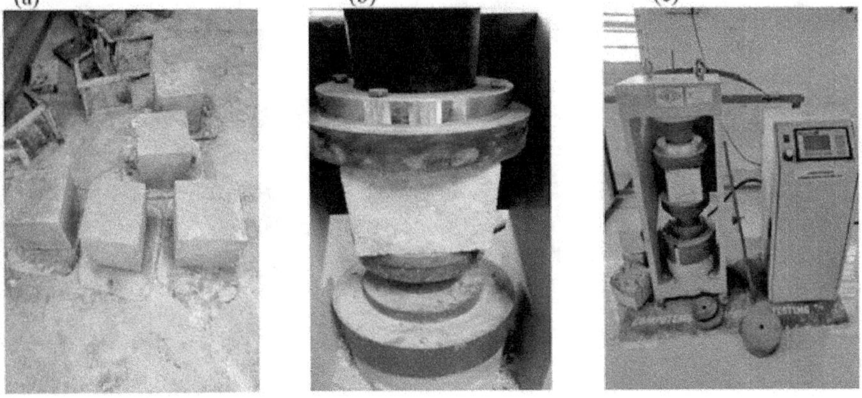

- **Ultimate Tension Stress- Rigidity of the concrete** is found by conducting ultimate tensile stress or strength test on universal testing machine.
    - In this test specimen of cylindrical size 150mm diameter and 350mm long has been used in the lab testing. Five samples were kept for testing and curing. Figure 6 is showing ultimate tension stress test on concreting cylindrical samples. Equation 2 is the formula for finding out ultimate stress of concrete cylinder specimens.

$$Fut = 2P / \Pi dl \qquad (2)$$

where, F_{ut} = Ultimate tensile stress on specimens [N/mm^2],
P = load application value [N],
l= length of cylinder specimen [mm],
d=diameter of cylinder specimen [mm].

To know the concreting characteristics Laboratory experiments were carried out in the lab. Results of ultimate tensile stress test and compression stress test tabulated and conclusions were made below and excel graphs are plotted to show the difference between increment and decrement of strengths.

Figure 6. (a) Molded cylinders; (b) Specimens; (c) Cracks developed in cylinders in Split Tensile testing

- **Concrete Design Mix Ratio:**

In present chapter authors described a general procedure for authentication work in which results were concluded by making a mix design ratio of ingredients or materials are taken as per IS 456:2000 i.e. 1:2.06:3.68 (cement: sand: aggregates) represent M-30.

- **Compressive Strength Results:**

The compression strength tests were carried out in laboratory on compression testing machine on square Indian standard size 150mm×150mm×150mm samples after proper curing.

Figure 7 showing the compression testing results got it from mix concrete with different quantities of Crushed-M-Sandy soil.

From Figure 8 it can be concluded that strength increases about 4.28% at 28 days for 55% replacement of crushed-M-Sandy soil.

When 100% replacement of crushed sand so evaluation is done with deviation of compression strength of with 65%, there is decrement in strength about 2.55%.

There is not much difference on the strength. So, it is concluded that 100% and 55% replacement of crushed-M-sandy soil are suitable with no harm for replacement of natural or river sandy soil in design mix of concreting for particular grade M-30.

Figure 7. Effect of % replacement of M-Sand including river sand on Compressive Strength of concrete cubes on 28 days

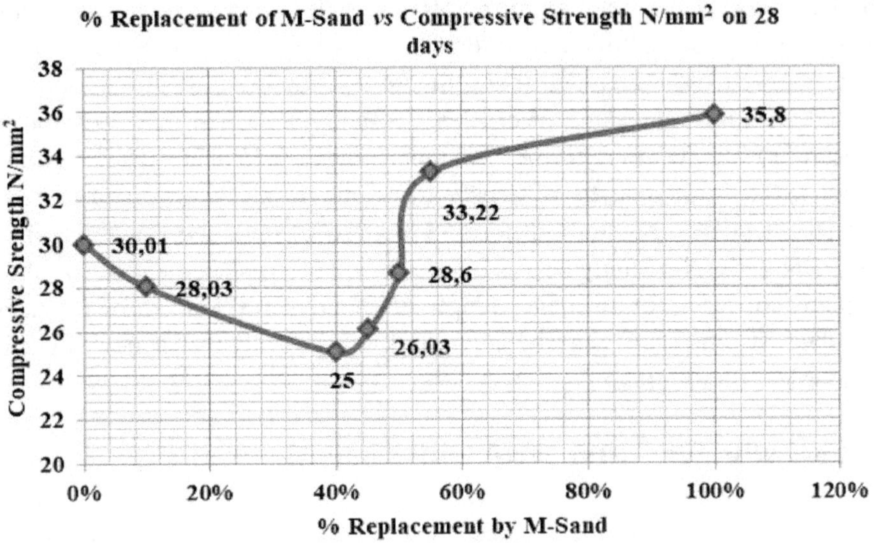

Figure 8. Effect of % replacement of M-Sand including river sand on Ultimate Split Tensile Strength of concrete Cylinders on 28 days

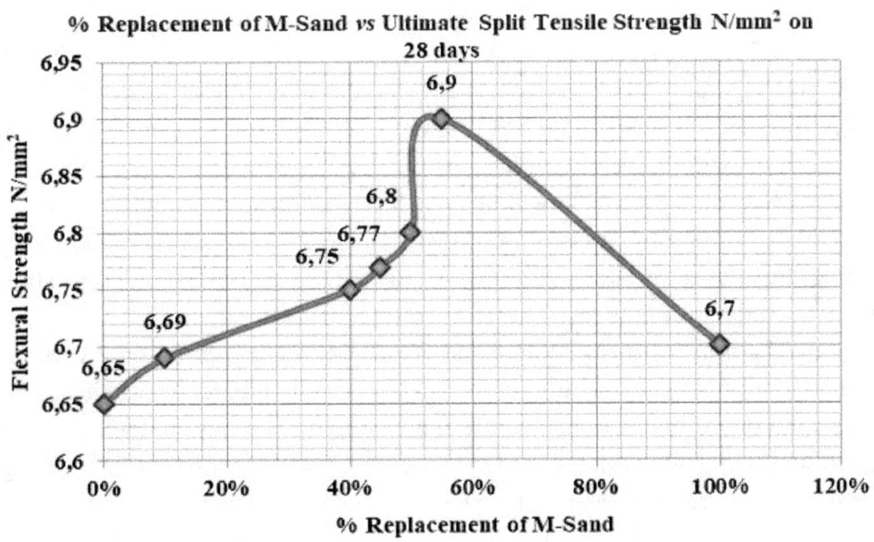

- **Ultimate/Split Tensile strength results:**

Indian standard size i.e. 150 mm× 350 mm × 70 mm cylindrical shape samples were used to find out Ultimate split/tensile strength on case-hardened cylinders. It can be observed and concluded that ultimate tension strength/stress continuously increased with the increasing of crushed-M-sandy soils. An increment of 4.55% is noticed in ultimate strength for 55% replacement in Crushed-M-Sandy soils as compare to 0.0% replacement of Crushed-M-Sand. It is also observed that there is an increment in flexural strength for 55% and 100% replacement of Crushed-M-Sandy soils with river sandy soil as compare to 0.0% replacement of Crushed-M-sand sandy soils. So, for ultimate strength of concrete 100% and 55% replacement of Crushed-M-sandy soils can be suggested in mix design of concreting mix for particular M-30 mix grade.

Analysis with RHA and WHA with Lime Material

Research was carried out by the authors in lab of SAMRAT ASHOK TECHNOLOGICAL INSTITUTE, Vidisha, M.P. INDIA during COVID-19 on agricultural waste materials like RHA-RICE HUSK ASH as well as on WHA-WHEAT HUSK ASH. The reason behind it that is absolutely match with soil index properties as of realistic natural soil like black cotton soil. As we know that not only in India even in abroad everywhere new areas of geotechnical engineering seems very much different i.e. UK, USA, JAPAN, CHINA, AUSTRALIA, so how often we can use RHA and WHA in partial replacement in concrete or in soil or in poor soil condition like cohesive soil? So many research studies have been conducted worldwide on this topic like agricultural waste materials which in dumped underground or in rivers. But, do we know how much it affects our ground water table/level? NO. We should do something creative in sense of neat and clean green technology field to reduce the environmental pollution. In this regard authors decided to show the parametric study in this area. After conducting parametric study it was observed that chemical composition of cement and WHA-RHA are approximately same so in future conventional material i.e. cement can be partially replaced with these wastage materials in concreting as demands are increasing day by day of cement and other conventional materials which are very costly. So to increase and to support green technologies is necessary even during COVID-19 we suggest that we should adopt these kinds of methodologies in daily construction activities in form of low cost housing.

To avoid exorbitant cost, the developers are forced to use low cost and speedy construction materials for major construction activities resulting in development of many innovative techniques in engineering. The various conventional techniques in practice include replacement of cement or aggregates with suitable materials which are social and environment eco-friendly. In the field of Civil Engineering there are so many construction challenges. It is a smart decision to build up mind for low cost construction materials with proper guidance or knowledge and efficiency skills with best construction technologies or methodologies. Low cost construction is based on three major factors and which are structural design, budget and cost reducing materials. Every year, worldwide approximately 600 million tons wheat are produced and in the mean time large amounts of agricultural waste known as "WHEAT STRAW" are produced by the de-husking/harvesting process of wheat. Wheat husk is an agricultural wastage material after harvesting process of wheat crops and mainly it is used for animals or paddy food purpose. Because of by-product of grains, husks are often burnt in the open air to produce energy and the ashes are transferred to landfills which cause significant environmental problems. Recovery of wheat husk brings economic benefits in the future as well as contributing to efforts of waste management by reducing environmental pollution.

WHEAT HUSK ASH (WHA): It is considered as potential and economical raw materials for preparing high-value products. Now-a-days sustainable infrastructure development is recommended as emerging trends and advanced methodology comprising of environmental, scientific, economic, social and political issues of global significance that doesn't compromise with needs.

RICE HUSK ASH (RHA): It is a carbon neutral green agent bi-product. So many ways are being observed for disposing them by making commercial use of this RHA. RHA is a good super-pozzolana material. It can be used in a large way for making special concrete mixes. There is a growing demand for fine amorphous silica in the production of special cement and concrete mixes, high performance concrete, high strength, low permeability concrete, for use in bridges, marine environments, nuclear power plants etc. This market is currently filled by silica fume or micro silica, being imported from Norway, China and also from Burma. Due to limited supply of silica fumes in India and the demand being high the price of silica fume has risen to as much as 500 millions / ton in India. From RHA 85% silica contents produced, in very small particle size of less than 30 microns – Silpozz for application in High Performance Concrete. To refine or to justify these statements that WHA and RHA are important replacement materials in concrete as well as in soil, few observations of cohesive soil and chemical parameters of RHA and WHA are shown in Table1, Table 2, Table 3 and Table 4.

As per above observation tables few results are here by the authors as from their own research work in the geotechnical engineering lab of SATI, Vidisha, M.P. INDIA.

Table 1. The composition of the rice husk: LAB TEST VALUES

Content, % wt.			
C	H	O	N
38.8-40.1	5.8-6.5	0.6-0.7	37.6-36.7

Table 2. The composition of the rice husk ash: LAB TEST VALUES

Content, % wt.						
SiO_2	Al_2O_3	Fe_2O_3	CaO	Na_2O	K_2O	Loss on Ignition
90.80	3.5	1.32	1.57	0.15	0.15	0.67

Table 3. Index Properties of Cohesive Soil (Vidisha, M.P., INDIA)

Soil Properties of Cohesive type Soil (without WHA & RHA)	LAB Test Results
Specific Gravity	2.5
Liquid Limit %	33.91%
Plastic Limit%	25%
Grain Size Analysis-% passing 75 micron sieve	99%
Optimum Moisture Content (OMC) %	12
Maximum Dry Density (MDD) g/cc	1.5
California Bearing Ratio (CBR)%	2.05

Table 4. Chemical Composition of Rice Husk Ash (RHA) & Wheat Husk Ash (WHA)

Chemical Parameters	RHA Composition Values % LAB TEST VALUES	WHA Composition Values % LAB TEST VALUES
Silica	90.80	90.24
Alumina	3.5	1.35
Ferric Oxide	1.32	4.47
Calcium Oxide	1.57	0.69
Sodium Oxide	0.15	0.13
Potassium Oxide	0.15	4.68
Loss on Ignition	0.67	1.03

Figure 9. Effects of rice husk ash content with 5-20% including 10% lime content on liquid limit

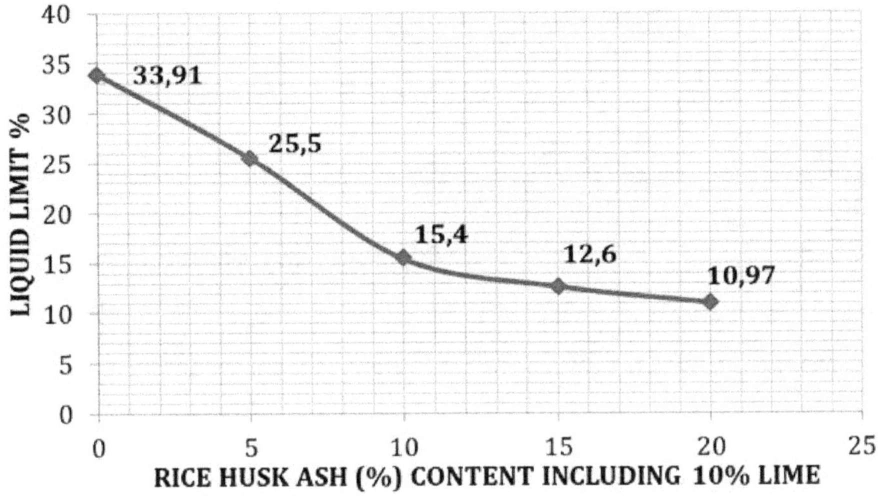

Figure 9 shows a Comparison Study With Different Percentage of RHA from 5-20% including 10% lime content mixed in cohesive Soils which shows the Impact on Liquid Limit. A series of experiments have been conducted in laboratory to compute mechanical properties of concrete like initial and final setting time of concrete, compaction factor test, workability test, rebound hammer test and compressive strength of reinforced cement concrete cubes including wheat straw ash and rice husk ash as a partial replacement materials of cement by varying percentage of WSA and RHA. By conducting this experiment and keeping in view of an R.C. pavement with low cost materials like wheat straw ash and rice husk ash it is noticed that this low cost materials will give higher strength because wheat straw ash and rice husk ash are having high silica contents like 90% to 95% respectively and silica content is much more beneficial for any types of cement for gaining higher strength.

Figure 10 shows a comparison study with different percentage of WHA from 5-20% including 10% lime content mixed in black cotton soils shows the impact on liquid limit of cohesive soil.

CONCLUSION

This entire chapter revels about an economical sustainable green technology type construction for all those people who are willing to construct their own shelters in this pandemic situation of COVID-19 even they are not very much stabilized even

Figure 10. Effects of Wheat Husk Ash with 5-20% including 10% Lime Content on Liquid Limit

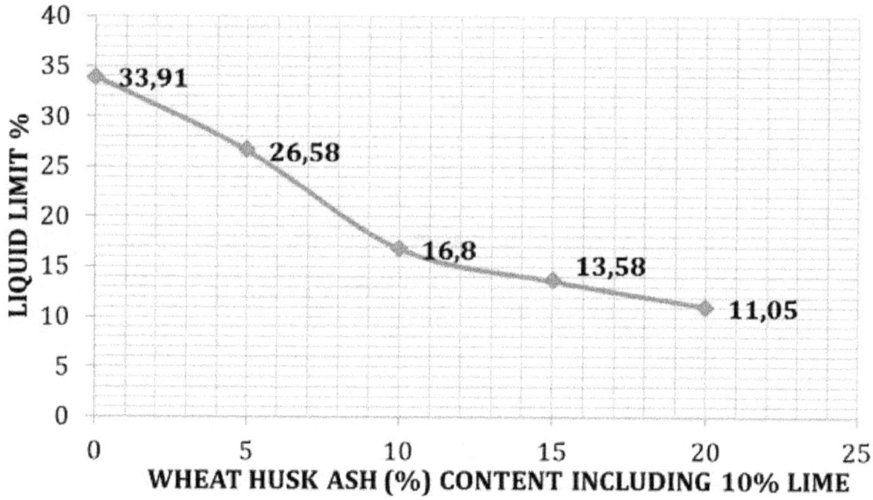

not able to buy completely conventional materials i.e. cement, sand and aggregates etc to construct a good structure. So, obviously each human being deserves a good life and first preference should be given to shelter/house, doesn't matter how much it is stylish, mainly matter is that it should have proper stability, strength, durability, sustainability, economical, smartly beautiful and most important thing low cost housing materials should be use such as agricultural waste materials or industrial waste materials. Authors of this chapter just tried research and development via their best of knowledge and gave contribution regarding new advanced green technologies because of experimental research work in the laboratory of Samrat Ashok Technological Institute, Vidisha, M.P.

It was really such a wonderful experience for us (authors) to do such kind of research work by developing advanced methodologies in neat and clean construction technology. Conclusions show the results summary and also required information of green technologies adopted particular in this chapter after conducting real life experimental studies by using agricultural waste materials as well as crushed stone sandy soil. The research works of this chapter in its final stage of conclusions are there as follows:

- From Figure 7 it can be concluded that strength increases about 4.28% at 28 days for 55% replacement of crushed-M-Sandy soil. When 100% replacement of crushed sand so evaluation is done with deviation of compression strength of with 65%, there is decrement in strength about 2.55%. There is not much difference on the strength. So, it is concluded that 100% and 55% replacement of crushed-M-sandy soil are suitable with no harm for replacement of natural or river sandy soil in design mix of concreting for particular grade M-30.

- From Figure 8, it can be observed and concluded that ultimate tension strength/stress continuously increased with the increasing of crushed-M-sandy soils. An increment of 4.55% is noticed in ultimate strength for 55% replacement in Crushed-M-Sandy soils as compare to 0.0% replacement of Crushed-M-Sand. It is also observed that there is an increment in flexural strength for 55% and 100% replacement of Crushed-M-Sandy soils with river sandy soil as compare to 0.0% replacement of Crushed-M-sand sandy soils. So, for ultimate strength of concrete 100% and 55% replacement of Crushed-M-sandy soils can be suggested in mix design of concreting mix for particular M-30 mix grade.
- From Figure 9 it can be observed that for only black cotton soil, liquid limit is found out as 33.91%. By increasing the rice husk ash content including 10% lime as an admixture liquid limit (LL) is found out to be decreased. When RHA is mixed 5%+10% lime so LL is decreased by 24.80% and when RHA is increased up to 20%+10%lime so LL value is fond out to be decreased up to 67.64% as compared to 5%+10% lime content.
- From Figure 10 it can be noticed that when in place of RHA, when WHA has been used from 5%-20% including 10% lime content so liquid limit is decreased but as compare to RHA its values are slightly increased because of heavy specific area and gravity of WHA as compare to RHA. When WHA is used only 5%+10% lime content so LL is found out to be decreased up to 26.58% from only black cotton soil LL which was 33.91% and difference is obtained as 21.61%. When WHA is increased from 5%WHA+10%lime to 20%WHA+10%lime content so liquid limit has been decreased continuously and the difference is found out to be 67.41%.

ACKNOWLEDGMENT

Authors are very much thankful to first of all GOD and our loving caring parents because while we were living in pandemic situation of fi-19 so their blessings were with us and their blessings are always with us. That's why today we are in this good situation that we are doing our work as it is as before. Thanks to those peoples also who helped us directly or indirectly during the entire research works and in chapter write-up work also.

REFERENCES

Akeke, G. A., Ephraim, M. E., Akobo, I. Z. S., & Ukpata, J. O. (2013). Structural properties of rice husk ash concrete. *International Journal of Engineering*, *3*(3), 8269.

Bahoria, B., Parbat, D. K., Nagarnaik, P. B., & Waghe, U. P. (2013, November). Effect of replacement of natural sand by quarry dust and waste plastic on compressive and split tensile strength of M20 concrete. *Proceedings of the International Conference on Engineering (NUiCONE 2013)*, 28-30).

Boopathi, Y., & Doraikkannan, J. (2016). Study on MSAND as a partial replacement of fine aggregate in concrete. *Int. J. Adv. Res. Trends Eng. Technol.*, *13*(1), 31–48.

Chauhan, J.S. (2019). Create Opportunity Victory In Disaster (COVID), SLOGAN-COVID-19, Samrat Ashok Technological Institute, Vidisha, M.P.

Cobîrzan, N., Muntean, R., Thalmaier, G., & Felseghi, R. A. (2022). Recycling of Mining Waste in the Production of Masonry Units. *Materials (Basel)*, *15*(2), 594. doi:10.3390/ma15020594 PMID:35057312

Habeeb, G. A., & Mahmud, H. B. (2010). Study on properties of rice husk ash and its use as cement replacement material. *Materials Research*, *13*(2), 185–190. doi:10.1590/S1516-14392010000200011

Jadhav, P. A., & Kulkarni, D. K. (2013). Effect of replacement of natural sand by manufactured sand on the properties of cement mortar. *International Journal of Civil and Structural Engineering*, *3*(3), 621.

Jaishankar, P. (2015). Experimental Investigation on Nano particles in High Performance Concrete. *International Journal of Chemtech Research*, *8*(4), 1666–1670.

Jayaraman, A. (2014). Experimental Study on Partial Replacement of Natural Sand with M-Sand and Cement with Lime Stone Powder. *International Journal of Chemtech Research*, *6*, 948.

Kankam, C. K., Meisuh, B. K., Sossou, G., & Buabin, T. K. (2017). Stress-strain characteristics of concrete containing quarry rock dust as partial replacement of sand. *Case Studies in Construction Materials*, *7*, 66-72.

Khatri, D. S. (2014). Impact of admixture and rice husk ash in concrete mix design. *IOSR Journal of Mechanical and Civil Engineering*, *11*(1), 13–17. doi:10.9790/1684-11141317

Khatri, S., & Jain, R. (2021). Improvement of Index and Engineering Properties Of Black Cotton soil by Using RHA & WHA including Lime. *44th Indian Social Science Congress (ISSC) in Collaboration with SATI Vidisha.*

Manikandhan, K. U., Kumar, N. S., & Sakthivel, R. (2015). Effect of replacement of river sand by M-sand in high strength concrete. *International Journal of Modern Trends in Engineering and Research*, *2*(02).

Meghashree, M., Gowda, A. K., & Roopa, K. (2016). Comparison of physical properties between natural sand and manufactured Sand. *International Journal for Innovative Research in Science and Technology*, *3*(7), 92–96.

Meisuh, B. K., Kankam, C. K., & Buabin, T. K. (2018). Effect of quarry rock dust on the flexural strength of concrete. *Case Studies in Construction Materials, 8*, 16-22.

Monika, N. R., & Sarankokila, S. (2015). Partial Replacement of M-Sand on Concrete by Self Curing and Standard Curing Method. *International Journal of Innovative Research in Science, Engineering and Technology*, *4*(12).

Mundra, S., Sindhi, P. R., Chandwani, V., Nagar, R., & Agrawal, V. (2016). Crushed rock sand–An economical and ecological alternative to natural sand to optimize concrete mix. *Perspectives in Science*, *8*, 345–347. doi:10.1016/j.pisc.2016.04.070

Munir, M. J., Kazmi, S. M. S., Gencel, O., Ahmad, M. R., & Chen, B. (2021). Synergistic effect of rice husk, glass and marble sludges on the engineering characteristics of eco-friendly bricks. *Journal of Building Engineering*, *42*, 102484. doi:10.1016/j.jobe.2021.102484

Nithyambigai, G., & Ash, F. (2015). Partial Replacement of Manufactured Sand and Fly Ash in Concrete. *International Journal of Emerging Technology and Advanced Engineering*, *5*(6), 166–170.

Santosh, P., & Shubha, K. (2018). Enhancement of index and engineering properties of black cotton soil using rice husk ash and wheat husk ash including lime. *International Journal for Research in Applied Science and Engineering Technology*, *6*(2), 1448–1455.

Shanmugapriya, T., & Uma, R. N. (2012). Optimization of partial replacement of M-sand by natural sand in high performance concrete with silica fume. *International Journal of Engineering Sciences and Emerging Technologies*, *2*(2), 73–80.

Shariati, M., Armaghani, D. J., Khandelwal, M., Zhou, J., & Khorami, M. (2021). Assessment of longstanding effects of fly ash and silica fume on the compressive strength of concrete using extreme learning machine and artificial neural network. *Journal of Advanced Engineering and Computation*, *5*(1), 50–74. doi:10.25073/jaec.202151.308

Subramani, A. K., Ramani, S. E., & Selvasembian, R. (2021). Understanding the microstructure, mineralogical and adsorption characteristics of guar gum blended soil as a liner material. *Environmental Monitoring and Assessment*, *193*(12), 1–17. doi:10.100710661-021-09644-4 PMID:34853926

Suresh, S., & Revathi, J. (2016). Effect of M-sand on setting time of high performance concrete. *Asian Journal of Research in Social Sciences and Humanities*, *6*(10), 1648–1654. doi:10.5958/2249-7315.2016.01118.7

ADDITIONAL READING

Das, S. K., Singh, S. K., Mishra, J., & Mustakim, S. M. (2020). Effect of rice Husk Ash and Silica Fume as strength-enhancing materials on properties of modern concrete—a comprehensive review. *Emerging Trends in Civil Engineering*, 253-266.

Eisenbach, P. (2017). *Processing of Slender Concrete Shells-Fabrication and Installation*. Kassel University Press GmbH.

Huseien, G. F., Khalid, N. H. A., & Mirza, J. (2022). *Nanotechnology for Smart Concrete*. CRC Press.

Hwang, C. L., & Chandra, S. (1996). The use of rice husk ash in concrete. In *Waste materials used in concrete manufacturing* (pp. 184–234). William Andrew Publishing. doi:10.1016/B978-081551393-3.50007-7

Luhar, S., Luhar, I., Abdullah, M. M. A. B., & Hussin, K. (2021). Challenges and prospective trends of various industrial and solid wastes incorporated with sustainable green concrete. In *Advances in Organic Farming* (pp. 223–240). Woodhead Publishing. doi:10.1016/B978-0-12-822358-1.00001-8

Mary, J. A., Gobinath, R., Shyamala, G., & Chary, K. R. (2020, December). Waste Products as an Alternative Construction Material-A Review. *IOP Conference Series. Materials Science and Engineering*, *981*(3), 032086. doi:10.1088/1757-899X/981/3/032086

Mokhtar, M. M., Morsy, M., Taha, N. A., & Ahmed, E. M. (2022). Investigating the mechanical performance of nano additives reinforced high-performance concrete. *Construction & Building Materials*, *320*, 125537. doi:10.1016/j.conbuildmat.2021.125537

Phonphuak, N., & Chindaprasirt, P. (2015). Types of waste, properties, and durability of pore-forming waste-based fired masonry bricks. *Eco-efficient masonry bricks and blocks*, 103-127.

Riach, G. K. (2022). 12 "Concrete fragments". *The Short Story in South Africa: Contemporary Trends and Perspectives*, 217.

Tripathi, D., Kumar, R., Mehta, P. K., & Singh, A. (2021). A Sustainable Concrete with Manufactured Sand in Different Aggressive Environments. In *Recent Advances in Structural Engineering* (pp. 3–10). Springer. doi:10.1007/978-981-33-6389-2_1

Wu, C., Li, J., & Su, Y. D. (2018). *Development of ultra-high performance concrete against blasts: from materials to structures*. Woodhead Publishing.

KEY TERMS AND DEFINITIONS:

Civil Engineering: To be at the forefront in the field of engineering by contributing towards the service and development of the society.

Clean Technology: To understand the impact of professional engineering solutions with technical skills and social values to develop quality infrastructure for the development of the mankind.

COVID: Create Opportunity Victory In Disaster.

Economical Construction: Appropriate consideration for the public health and safety and environmental consideration including prediction with an understanding of the limitations.

Green Technology: To apply appropriate techniques, resources, and modern smart materials relevant to the professional engineering practices for sustainable development.

M-Sand: Crushed stone sand as manmade materials as a smart material in civil engineering construction for economical and pollution free sustainable development.

Sustainable Development: To contribute towards service and development of the mankind through quality education and research in the area of science, technology, and management.

Chapter 5
Reliability and Sustainability of Water Transport Systems

Ana Diana Ancas
Technical University of Iasi, Romania

Florin Emilian Turcanu
Technical University of Iasi, Romania

Marina Verdes
Technical University of Iasi, Romania

Valeriu Sebastian Hudisteanu
Technical University of Iasi, Romania

Nelu Cristian Chereches
Technical University of Iasi, Romania

ABSTRACT

The development of populated centers has as a consequence the corresponding increase of the consumption of resources, first of all of the consumption of energy necessary for the transport of the additional flows of water. Their often-limited character materializes the desire to redistribute existing resources in time and space, in accordance with the requirements of the uses served. The widespread introduction, due to competitive prices, of composite pipes, but for which there is insufficient information on their behavior over time, has led to the need to research the sustainability and reliability of new systems in development for water transport. This chapter aims to provide an efficient tool for those interested in managing water supply systems with direct reference to the reliability and sustainability of water transport systems: supply and distribution networks.

DOI: 10.4018/978-1-7998-9810-8.ch005

INTRODUCTION

Civil Engineering is a very wide field that includes not only buildings of any kind, but also the transport infrastructure and the installation infrastructure. Groundwater transport systems can be considered as true "arteries" or vital lines for urban settlements and for these reasons are considered critical infrastructure *(Ancas, A.D., & Atanasiu, G.M.,2011)*. Sustainable development in civil engineering also requires the sustainability of water transport systems.

The socio-economic development of urban and rural localities is decisively conditioned by the water supply. Taken from surface or deep sources, the raw water is subjected to appropriate treatment processes to meet the quality requirements, transported and distributed to consumers and, after use, returned to the natural environment.

This closed circuit presupposes the existence of complex water supply and sewerage systems, capable of ensuring, on the one hand, the qualitative and quantitative requirements of the users, and on the other hand the measures imposed for the protection of the receiving environment.

From a technical and economic point of view, an important share in these systems belongs to the works for water transport. A water transport system consists of the supply pipes and the distribution network.

The complexity of the factors that contribute to the overall efficiency of a water transport and storage system, with a certain degree of operational safety, raises a number of technical and economic issues: choice of material, delimitation of economic operation, determination of economic diameter, staging making the adduction, etc. *(Ancaş, A. D., et.al., 2021)*

Water is transported from tanks or pumping stations to consumer connections via the distribution system. The distribution network must ensure the maximum hourly flow and the minimum service pressures, in order to satisfy all the uses served - household and public consumption, industrial and consumption for fighting fires.

The pipe, as the main element of a water transport network, must have sufficient strength and / or rigidity to perform its intended function in good condition.

Background

The term *strength* is the ability of the pipe to withstand the stresses that occur in the wall material. These are caused by the *actions* exerted on it such as the internal pressure of the transported fluid, the action of the soil around the pipe, the mobile actions caused by the surface traffic, differential settlements etc *(Ungureanu, N. & Vrabie, M., 2004)*.

Any direct or indirect cause capable of producing destructive effects on building elements or structures shall be deemed to be action.

The term *stiffness* refers to the ability of the pipe's constituent material to resist movement. *Bending stiffness* is the product of the modulus of elasticity of the material and the abrasion resistance (*Ungureanu, N. & Vrabie, M., 2004*).

In addition to all these requirements of strength, rigidity and stability, which must be met by the materials and components of a water transport system, the system as a whole must be **reliable** and **durable**.

The term *reliability* refers to components, appliances, installations, equipment, which present safety in operation, under given conditions and for a determined period of time. Reliability is therefore the property of being reliable, associated with components, appliances, installations and equipment. In a more compact expression, reliability is safety in operation *(Catuneanu, M.V., 1990).*

According to the definition of the term sustainable, it means something that lasts a long or long time, lasting, resilient, viable. Durability is the quality of what is durable, durability, strength, viability. Another acceptance of durability refers to the period during which a good can be used.

Sustainability also refers to the system's ability or ability to cope with environmental effects over time. Notions such as design life, corrosion resistance and abrasion resistance are durability factors.

In other words, *sustainability* refers to the ability or ability of the pipeline to cope with environmental effects over time.

The notions of durability and reliability, associated with a water transport system, are interdependent and related to the life of the system *(Cioc, D., Anton, A., 2001).*

The need for and timeliness of reliability and sustainability studies in the water sector also stems from the assumption of environmental protection responsibilities *(UE 2000/60/CE).*

RELIABILITY OF WATER TRANSPORT SYSTEMS

General Notions.Reliability Criteria

The reliability of a distribution system is its ability to provide consumers with the necessary flows, at predetermined parameters, under normal operating conditions and over a period of time *(Manescu, A., 2010).*

The characteristic feature of distribution networks is that they behave as long-acting systems.

Distribution systems are developing simultaneously with the development of an urban settlement, expanding and rehabilitating - the usual pipes are replaced with new ones, whenever the degree of wear requires it. The calculation time taken to determine the reliability of the system must be chosen in such a way that, during this period, no reconstruction work is required, which would alter the initial reliability.

For example, for a developing locality, water distribution systems must have different reliability indicators, in different successive stages of their realization. In the first period, a branched scheme can be adopted, for which the reliability corresponding to a short duration can be considered satisfactory. In the next stages of development it is necessary to introduce backup solutions, ringing the network or doubling the elements, which will increase the reliability for a longer period.

At a new stage of system expansion, its reliability is to be specified by calculations that take into account the introduction of a new reservation, the rehabilitation and modernization works performed in the system, as well as the modification of the reliability parameter of the components.

The substantiation of the level of reliability can be done by technical-economic calculations, by comparative evaluation of the damages caused by the system failures and of the additional expenses, performed for the increase of the level of reliability *(Stoica, S, 2007).*

In principle, distribution systems must operate continuously and without defects. As a result, the level of reliability must be adopted as high as possible, which requires the structural reservation of the distribution scheme.

In case of failure of the elements, the system goes into a hydraulic failure mode, in which it is accepted that, for a short period of time, the distributed flow is lower than the calculation.

The permissible limit value of the flow distributed in the event of a failure determines the capacity reserve of the system.

In the theory of reliability of technical systems, the **basic notion** for assessing the working capacity of the element or system is the **value of the probability of operation without defects**, over a period of time t, the execution and commissioning of the system, as well as factors that influence the state of operation of the element in the operation process. The action of random factors cannot be determined precisely and, therefore, the value of the time of operation of the element is adopted according to a probabilistic law *(Haghighi, A., & Samani, S.M.V. 2011).*

The basic criterion for the reliability of non-renewable systems is the probability $P(t)$ of malfunctioning in the time interval t. The reliability of composite technical systems is assessed by a probability indicator, which more fully characterizes the system in terms of output function, namely the performance indicator. It is adopted as the mathematical hope of the characteristic of the quality of operation. These indicators determine the necessary reservation structure of the system.

In the case of hydraulic failure regimes of backup systems, when the operation of a certain element is interrupted, the cutting capacity is reduced and the supply of consumers in unfavorable positions may become unsatisfactory. Such a situation is a malfunction of the system.

In order to avoid the occurrence of such system failures in operation, the network calculation is performed for different hydraulic loading regimes, corresponding to the failure of the most important elements in the system.

In the case of hydraulic damage checks, a minimum permissible flow rate shall be established for each consumer. This limit flow is the second determining indicator for the reliability calculation.

The minimum flow rate is established according to the particularities of all consumers and determines the transport capacity required in the event of a breakdown - the capacity reserve of the system.

The basic possibilities for booking the system are: the use of ring-type networks, the doubling of the network on some main sections and the increase of the accumulation capacity for damages.

A judicious organization of the network can essentially improve the quality of operation of composite technical systems.

There are two basic ways to increase the reliability of distribution systems:

a.A first way is to increase the reliability and quality of the system components. In this sense, pipes, fittings and equipment made of high quality materials will be used, more attention will be paid to the execution and preparation of parts (pipes, fasteners) and, of course, the aim will be to raise the quality of construction-assembly works, by increasing the requirement for quality of execution.

In order to increase reliability, priority should be given to measures relating to the preparation, execution, reception and commissioning of system elements.

When the technical possibilities of raising the quality of the system elements are exhausted, or when the raising of the quality becomes uneconomical, the second way, that of the reservation, is resorted to.

b. The second way, reservation, is required when the reliability of the system is required to be superior to the reliability of the components. The reliability of the elements and systems is assessed using the reliability-time characteristics.

The probabilistic character of random quantities is expressed by the distribution law, which determines the correspondence between the possible quantities and the corresponding probabilities. The distribution law of the random variable ξ, is expressed more fully by the distribution function:

$$F\xi(x) = P(\xi<x)$$

The distribution function of the random variable is precisely the probability that the variable ξ takes values less than x.

The distribution function $F\xi(x)$ is an increasing function, with the argument x.

$$x_2 > x_1 \rightarrow F\xi(x_2) \geq F\xi(x_1)$$

At the value $-\infty$, the function becomes null, and for $+\infty$ it takes a unit value.

$$F\xi(-\infty)=0$$

$$F\xi(+\infty)=1$$

For discrete random values, the distribution function is discontinuous and varies in steps. Each step of the function corresponds to the possible value of the random quantity and has a height equal to the probability of these values. The sum of all values of the function is equal to unity. For continuous random variables, the distribution function is continuous. The distribution function completely characterizes the probabilistic random size. An example of a continuous distribution function can be the period of operation of an irreparable element, until failure. The average operating time, which characterizes the quality level of the element, is calculated as the average v**alue of the operating times** of the elements of the same type until failure - service life. But, by its random nature, the lifespan of each concrete element will be different from the average value. This deviation expresses precisely the law of distribution.

It is possible that in the first working period, the probability of defects is low and the value of the distribution function increases slowly. The higher the slope of the curve, the smaller the difference in quality indicators regarding the manufacture of the elements. In the final part, the curve can have a very small slope, as the probability of long operation of the element above the average value is lower. The function $F_\tau(t)$ characterizes the reliability of the element, as it represents the probability of failure up to a given moment t.

$$F_\tau(t) = P(\tau < t)$$

The probability of trouble-free operation over time t is determined by the expression:

$$P_\tau(t) = 1 - F_\tau(t) = P(\tau > 1)$$

This probability is called the probability function. It is a basic criterion of reliability, characterizing the trouble-free operation of the element (system) over a given period of time. The trouble-free operation of composite technical systems, which also includes water distribution systems, is characterized by a more complex probability indicator, namely the qualitative indicator of operation.

Often the law of distribution of a continuous random quantity is expressed in the form of the distribution density, $f(t)$, which is related to $F(x)$ by the relations:

$$f(x) = F'(x)$$

$$F(x) = \int_{-\infty}^{x} f(x)\,dx$$

The average lifetime value (trouble-free operation) of the element is determined as the mathematical expectation M of the random quantity τ:

$$T_0 = M\tau = \int_0^{\infty} tf(t)\,dt \text{ and } T_0 = \int_0^{\infty} P(t)\,dt$$

Reliability is therefore characterized by durability - that is, the maintenance of the system's operating capacity up to the limit value, with or without permissible interruptions, in order to ensure the necessary technical assistance and repairs.

It is found that water distribution systems behave as long-term systems, as long as the prospective operating time is not fixed. They exist and develop, as long as the locality exists or as long as elements that require another distribution system will not contribute.

As a result, durability cannot be used to characterize the reliability of distribution systems.

Durability is determined by the reliability of the system: pipes, fittings, equipment, etc. The service life of the system elements is chosen so as to avoid the aging effect. This results from the need for the safe operation of the system, but in the composition of the distribution systems there are also elements subject to aging phenomena.

The durability of the assembly is determined based on the service life of the element with the lowest durability. The issue of sustainability does not influence the choice of system distribution scheme. Durability determines the service life between major repairs.

Water distribution systems are systems that can be repaired. Therefore, they are also characterized by the property of repairability, which consists in adapting the system to the prevention, location and removal of defects, based on continuous monitoring and maintenance. A basic indicator of system repairability is the repair time of the defective element τ_p.

In order to assess the reliability of the system, it is necessary, first of all, to specify the notion of defect of the element and to analyze the physical and probabilistic conditions of occurrence of defects.

In general, the failure of the element represents the cessation of the operating capacity, for the restoration of which it is necessary to repair by removing it from the system. But, not every malfunction of the elements leads to the interruption of the system. If its decommissioning occurs during the reduced demand period (minimum consumption), then this operation has no essential consequences on the supply of consumers. As a result, the failure of an element means an untimely defect, which involves its decommissioning. Such a defect will cause the system to malfunction and lead to material and functional damage.

The causes of failures, related to the loss of durability of the elements, are random actions of overloading the weaknesses of the elements and therefore, the nature of the defects has a probabilistic character.

The robustness of the element is determined by the values of independent random quantities, each with its own distribution law. Often these distribution laws are of the normal type, because the value of the size depends on a series of random factors, with similar effects. For example, the reduction of the strength of the weld bead can be caused by insufficient penetration, failure to ensure welding times and technological pressures or other causes that depend, in turn, on the qualification of the welder, the quality of electrodes, welding conditions, compliance with technology, etc.

Another peculiarity of the defects is that they fall into the category of rare events.

These properties characterize not only defects related to loss of durability, but also all other types of defects. If the loss of the operating capacity of the element is not a random phenomenon, it can be predicted and taken into account in the calculations.

One of the most important characteristics of the reliability of the elements is the intensity of the defects (λ), defined as the probability that an element that works without defects a time t, refuses to operate at the next moment.

From probability theory it follows that the probability of the simultaneous occurrence of two events or the probability of their product is equal to the product of the probability of one of the events with the conditional probability of the other event.

$$P(AB) = P(A).P({}^{B}/_{A})$$

Event A will be considered as representing the defect-free operation of the element in the interval $(0,t)$.Its probability will be:

$$P(AB) = P(0,t)$$

For event B, we consider the operation without defects in the interval (t,t_1), provided that it has operated without defects in the interval $(0,t)$, so provided that event A is performed. The probability will be:

$$P({}^{B}/_{A}) = P(t,t_1)$$

Probability of occurring both events:

$$P(AB) = P(A).P({}^{B}/_{A})$$

$$P(0,t_1) = P(0,t).P(t,t_1)$$

$$P(t_1) = P(t).P(t,t_1)$$

$$P(0,t_1) = P(t_1) = P(0,t).P(t,t_1) = P(t).P(t,t_1)$$

It follows that the probability of the element functioning flawlessly in the range (t,t_1), provided that it has functioned normally in the range $(0,t)$, is:

$$P(t,t_1) = {}^{P(t_1)}\!/_{P(t)}$$

The probability of failure in the interval (t,t_1), is determined with the expression:

$$F(t,t_1) = 1 - .P(t,t_1) = \frac{P(t) - P(t_1)}{P(t)} = \frac{P(t) - P(t+\Delta t)}{P(t)}$$

$$F(t,t+dt) = -\frac{dP(t)}{P(t)} = -\frac{P'(t)}{P(t)}dt = \lambda(t)dt$$

Therefore, the probability of failure in a very small interval, dt, in the situation where until this elementary moment it functioned normally, will be $\lambda(t)dt$, where:

$$\lambda(t) = -\frac{P'(t)}{P(t)} \approx \frac{P(t) - P(t_1 t+\Delta t)}{\Delta t.P(t)} \approx \frac{\dfrac{n(t) - n(t+\Delta t)}{N}}{\Delta t \dfrac{n(t)}{N}} = \frac{\Delta n}{\Delta t.n(t)}$$

where Δn represents the number of defects that occur in the range Δt.

The size $\lambda(t)$, determined experimentally, expresses the specific frequency of failure. The experimental results show that the function $\lambda(t)$, for most elements in the systems, has three characteristic periods.

The first period is the running-in period, in which those elements that had hidden defects fail. This period is characterized by a high intensity of defects, an intensity that decreases significantly after the moment T_{τ}, remaining constant.

The second area is the period of normal operation and represents the most important working regime and is characterized by a constant intensity of failures.

After this period, determined by the operation, the defects begin to influence wear and aging, and the elementary enters the third stage, the aging stage. In this area, the intensity of faults increases.

Such a time dependence λ is appropriate for irreparable parts. The water distribution elements are repairable. The analyzed dependency is correct for the considered system until the first failures of the component elements. All elements of the distribution systems, before commissioning, are subjected to tests and adjustments. During this period, all defects are detected and removed. Therefore, the period $(0,T_{\tau})$ can be neglected for distribution systems.

From the point of view of the dangers that appear at the failures, in the calculation of the water distribution systems, the intensity of the failures can be considered constant.

For constant λ, the probability of malfunction of the element up to time t is determined as follows:

$$\lambda dt = -\frac{dP(t)}{P(t)}$$

Integrating the above equation results:

$$\lambda t = \ln P(t)$$

at $t=0$, $P(t)=1$.

It follows that the probability of trouble-free operation for time t will be:

$$P(t) = e^{-\lambda t}$$

In this way, we can consider that the function of the reliability of the elements of the distribution systems has a law of exponential variation.

The probability of failure of the element during t (unreliability) is:

$$F'(t) = f(t) = \lambda e^{-\lambda t}$$

The characteristics considered refer to irreplaceable elements and are used to assess the reliability of repairable components.

In the case of locking fittings, faults fall into two categories:

- A first category, determined by the impermanent phenomena of overpressures and / or underpressures, generated by the sudden maneuvering of the closing elements, which disturb the normal operation by damaging the pipe or fittings.
- The second category refers to the situation in which there are losses due to leaks in the fittings.

Not in all cases, which influence the reliability of the systems, it is necessary to remove the element from the network in order to fix it. Simple situations are solved by keeping the network running. They do not affect distribution to consumers and are therefore not considered defective.

If it is necessary to disassemble the element from the network in order to remedy the element, then such a situation constitutes a failure of the element. Defects in items in unreserved systems lead to system failure. It depends on the degree of reservation. Therefore, the situations encountered in distribution networks must be divided into two groups:

- Situations that lead to the failure of the element and require its disassembly for repair: welded joints, cases of penetrating corrosion with dimensions over 5 mm, cracks, flange opening, damage due to mechanical causes;
- Simple, non-defective situations that can be remedied without reducing the mains pressure and without disconnecting the pipe section: cases of non-penetrating corrosion, small cracks, puncture situations, etc.

Reliability Indicators for Distribution Systems

The state of the system at time t is determined by the state of its elements. If, as a whole, the elements will be in working order, then the system will be completely in good working order.

Complex technical systems are characterized by the fact that, in addition to the two extreme states: operation / defect, they may be in other decisive states, in which they will be in a partially good state of operation.

The transition of systems from one state to another is related to the change of the state of the component elements. The state of the system is determined by the vector $x(t)$, which is a mathematical model of system operation.

For a water distribution system:

$$\bar{X} = \begin{Vmatrix} x_1(t) \\ x_2(t) \\ \dots \\ x_n(t) \end{Vmatrix}$$

were:

$x_1(t)$ is the function describing the state of element i of the system;
$x_2(t)=1$ - if element i of the system is in working order;
$x_2(t)=0$ - if element i of the system is not in working order;
n - the number of elements that determine the reliability of the system.

The operation process of the system composed of two elements is analyzed. The status vector will be:

$$\bar{X} = \begin{Vmatrix} x_1(t) \\ x_2(t) \end{Vmatrix}$$

Four states of the system are possible, determined by the various states of the operating and defective elements:

$$\overline{X_0} = \begin{Vmatrix} 1 \\ 1 \end{Vmatrix} \quad \overline{X_1} = \begin{Vmatrix} 1 \\ 0 \end{Vmatrix} \quad \overline{X_2} = \begin{Vmatrix} 0 \\ 1 \end{Vmatrix} \quad \overline{X_2} = \begin{Vmatrix} 0 \\ 0 \end{Vmatrix}$$

State 0 corresponds to the functional elements.

The complete or partial failure of the system can be assessed only if its structure is known.

There are two characteristic structures of the systems: series structure and parallel structure.

The series assembly of the elements, from the point of view of reliability, refers to the fact that the defect of any element causes the failure of the system.

For trouble-free operation at time t, it is necessary for each element to work without defect in time period t. So, for independent elements, the probability of trouble-free operation will be given by the product of the probabilities of each element:

$$P(t) = P_1(t), P_2(t), \ldots, P_n(t)$$

The probability of operation without defects is given by an exponential law:

$$P(t) = e^{-\omega_1 t} e^{-\omega_2 t} \ldots\ldots e^{-\omega_n t}$$

were:

$$\omega_0 = \omega_0 + \omega_1 + \ldots + \omega_n$$

The flawless operation of the system is also expressed by an exponential law.

In case of parallel connection of the elements, the failure of the system is determined by the failure of all elements. Therefore, for the characterization of the good functioning they are considered sterile (0), (1), (2). System failure corresponds to condition (3).

If the elements taken independently do not influence the operation of the system, then only for state (0) the system is in working order, and in tailings (1), (2) the corresponding indicators will be lower.

When analyzing the probability of failure of repairable systems, consisting of elements in parallel, it is found that the failure of the system is caused by the simultaneous failure of all elements, as the recovery time is much shorter than the working time. The simultaneity of two defects is an exceptional event, with a low probability. For this reason, the parallel coupling of the elements results in a high reliability of the distribution systems.

Simultaneous failure of two components in the network can occur by overlapping two situations: failure of one component during the repair period of the other.

If at some point t_i the elements fail, they are disassembled and the repair operations begin in the time period τ_p, after which the system re-enters the operation process.

We consider that the first item has failed. The probability of such an event is $\left(1 - e^{-\omega_1 t}\right)$.

Simultaneous failure of two elements is possible in the following two situations: either the second element failed during the repair of the other, or it failed before the first, in a time interval, from t_1 to t_2, for example.

In the first case, the start of the repair of the second element is in the repair period of the first element. The extreme case of such a situation corresponds to the coincidence between the beginning of the repair of the second element and the end of the repair of the first.

In the second situation, the first element fails with the second, and the end of its repair enters the period τ_i. The limit of the situation will be the moment t_1 when the end of the repair of the second element coincides with the beginning of the repair of the first.

If the defect of the second element occurs in the interval $\tau_1+\tau_2$ the probability will be: $\left(1-e^{-\omega_2(\tau_1+\tau_2)}\right)$.

The probability of simultaneous failure of the two elements is equal to the product of the probabilities of the two states:

$$\left(1-e^{-\omega_1 t}\right)\left(1-e^{-\omega_2(\tau_1+\tau_2)}\right) \approx \omega_1\omega_2\left(\tau_1+\tau_2\right)$$

For example, if t=10years, $\tau_1+\tau_2=10^{-3}$years and $\omega_1=\omega_2=5.10^3 l$/year, then

$$\omega_1\omega_2\left(\tau_1+\tau_2\right)t=\left(5.10^{-3}\right)^2.10^{-3}.10=25.10^{-6},$$

ie 5 times smaller than in the case of failure of a single element. Therefore, the simultaneous failure of two network elements in terms of reliability can be neglected.

The set of all the states of the system constitutes the phase plan of the states. When considering the effective operation of a complex technical system, for each condition the characteristic of the quality of operation is determined accordingly.

Noting this characteristic with $\Phi_x\left(t\right)$,the function will be written as:

$$\Phi_x\left(t\right)=\Phi\left[\bar{X}\left(t\right)\right]$$

The vector $\bar{X}\left(t\right)$ has a random variation and therefore the indicator of the quality of operation of the system is the mathematical expectation of the random function $\Phi_x\left(t\right)$ during t, ie the average of the set of realization of the process: $\Phi\left(t\right)=\Phi_x\left(t\right)$.

In order to be able to calculate the system reliability indicators, it is necessary to specify the concept of system failure.

Failure of the system will be considered as the situation in which at least one consumer is disconnected from the network and is not powered.

On unreserved systems, failure of any item will result in failure of the system.

For reservation systems, system failure occurs only for the following types of faults:

- Failure of the part of the network to which a consumer is connected by means of separation devices;
- Failure of separation devices;
- Simultaneous failure of two network elements.

Only the first two cases will be taken into account in the calculation, as the third has a low probability of occurrence.

The reliability indicator of the distribution system, $R_{sist}(t)$ will be determined as the ratio between the indicator of the quality of operation of the real system $\Phi(t)$ and that of the operation of the ideal system $\Phi_0(t)$.

The ideal system will always be in good working order $\overline{x_0}(t)$, and all components of the vector $\bar{X}(t)$ are equal to the unit:

$$R_{sist}(t) = \Phi(t) / \Phi_0(t)$$

The characteristic of the quality of operation is determined in relation to the function of the system. The main function of distribution systems is to permanently supply consumers in accordance with their requirements, or on the basis of predetermined supply schedules. Therefore, according to the quality characteristic of the operation of the system, the necessary hourly flow for consumers must also be adopted.

Each state of the distribution system $\bar{X}(t)$ will be compared with the hourly consumption of the system $\Phi\bar{X}(t)$.This consumption will depend only on the state of the system and determines the model of assessment of the performance of the system function.

In the case of certain states of the system $\overline{X}(t)$ a certain number of consumers remain without power, which causes the quality of operation indicator to decrease.

If the calculated consumption is considered for the system in good working order, Q_0 and the flow reduction corresponding to the affected consumers in state $\bar{X}(t)$, ΔQ_x then:

$$\Phi_x(t) = Q_0 - \Delta Q_x$$

The quantity ΔQ_x is determined without hydraulic calculations, directly from the diagram corresponding to the considered emergency condition.

The transitions of the system from the good operating state to the defective one are characterized by the parameter of the failure frequency of the sections, $\omega_1, \omega_2, \omega_3$.

From the fault state, after repair, the system returns to the good operating state with the return frequency μ_1, μ_2, μ_3. The probability of the system passing from one state to another, during *dt*, will, consequently, be equal to $\omega_i dt, \mu_i dt$.

Note on the state graph, under each arrow, the frequency parameters of the elements (ω_i or μ_i). It can be considered that the current state of the system does not depend on its past and then the transitions from one state to another will be Markov type events. The intermediate probabilities corresponding to the transition from state j to state ; k(j[1]k) will be:

$$P_{jk} = \omega_j dt \quad P_{jk} = \mu_i dt \quad P_{jk} = Y_{jk} dt$$

The probability that the system remains in its state, without having any transition to another state during *dt*, is determined as the probability of the event resulting from the sum of all possible transitions from this state to others, ie:

$$\mathrm{P}_{ij} = 1 - \sum_{j=k} Y_{jk} dt$$

So it turns out:

$$\mathrm{P}_{ij} = \begin{cases} Y_{jk} dt, & j \neq k \\ 1 - \sum_{j=k} Y_{jk} dt & j = k, \Sigma P_{jk} = 1 \end{cases}$$

The probability will be made between the probability of the state of the system at time $t^1 dt$ and the probability of the state at time *t*, as well as the transient probabilities.

At time $t+dt$ the system can reach state *k* with the probability $P_k(t)$, and after the duration *dt* no transition to another state can occur.

The probability will therefore be:

$$\mathrm{P}_k(t)\left(1 - \sum_{k \neq j} Y_{kj} dt\right)$$

The quantity $\left(1 - \sum_{k \neq j} Y_{kj} dt\right)$ represents the conditional probability of not leaving state *k* during *dt*, provided that the system was in state *k*.

The second path corresponding to the situation when at time t the system is in state j^1k and after time dt the transition to state k took place. The probability of such an event will be:

$$\sum_{k\neq j} P_j(t) Y_{jk} dt$$

Then the quantity $Y_{jk}dt$ represents the conditional probability of passing from state j to state k, provided that the system has been in state j. You can write the relation for the probability $P_k(t+dt)$ in the form:

$$P_k(t+dt) = P_k(t)(1-\sum_{k\neq j} Y_{kj}dt) + \sum_{j\neq k} P_j Y_{jk} dt$$

We transform the relationship by derivation:

$$\frac{dP_k(t)}{dt} = P_k(t)\sum_{k\neq j} Y_{kj}dt + \sum_{j\neq k} P_j(t) Y_{jk} \quad and\ k, j = 0,1,2.....$$

So the relationship can be written for all system states.

To solve the system of differential equations, it is necessary to express the initial conditions in the form of the probability $P_j(0)$ of the state of the system at time t=0. Let, for example, be the equations for the analyzed system. The k states of the system will be (0), (1), (2), (3). So:

$$\begin{cases} P_0^{'}(t) = -P_0(t)(\omega_1+\omega_2+\omega_3) + P_1(t)\mu_1 + P_2(t)\mu_2 + P_3(t)\mu_3 \\ P_1^{'}(t) = -P_1(t)\mu_1 + P_0(t)\omega_1 \\ P_2^{'}(t) = -P_2(t)\mu_2 + P_0(t)\omega_2 \\ P_3^{'}(t) = -P_1(t)\mu_3 + P_0(t)\omega_3 \end{cases}$$

Considering that by failure the system no longer fulfills its role (for irreparable system), it results μ_i=0. The equations will then become:

$$\begin{cases} P_0^{'}(t) = -P_0(t)(\omega_1 + \omega_2 + \omega_3) \\ P_1^{'}(t) = P_0(t)\omega_1 \\ P_2^{'}(t) = P_0(t)\omega_2 \\ P_3^{'}(t) = P_0(t)\omega_3 \end{cases}$$

The initial conditions are considered: t=0, $P_0(0)$=1, $P_2(0)$, $P_3(0)$=0.

Integrate the first equation:

$$\frac{dP_0(t)}{dt} = -\Sigma\omega_i P_o(t)$$

$$\int \frac{dP_0(t)}{P_o(t)} = -\int \Sigma\omega_i(t)dt$$

or:

$nP_o(t) = -\Sigma\omega_i T + C$

for: t=0, $P_0(t)$=1, and C=0,the solution will be:

$$P_0(t) = e^{-\Sigma\omega_i(t)}$$

Solve the second equation using the solution of the first equation:

$$\frac{dP_1(t)}{dt} = -\omega_1 P_o(t) \rightarrow dP_1(t) = \omega_1 e^{-\Sigma\omega_1(t)}dt$$

or:

$$P_1(t) = \omega_1 \int e^{-\Sigma\omega_i(t)}dt + C = \omega_1 \frac{e^{-\Sigma\omega_i(t)}}{\text{In}\left(e^{-\Sigma\omega_i(t)}\right)} + C$$

$$P_1(t) = \frac{\omega_1}{\Sigma\omega_i} e^{-\Sigma\omega_i(t)} + C$$

for $t=0, P_1(0)=0 \rightarrow C=\frac{\omega_1}{\Sigma\omega_i}$

The solution for the second equation is:

$$P_1=\frac{\omega_1}{\Sigma\omega_i}\left(1-e^{-\Sigma\omega_i(t)}\right)$$

The third and fourth equations of the system are solved similarly. In this way, the general solution of the system of equations will be:

$$P_0(t)=e^{-\Sigma\omega_i(t)},\quad P_1(t)=\frac{\omega_1}{\Sigma\omega_i}\left(1-e^{-\Sigma\omega_i(t)}\right)$$

$$P_2(t)=\frac{\omega_2}{\Sigma\omega_i}\left(1-e^{-\Sigma\omega_i(t)}\right),\quad P_3(t)=\frac{\omega_3}{\Sigma\omega_i}\left(1-e^{-\Sigma\omega_i(t)}\right)$$

Since the system can be in one of the tailings (0), (1), (2), (3), the sum of the probabilities that the system is in these states is equal to the unit, so å$P_j(t)$=1.

The mathematical expectation of the quality characteristic for the operation of the system is determined by the formula:

$$Q(t)=\overline{Q_x}(t)=Q_0e^{-\Sigma\omega_i(t)}+\sum_{j=1}^{l}Q_j\frac{\omega_1}{\Sigma\omega_i}(1-e^{-\Sigma\omega_it})$$

Using the relation $Q_i=Q_0-\Delta Q_j$, the previous formula becomes:

$$Q(t)=Q_0-\sum_{j=1}^{l}Q_j\frac{\omega_1}{\Sigma\omega_i}(1-e^{-\Sigma\omega_it})$$

Accordingly, the formula for determining reliability will be:

$$R_{sist}(t)=1-\sum_{j=1}^{l}\frac{\Delta Q_j\omega_i}{Q_0\Sigma\omega_i}(1-e^{-\Sigma\omega_it})$$

For larger isolated nodes, it may be necessary to calculate the reliability of the power supply. The reliability indicator of the node is obtained from the reliability indicator of the system, by considering the following particularizations: only those fault situations are analyzed that lead to the interruption of the node supply and therefore the number of such situations will be equal to the number of elements whose failure determines emergency situation.

Since for the disconnected node $Q_j=0$, the term of the relation $\sum_{j=1}^{l} Q_j \frac{\omega_1}{\sum \omega_i}(1-e^{-\sum \omega_i t})$

tends to zero and the indicator of reliability of the node will be: $R_{nod}\left(t\right) = e^{-\sum_{1}^{l}\omega_i t}$) where 1 is the number of elements whose failure causes the interruption in the supply of the node.

Therefore, the node reliability indicator represents the probability of the node being supplied without defect. The size $R_{sist}(t)$ completely characterizes both the reliability of systems with reservation and those without reservation. But for reservation systems such an indicator is not enough. As indicated above, specific to distribution systems is the fact that during breakdowns, when repairing the disassembled element, it is possible to reduce the quality of the system, expressed by limiting the flow distributed to consumers Q_{lim}, which is lower than the calculation value . The Q_{lim} size is determined by the character of the consumer.

The obtained reliability indicator, Q_{lim}, determines the fault situation for the reservation systems. If the water flow in the event of a Q_{av} fault is less than Q_{lim}, then the system fails. The size of the determined indicator indicates the capacity reserve of the reservation system. This is determined by calculating the distribution frequency in the system, in the case of very intense hydraulic regimes.

Under these conditions, the reliability of reservation systems is determined using:

- The reliability indicator $R_{sist}(t)$, which determines the structure of the system, and
- The determined indicator Q_{lim}, which evaluates the reserve regarding the transport capacity of the pipelines.

CALCULATION OF THE RELIABILITY OS DISTRIBUTION NETWORKS WITHOUT RESERVATION

The main problem for calculating the reliability of branched networks is to determine the reliability indicator for the considered scheme and the indicator that expresses the need to reserve the network.

The failure of any element of the branched scheme causes the failure of the system, but with another importance of the defect. So, for the same arrangement of the nodes, the various schemes of their connection through pipes will be distinguished by appropriate reliability.

Also, the different redistribution of flows in defective systems (ie different sectioning of the system) is characterized by a different reliability.

Depending on these elements, there is the problem of optimizing the branched scheme of the distribution networks and the sectioning mode that allows the design of networks with high values of the reliability indicator. If this indicator is lower than the required level, then the network reservation is required. The calculation of the reliability of the annular networks is done in two stages: the necessary structure reserve is established, then the transport capacity reserve.

Calculation Algorithm for Network Reliability without Reservation

We have in mind a branched scheme, for which the reliability indicator must be determined. Consumers are connected to the ends of the terminal nodes of the branches. The length of the terminal branch can be 0. All nodes are numbered. The first node is considered the power point of the network. A valve is mounted in front of each consumer, bearing the number of the node to which the consumer is attached. The sectioning valves are provided at the entrance to the sections and, therefore, bear their number.

The calculation of the reliability indicator for a branched scheme of a distribution network, supposes the knowledge of the following data:

- The value of the element failure frequencies: $\omega_1(^1/_{km.year})$ for the pipe and $\omega_2(^1/_{year})$ for the valves;
- Duration considered, t(years);
- Network topology given by the incidence matrix (Wind Matrix) MV.

The Wind Matrix has on the first line the origin of the sections, and on the third line the information on the presence of sectioning valves: 1 - if the valve is mounted on the section and 0 otherwise.

The lengths of the sections L(m) are inscribed in the matrix of the lengths ML (I). The numbering of the sections is in accordance with the one in the wind matrix and with the number of sections in the network.

The flows transited on sections are entered in the MQ (I) matrix.

The flow on the branched sections is equal to that of the final node and therefore, the consumption concentrated in these nodes is recorded in the matrix.

For the other sections, the value 0 is entered at the beginning, which will change with the consumptions resulting from the calculation.

The flow on the branched sections is equal to that of the final node and therefore, the consumption concentrated in these nodes is recorded in the matrix.

For the other sections, the value 0 is entered at the beginning, which will change with the consumptions resulting from the calculation.

In carrying out the calculations, the first step is to set up equivalent network areas, adapted for the calculation of reliability. Network sections equipped with special fittings or valves will be logically grouped, if the effect of a certain element determines the need to disconnect the same consumers. These elements constitute an equivalent area, which is delimited by the sectioning valves. The area will include: distribution pipes, fittings and valves in front of consumers. Disconnecting valves enclosing the fluid flow zone and the disconnecting valve starting from the equivalent zone shall not be included in the zone. For repair or replacement of any element in the equivalent zone, it is necessary to close all sectioning valves delimiting the zone. The number of equivalent zones results depending on the number and arrangement of the sectioning valves.

The failure frequency of zone ωJ is determined as the sum of the failure frequencies of all elements within that zone, i.e ω_J=åω_J.

The flow rate ΔQ_J which is reduced to the failure of the zone is equal to the consumption afferent to the end section Q_C of the equivalent zone, i.e $\Delta Q_J = Q_C$.

The delimitation always starts with the end valve of the first portion included in the area. Only sections not equipped with section valves and all branches starting from them are included in the area. Any section included in the area that has a valve in front of the consumer is considered the end. Following the formation of the area, the lengths of the sections and the number of valves included in the area are added together. Consequently, for a delimited area we will have:

$$\sum^{nj} L_k; \quad k_{zj}; \quad \Delta Q_j.$$

where:

- ΔQ_j – the flow of the entrance section in zone j;
- L_k - length of section k;
- n_j, K_{zj} - correspond to the number of linear sections, respectively the number of valves in area j.

Based on the characteristics obtained for the areas and the data considered, the reliability indicators can be calculated as follows:

- The contribution of each zone to the unreliable operation is determined:

$$\sigma_j = \left(\omega_a \sum_{k=1}^{nj} L_k + \omega_z k_{zj}\right) \Delta Q_j$$

- The contribution of all areas to the unreliable operation of the network shall be determined:

$$\sigma = \sum_{j=1}^{lj} \sigma_j$$

- Determine the sum of the failure frequency of all network elements:

$$\Omega = \sum_{j=1}^{lz} \left(\omega_a \sum_{k=1}^{nj} L_k + \omega_z k_{zj}\right)$$

- Determine the reliability indicator:

$$R_{cr}\left(t\right) = 1 \frac{1 - e^{-T\Omega}}{Q_0 \Omega} \sigma$$

where:

- Q_0 – normal network load;
- T - calculation period.

If the branched network is independent and does not represent a bar in an annular structure, then its reliability is assessed based on the result of the calculation of the indicators $R_{cr}(t)$. If the terminal network is a branch of the ring network, then it is replaced by a branch equivalent to ω_{rkJ} and Δ_{ekJ}.

These parameters are used to calculate the ring network. Therefore, the calculation scheme of the ring network is presented in the form of a network with flow rates concentrated in nodes and without branches. Each demand concentrated in nodes is characterized by an equivalent value of the failure frequency and by determined consumptions. If this is a simple branch with a valve then:

$$\omega_i = \omega_a L_{tr} + \omega_z$$

If the network is multiple branched, then the quantities $R_{cr}(t)$, the equivalent quantity ω_{eJ} and ΔQ_{eJ} .are considered. The quantity ω_{eJ} results as the sum of the values ω_i specific to a terminal branch, ie: $\omega_{eJ}=\Sigma\omega_{i.}$

The quantity $\Delta Q_{e.J}$ results from the expression obtained for $R_{cr}(t)$:

$$\frac{1-R_{cr}\left(t\right)}{1-e^{-\Sigma\omega t}} = \Sigma\frac{\Delta Q_j \omega_j}{Q_0 \Sigma\omega_j} = \frac{1}{Q_0}\frac{\Sigma\Delta Q_j \omega_j}{\Sigma\omega_j} = \frac{\Delta Q_{ej}}{Q_0}$$

So:

$$\Delta Q_{ej} = \frac{\Delta Q_j \omega_j}{\omega_{ei}} \quad or \quad \Delta Q_{ej} = \frac{1-R_{cr}\left(t\right)}{1-e^{-\Sigma\omega t}} Q_0$$

The wide range of materials for water transmission and distribution pipes that currently exist, as well as the techniques of excision and commissioning, make their choice a complex problem. The choice of pipe material for networks must be based on a comparison of the costs of execution and operation.

In conclusion, the reliability of the elements and systems is assessed with the help of the probability characteristic - time, the reliability being characterized by durability - respectively maintaining the capacity of the system to the limit value, with or without interruptions, in conditions of technical assistance. and necessary repairs.

- repairability property: adaptation of the system to the prevention, location and removal of defects based on continuous supervision and maintenance;
- intensity of defects;
- defects of replaceable elements and frequency of defects, obtaining the defect characteristic, types of defects;
- reliability indicators for distribution systems.

SOLUTIONS AND RECOMMENDATIONS

The wide range of materials for existing water transport pipelines, the execution techniques, the implementation, make their choice a complex issue. The choice of pipe material must be based on a comparison of the costs of execution and operation. In order to have an image on the materials used, it is necessary to make some comparisons from the point of view of:

- Behavior of the pipes in terms of the quality of the material from which they are made;
- Behavior in operation of pipes made of different materials.

The design of networks with high reliability indicator values must always be considered. The choice of the calculation algorithm depends on the value of this indicator.

This chapter presented the calculation model when network reservation is not required (obtaining a high reliability indicator) but it should be borne in mind that after designing the network a small value can be obtained for this indicator, which necessarily leads to network reservation. . In this case, the calculation of the network reliability must be done in two stages: in the first, the necessary structural reserve is established, and in the second, the transport capacity reserve.

FUTURE RESEARCH DIRECTIONS

As a next step, we intend to develop, based on the method of calculating the reliability of reservation transport networks, a calculation algorithm for ring distribution networks with reservation.

Thorough knowledge and deepening of the issues related to the sustainability of new and evolving water transport networks contribute to the development of technical documentation in accordance with the requirements imposed internationally.

CONCLUSION

The reliability of the elements and systems is assessed by means of the probability-time characteristic, the reliability being characterized by durability - respectively maintaining the capacity of the system to the limit value, with or without acceptable interruptions, in conditions of technical assistance and repairs necessary. It is imperative to know the reliability criteria and the basic ways to increase the reliability of water transport systems: either to increase the reliability and quality of the system components or to "reserve" when the reliability of the system is required to be superior to the reliability of the components.

REFERENCES

Ancaş, A. D., Aşchilean, I., Profire, M., Ţurcanu, F. E., & Felseghi, R. A. (2021). Experimental Study on the Behaviour of Seismic Actions on a Flexible Glass-Reinforced Plastic Structure Used in Water Transport Pipes. *Materials (Basel), 14*(11), 2878. doi:10.3390/ma14112878 PMID:34072081

Ancas, A. D., & Atanasiu, G. M. (2011). Seismic Risk Management considering the Urban Lifeline Existing System. *Business Excellence* vol. I, *Proceedings of the 6th International Conference on Business Excellence*, 20-23.

Catuneanu, M. V. (1990). *Optimizarea fiabilităţii sistemelor* Ed. Academiei.

Cioc, D., & Anton, A. (2001). *Reţele hidraulice: calcul, optimizare, siguranţă*. O. Universitare.

Directiva cadru 2000/60/CE. *Cadrul comunitar de acţiune în domeniul strategiei apelor*.

Haghighi, A., Samani, S. M. V., & Samani, Z. M. V. (2011). GA-ILP Method for Optimization of Water Distribution Networks. *Water Resources Management, 25*(7), 1791–1808. doi:10.100711269-011-9775-4

Manescu, A. (2010). *Aspecte legate de durabilitatea retelelor de canalizare realizate din mase plastice*. https://zdocs.ro/doc/durabilitatea-retelelor-de-canalizare-din-materiale-plastice

Stoica, S. (2007). *Managementul retelelor de alimentare cu apă şi canalizare urbane*. Technical University of Civil Engineering Bucharest.

ADDITIONAL READING

Alagirisamy, B., & Ramesh, P. (2022). Smart Sustainable Cities: Principles and Future Trends. In *Sustainable Cities and Resilience* (pp. 301–316). Springer. doi:10.1007/978-981-16-5543-2_25

Chowdhury, P. R. (2022). Variation in Seismic Behaviour of RC Shaft Supported Elevated Water Towers with Change in Proportion of the Shaft Staging. In *Recent Advances in Earthquake Engineering* (pp. 53–68). Springer. doi:10.1007/978-981-16-4617-1_5

Federico, V., Cesare, C., & Alessandro, A. (2022). Evaluation of different methods for deriving geotechnical parameters from electric and seismic streamer data. *Engineering Geology*, 106670.

Halkijević, I., Vouk, D., Posavčić, H., & Mostečak, H. (2021). Damage assessment of water supply networks due to seismic events using vulnerability functions. *Građevinar, 73*(7), 737-749.

Han, L., Zhao, X., Chen, Z., Gong, H., & Hou, B. (2021). Assessing resilience of urban lifeline networks to intentional attacks. *Reliability Engineering & System Safety*, *207*, 107346. doi:10.1016/j.ress.2020.107346

Huiquan, M., & Jie, L. (2022). Serviceability evaluation of water supply networks under seismic loads utilizing their operational physical mechanism. *Earthquake Engineering and Engineering Vibration*, *21*(1), 283–296. doi:10.100711803-022-2084-5

Marasco, S., Cardoni, A., Noori, A. Z., Kammouh, O., Domaneschi, M., & Cimellaro, G. P. (2021). Integrated platform to assess seismic resilience at the community level. *Sustainable Cities and Society*, *64*, 102506. doi:10.1016/j.scs.2020.102506

Marasco, S., & Cimellaro, G. P. (2022). A New Finite Element–Based Methodology for Earthquake Simulation of Large-Scale Urban Areas. *Journal of Structural Engineering*, *148*(4), 04022019. doi:10.1061/(ASCE)ST.1943-541X.0003264

Munasinghe, M. (2019). *Water supply and environmental management*. Routledge. doi:10.4324/9780429267567

Schelwald-van der Kley, A. L., & Reijerkerk, L. (2009). *Water: A way of life: Sustainable water management in a cultural context*. CRC Press.

KEY TERMS AND DEFINITIONS

Bending Stiffness: The product of the modulus of elasticity of the material and the abrasion resistance.

Probability: The probability of an event occurring is the ratio between the number of favorable casese and the number of possible cases.

Reliability: Safety in operation (refers to components, appliances, installations, equipment).

Stiffness: Ability of the pipe's constituent material to resist movement.

Strength: Ability of the pipe to withstand the stresses that occur in the wall material.

Sustainability: To the ability or ability of the pipeline to cope with environmental effects over time.

System Capacity Reserve: Transport capacity required in an emergency situation.

Chapter 6
Novel Process of Using Solar Thermal Energy for Improving the Mechanical Properties of Bricks

Rajmunni Hombal
Sahyadri College of Engineering and Management, India

Rathishchandra Ramachandra Gatti
https://orcid.org/0000-0002-4086-2778
Sahyadri College of Engineering and Management, India

ABSTRACT

The proposed process is a novel, inexpensive, and eco-friendly method to convert the sunburnt bricks into second-class bricks by increasing their compressive strength and other mechanical properties through the use of solar thermal energy. This is achieved by applying black lead-free paint on the whole surface of the brick and exposing it to solar radiation for about 28-30 days. As the black body absorbs radiation, there is an increased flux of radiation energy into the pores of the clay brick, and thus it increases the compression strength of brick and other its mechanical properties. Experimental samples were prepared and tested as per the prevailing industry standards. The results obtained proved that the tested samples manufactured by this novel process qualified as second-class bricks with a compressive strength of 65-70 kg/cm2. The proposed novel process is simple, cost-effective, and environmentally friendly as there is no use of fossil fuels in the conversion process. The current invention is filed for Indian patent with application number 201941036244 as in IPO publications.

DOI: 10.4018/978-1-7998-9810-8.ch006

INTRODUCTION

There is an increasing demand for building materials owing to the rapid rise in population and urbanization. Out of building materials, demand for bricks has increased significantly as it continues to be one of the oldest forms of masonry units used in construction. Globally, about 1500 billion bricks are produced every year wherein about 1300 billion bricks (87% of the global market share) are produced in Asia alone. Although bricks use eco-friendly materials, their product is highly energy intensive and uses fossil fuels. Typically, about 0.54MJ to 3.14MJ of energy is required for every 1kg of brick production. At the global level, about 375,000,000 tons of coal is burnt every year to meet the demand of bricks. With rapid urbanization at approximately 6%, this demand will continue to rise with subsequent increase in fossil fuel consumption. While this is one major concern, the other repercussions based on these consumptions include global warming and air pollution caused by the brick kilns. It is estimated that production of 1kg of brick yields about 0.282 kg of carbon dioxide, 5mg of CO, 1mg of particulate matter. On top of that, many small bricks manufacturers resort to other non-conventional fuels (other than coal) such as old tyres, wooden logs, saw dust, dry dung, and battery cases etc. which yield more harmful gases. Hence, there is a dire necessity to develop safe, ecofriendly and sustainable brick production processes.

Traditional bricks used are usually made of red clay. Methods of production, composition and firing temperature are the key factors, which influence the properties of bricks. The two main categories of bricks are unburnt and burnt bricks. Non-fired or unburnt bricks are the clay or mud molded bricks usually exposed directly to the sun for drying and thus increasing their strength properties. Hence, these are generally called sun-dried bricks. These are typically used for constructing temporary structures and in non-load-bearing structures. Fired bricks are the clay molded bricks exposed to high temperature in the kiln to attain the required strength, often used for load-bearing and seldom used for non-load bearing wall structures.

Red bricks are usually preferred owing to their superior strength properties, but the process involves the burning of fossil fuels or wood. This aggravates the existing increase in pollution due to non-sustainable production processes. Contrary to this, sunburnt bricks are eco-friendly as they utilize only solar thermal energy, but their strength and other properties are inferior compared to fired bricks. Sunburnt bricks are inexpensive due to a simpler method of production. However, the compressive strength of the sunburnt bricks is about 35-40 kg/cm^2, whereas the compressive strength of the fired bricks is 65-105 kg/cm^2. Thus, during the construction of buildings as per the prevailing building codes, these sunburnt bricks are currently restricted to only non-load-bearing structures.

The evolution of blocks or bricks says that they are replaceable, non-poisonous and heatproof after the use of some years. However, bricks can be utilized all through one thousand years from the day of manufacturing *(Fiala et al., 2019)*. The study of mud architecture of building materials considers clay as locally available, economical and biodegradable. But, the material like clay used in bricks as inexpensive building materials allowed for rapid production and construction, influencing the economical housing *(Tarrad, 2020)*. Bio bricks made of agro-waste with compressive strength of 9.36 kg/cm^2 and a good carbon foot-print was suitable for building non-loading structures *(Rautray et al., 2019)*. The production of brick involving the addition of reeds for brick strength for about 5 -20% by weight of the brick before burning the brick was reported *(Salman & Salih, n.d.)*. Fly Ash and bentonite, mechanical waste materials, were utilized as crude materials in manufacturing blocks without burning operation. These blocks were typically used for non-loading structures *(Philip et al., 2019)*. Another alternative, such as the red mud blocks, expands at a range of ðring temperature due to the expansion of the slagging process. The quality of red mud blocks is equivalent to third-class bricks with a compression strength of 30-35 kg/cm^2 and thus suitable for non-load bearing structures *(Arroyo et al., 2020)*. Another eco-friendly brick explored was biomass block that was manufacturing using a dehumidification-based drying strategy to get compressive strength equivalent to third-class bricks. In general, most of the eco-friendly manufactured bricks using eco-friendly materials were able to achieve mechanical properties equivalent to third class bricks and hence suitable for non-load bearing constructions. Hence, there is a dire need to explore a process to convert these economically made third-class bricks to second-class bricks using an economical manufacturing process.

Although, recent developments in the bricks manufacturing have adopted newer sustainable building materials, the process of heat treatment of bricks for increasing the mechanical properties still involves the utilization of non-ecofriendly fuel such as firing wood and coal *(Huarachi et al., 2020)*. For example, Brick kilns in Bangladesh have inefficient in the coal-burning methods which eventually produce air pollution *(Luby et al., 2015)*. Thus, there is a need to extend the framework limits in life cycle assessment of the produced bricks. This is possible by exploring renewable sources of energy for heat treatment of bricks.

According to section 4.1 of IS: 1077 – 1992 (R2002) *(Indian Standard, 2005)*, the classes 3.5, 5, 7.5 and 10 have compressive strengths 35 kg/cm^2, 50 kg/cm^2, 75 kg/cm^2 and 100 kg/cm^2 respectively are commonly used in residential and commercial buildings. Commercially, there are three classes of bricks in burnt bricks that are classified according to their quality and mechanical properties, particularly compressive strength. Compressive strength possessed by first class bricks (Class A) are greater than 70 kg/cm^2, second class bricks (Class B) are 65-70 kg/cm^2 and third-class bricks (Class C) are 35 - 64 kg/cm^2. These commercial classifications may deviate from the above-mentioned values based on the regions and applications, but the IS classes remain same. The above classification assumes that other quality parameters such as squared vertices, sharp edges without any damage or cracks are already achieved in the bricks.

To address the above two needs of an economical as well as ecofriendly way of manufacturing bricks, a novel method is proposed in this paper. The proposed method involves increasing the compressive strength of sun burnt bricks (third class bricks) to second class bricks by painting the whole surface area of the brick using black lead free paint and then sun drying it at normal temperature (even 33-37 °C is sufficient). The proposed method aims to eliminate the process of brick burning using fuels such as firewood or coal that was traditionally used to attain the required strength of bricks used for construction of buildings.

MATERIALS AND METHODOLOGY

The method of experimental study involved four stages - preparation of the prototype bricks by the proposed novel solar thermal method, experimental testing of the specimen as per the IS standards, comparison with the standard bricks followed by analysis and conclusions.

The standard followed for testing of the bricks was IS: 1077 – 1992 (R2002). The material used for making the brick was composed of 20-30% of alumina, 50-60% of silica, 5% of lime, 5-6% of oxide of iron and small quantity of magnesia.

EXPERIMENTATION

As discussed in methodology, the experimentation involved two parts, namely, the preparation of the prototype bricks and testing of the bricks as per the brick testing standards IS: 1077 – 1992 (R2002). The preparation of the samples took about 35 days for prolonged exposure of the bricks in the sunlight a followed by tests that were done within one day.

Planning of the Prototypes

As per the IS: 1077 – 1992 (R2002), five types of brick tests need to be conducted to determine the essential mechanical properties and classify them as first class, second class and third-class bricks. These tests are visual inspection (shape and size) test, compression strength test, soundness test, hardness test and water absorption test.

Except for the first test, the rest of the four tests are destructive tests and thus involves the consumption of the specimen. Hence, the first test namely the visual inspection was conducted for all the bricks. For the remaining four tests, minimum three bricks were required to get the average value. Hence, 12 painted and 12 unpainted bricks was planned in each set of bricks as in Table 1.

The experiment was designed for five weeks. This variation of solar thermal exposure for different sets of prototype bricks was intended to study the influence of solar thermal exposure with the compressive strength of the brick samples. The duration of the project was scheduled for 35 days.

The schedule of sun drying was carried in five sets of bricks with a time difference of seven days (7 days, 14 days, 21 days, 28 days and 35 days) as shown in Figure 1. As the bricks were exposed to sunlight, the painted bricks being black body absorbs sunlight and gets hardened in order to achieve durability.

Table 1. Test configuration planned for each set of prototype bricks

	Test as per IS: 1077 – 1992 (R2002)	Test type	No. of bricks planned	
			Painted	**Unpainted**
1	Visual inspection (Shape and size)	Non-destructive	0	0
2	Compression test	Destructive	3	3
3	Water absorption test	Destructive	3	3
4	Hardness test	Minor destructive	3	3
5	Soundness test	Destructive	3	3
	Total no. of bricks planned per set		12	12

Preparation of the Prototype Brick Specimen

As mentioned in Table 1, 120 samples (24 bricks per set x 5 sets) of prototype bricks were molded as per the standard IS: 1077 – 1992 (R2002). The preparation of the specimen includes the building materials like clay, potable water and fly ash and molded to rectangular blocks of standard size of 20 cm x 10 cm x 10 mm including the thickness of the mortar and actual size of brick is 19 cm x 9 cm x 9 cm.

This was followed by a hack drying process, which involves sun drying of molded bricks for about two weeks, by placing the molded bricks in rows on their edges. This hack drying was a basic essential part for manufacturing any type of brick to increase its hardness, compressive strength and density. After the hack drying process, black paint was applied on the whole surface of brick and exposed to the solar radiation for a specified number of days as per the schedule described in Figure 2.

Figure 1. Schedule of solar-thermal exposure for planned sets of bricks

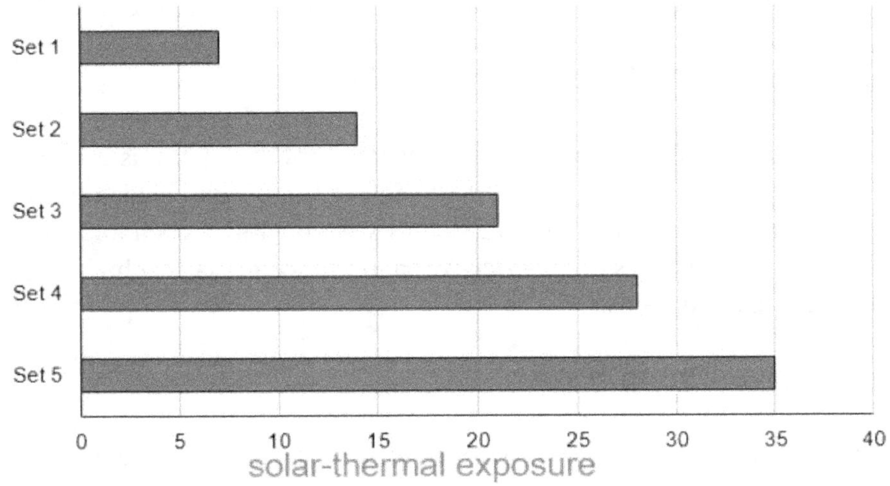

Figure 2. Sunlight exposure of Prototype sets

Testing of the Specimen

The physical and mechanical properties, namely, compressive strength, water absorption, hardness, soundness, structural integrity (shape, size and sharpness) was tested using the five standard brick tests as prescribed in the standard of IS: 1077 – 1992 (R2002). The Set1 was initially taken out and tested after Week1 for the above five tests. This was followed by taking out Set 2 at the end of Week 2, Set 3 at the end of Week 3, Set 4 at the end of Week 4 and Set 5 at the end of Week 5.

Visual Inspection (Shape and Size) Test

All bricks must be of the same size. The shape of bricks as per this test, must be purely rectangular with sharp edges. It was inspected to have standard brick size of length x breadth x height as 19 cm x 9 cm x 9 cm with an open tolerance of 5 mm. Variation in the edges of the brick was allowed within narrow limits of, ±8% of the specified dimensions.

Compressive Strength Test

The bricks were placed in between plywood sheets. The brick was sandwiched between the plywood sheet and centered between the compression testing machine as shown in Figure 3. Then uniform axial load of 140-150 kg/cm^2 per min was applied till it fails. Then the compression strength was noted as the maximum load at failure divided by the average area of the bed faces.

Figure 3. Compression Strength using UTM for (a) painted brick and (b) unpainted burnt bricks

(a) (b)

Water Absorption Test

Initially, the brick was weighed under dry condition in its room temperature as W1. This was followed by the water absorption test by immersing the prototype bricks in water for 24 hrs as shown in the Figure 4. The brick was then damped by a dry cloth and weighed to get W2. The water absorption coefficient was then calculated using the formula,

$$(W2-W1)/W1*100 \tag{1}$$

Figure 4. Water absorption test

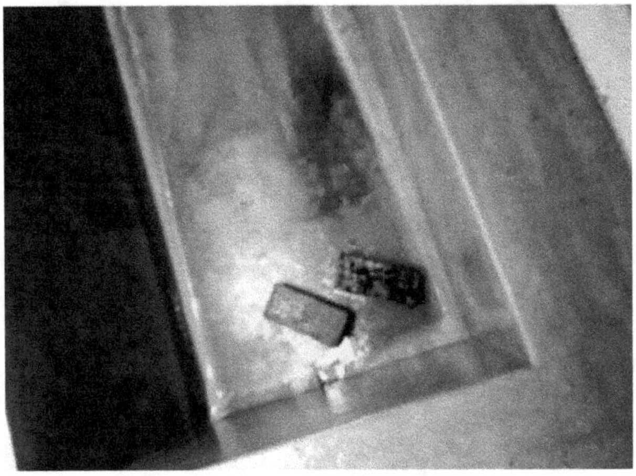

Hardness Test

A good brick should resist scratches against sharp things. The prototype bricks were tested against the sharp tool or finger nail to make scratch on the brick. Since there was no significant deep scratch as in Figure 5, it was considered as a hard brick.

Figure 5. Hardness test

Soundness Test

Soundness test of bricks shows the nature of bricks against sudden impact. In this test, 2 bricks were chosen randomly and struck with one another as shown in Figure 6. If the sound produced was a clear bell ringing sound, and if the brick does not break, it qualifies the soundness test.

Figure 6. Soundness test

RESULTS AND DISCUSSION

The visual inspection after each drying, the prototype was carefully selected and those who were out of shape were rejected. Therefore, all the 120 bricks had proper shape. The size was within the tolerance of ±8% of the specified dimensions as permitted by the IS: 1077 – 1992 (R2002).

The compressive strength tests conducted on the sample sets of painted and unpainted bricks are as shown in the Figure 7. It is observed that the prolonged exposure to the solar radiation significantly increases the compressive strength from 7 to 35 days and then starts to flatten. This may be due to decrease in the moisture content and more bonding of the clay microstructures due to slow and steady flow of thermal radiation from the sun.

The maximum compressive strength observed during the optimal period of 28 days was about 32.5 kg/cm^2 for unpainted brick samples and 65 kg/cm^2 for painted brick samples. Thus, the compressive strengths of painted brick samples come within standard qualifying compressive strength of 65-70 kg/cm^2 for second class bricks of prevalent commercial standards and between class 5 and class 7.5 of section 4.1 of the IS: 1077 – 1992 (R2002).

Figure 7. Compression Strength test of five sets of brick prototypes

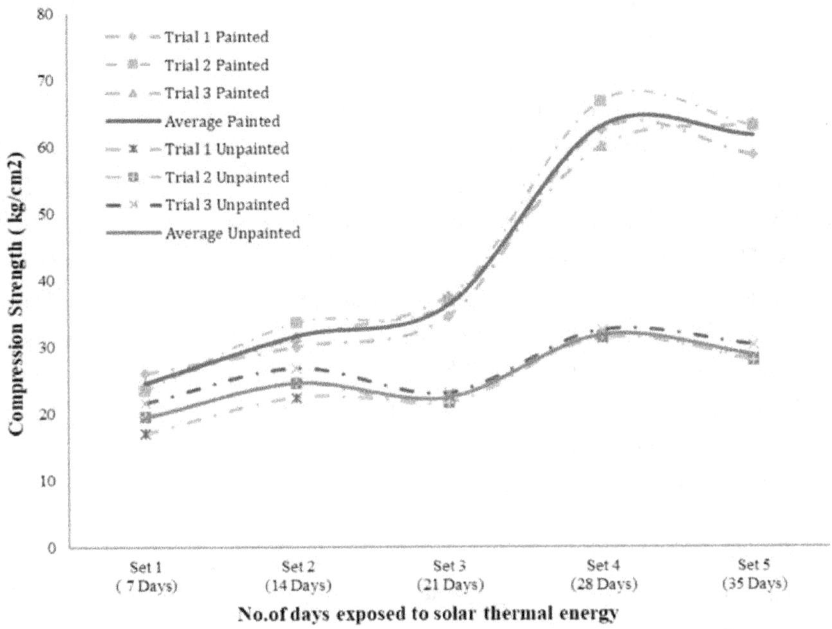

In the water absorption test results as shown in Figure 8, it was observed that the painted bricks have 13.05% -19.8% by weight. The unpainted bricks have 21% - 27.75% by weight. It can be observed that the water absorption coefficient was within the 12.5 -20% by weight as specified in the standards. It was also observed that sunburnt unpainted bricks were outside the maximum limit of water absorption coefficient as above.

Figure 8. Water absorption test results

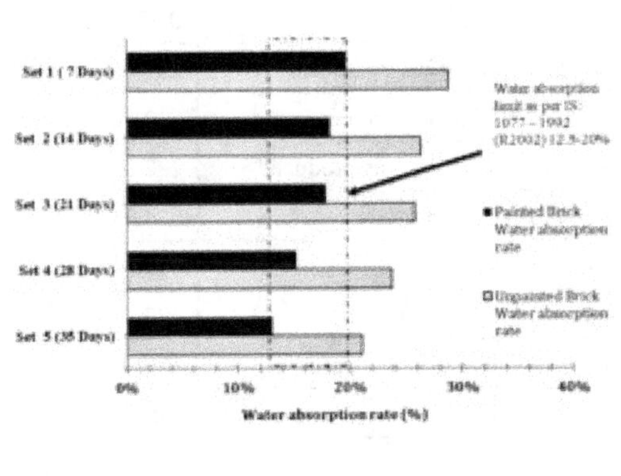

A good brick should resist scratches against sharp things. In the scratch test, it was observed that there was no deep scratch impression on painted brick thus passing the hardness test.

Overall, the painted bricks passed all the five tests to qualify as second-class brick thus proving that the proposed process can be used to convert third class bricks or sun burnt bricks to second class bricks. In India, the average cost is Rs. 6 per brick for second class brick and Rs.3 per brick for third class (sun burnt brick). In some regions, third class bricks are even given as free of cost with purchase of second-class bricks. The painting cost for 120 samples in this experiment was an additional cost of about Rs. 0.50 per brick inclusive of labor costs at the rate of Rs.800 per day. Therefore, there is an economic saving of Rs. 2.5 per brick. For an average 60 X 40 house that consumes about 20,000 bricks, the savings would be approximately Rs.50, 000. In addition to this economic advantage, the process involved in making these black bricks is ecofriendly.

The global production of bricks is about 1500 billion bricks, At a cost saving of approximately Rs.2.50 per brick, the overall cost savings is about 3750 billion rupees or 53 billion USD. Another great advantage of using this technology is the environmental impact as in Table 2 below. A typical brick produces about 280g of CO_2 emission per kg of brick. This typically amounts to reduction of 227.5 billion kg of CO_2 globally and 227.5 billion kg of CO_2 in India. The third advantage of the proposed technology is in terms of savings of coal which amounts to about 3750 billion kg for the world and 625 billion kg in India. Therefore, at a large scale, the environmental impact is tremendously beneficial.

Table 2. Environmental Impact of the proposed brick production technology

	Average Weight of the Brick	3.25	kg
	Amount of Co2 produced per kg of brick	0.28	kg
	Amount of coal burnt per kg of brick	0.77	kg
India	No of bricks produced in India (2017 estimate)	250	bricks in billions
	Mass of total amount of bricks produced in India	812.50	kg in billions
	Mass of total amount of Co2 emissions produced	227.50	kg in billions
	Mass of total amount of coal burnt	624.81	kg in billions
World	No of bricks produced in the world	1500	bricks in billions
	Mass of total amount of bricks produced	4875	kg in billions
	Mass of total amount of CO_2 emissions produced	1365	kg in billions
	Mass of total amount of coal burnt	3750	kg in billions

CONCLUSIONS AND FUTURE DIRECTIONS

In this research, a unique technique is being proposed in to convert sun burnt bricks to second class bricks by painting the whole surface area of the brick through solar drying. The proposed method of manufacturing the clay bricks is very economical and ecofriendly.

The compressive strengths of painted brick samples come within standard qualifying compressive strength of 65-70 kg/cm^2 for second class bricks of the IS standard IS: 1077 – 1992 (R2002).

The other mechanical tests add to the validation of the manufactured brick as second-class brick as per the IS standards.

This adoption of this process of manufacturing bricks does not use non-renewables such as wood and kerosene and substitutes them with the efficient solar thermal brick burning process.

The intent of this invention is to develop an economical cum ecofriendly process of manufacturing bricks for small scale industries involved in brick manufacturing.

The future scope of the proposed project is to further increase the compressive strength to 70-75 kg/cm^2 by improving the economical passive heating method of black painted bricks by optimizing the heat dissipation.

ACKNOWLEDGMENT

This research is supported by Hombal Bricks company and Sahyadri College of Engineering & Management. We are thankful to our co-inventors of this proposed invention Akshay P.N and Pooja Kumble for their contribution in developing this process. The authors are thankful to Director-Research, Dr. Manjappa Sarathi and Dr. Nalini Elizabeth Rebello, HOD Civil Engineering (then) for identifying the IPR in this novel process and helping us to patent this project. We would like to extend our gratitude to Sahyadri college of engineering and Management for supporting and patenting this project.

REFERENCES

Arroyo, F., Luna-Galiano, Y., Leiva, C., Vilches, L. F., & Fernåndez-Pereira, C. (2020). Environmental risks and mechanical evaluation of recycling red mud in bricks. Environmental Research, 186, 109537. PubMed PMID:32315825

Fiala, J., Mikolas, M., Junior, J. F., & Krejsova, K. (2019). History and Evolution of Full Bricks of Other European Countries. IOP Conference Series. Materials Science and Engineering, 603(3), 32097.

Huarachi, D. A. R., Gonçalves, G., de Francisco, A. C., Canteri, M. H. G., & Piekarski, C. M. (2020). Life cycle assessment of traditional and alternative bricks: A review. Environmental Impact Assessment Review, 80, 106335.

Indian Standard. (2005). Common burnt clay building bricks - Specification (IS 1077 (1992), Fifth Revision. Retrieved from https://law.resource.org/pub/in/bis/S03/is.1077.1992.pdf

Luby, S. P., Biswas, D., Gurley, E. S., & Hossain, I. (2015). Why highly polluting methods are used to manufacture bricks in Bangladesh. Energy for Sustainable Development, 28, 68–74.

Philip, S., Ajin, A., Shahul, F., Babu, L., & Tina, J. (2019). Fabrication and performance evaluation of lightweight eco-friendly construction bricks made with fly ash and bentonite. International Journal of Engineering and Advanced Technology, 8(6), 2090–2096.

Rautray, P., Roy, A., Mathew, D. J., & Eisenbart, B. (2019). Bio-Brick-Development of sustainable and cost effective building material. Proceedings of the Design Society: International Conference on Engineering Design, 1(1), 3171–3180.

Salman, G. A. S., & Salih, S. A. (n.d.). Effect of Firing Temperatures on Physico-Mechanical Properties of Clay Bricks Containing Reeds. Academic Press.

Tarrad, M. (2020). A Vision to Revive Mud Architecture, a Community Heritage Architecture in Jordan, for Low Income. Journal, 15(2), 269–275. Http://Iieta. Org/Journals/Ijdne

Chapter 7
Technologies for the Clean and Renewable Energy Production for the Sustainable Environment

Amit Kumar Chaurasia
https://orcid.org/0000-0003-1786-3745
Indian Institute of Technology, Roorkee, India

Swati Mohapatra
Wonkwang University, South Korea

Ravi Shankar
Madan Mohan Malaviya University of Technology, India

Lokendra Singh Thakur
Department of Chemical Engineering, Ujjain Engineering College, India

ABSTRACT

Energy generation or green energy generation from waste materials is a necessity for the sustainable development of our society. In the last two decades, there is substantial scientific and technological development in waste to energy generation research. This chapter discussed the various aspects of the waste to energy generations technologies such as the hydrogen/energy generation from the waste materials. The chapter enlisted the advanced in the technology's development of waste to energy generation. It emphasized the most sustainable way to treat waste and their current state of the art in the waste to energy generation. This chapter also includes the biological, biochemical, and various fuel cells technologies for waste to energy generation. This chapter summarized the various state of the art of hydrogen production technologies from the waste materials or wastewater and emphasized the applications of nanotechnology in the cathode development for the bio-electrochemical system.

DOI: 10.4018/978-1-7998-9810-8.ch007

INTRODUCTION

Increasing population, urbanization and industrialization has several adverse impacts over our society such as increasing waste generation, degradation of the natural resources and saviour critical health issue in the living creatures. The amount of waste generation across the globe is increasing rapidly and it is estimated that 2 billion tonnes of municipal solid waste generated annually that expected to increases by 70 percentage by 2050. In which ~33 percentage of that are not disposed properly or environmentally friendly manner. Worldwide, waste generated per person per day averages 0.74 kilogram but ranges widely, from 0.11 to 4.54 kilograms (Chaurasia & Mondal, 2022).

On the other hand, the volume of wastewater generated per year in North America and Europe are 85 cubic meters, where as in Asia around 160 cubic meters. In which more than 80% of wastewater flows back into the ecosystem without being treated (Climate News Network; United Nations University, 2020). In India 53 cubic meter wastewater generated annually and more than 86% of untreated wastewater is discharged into the ecosystem (CPCB, India 2017).

A recent literatures suggest that high-income countries on average treat 70% of the generated wastewater, followed by upper-middle-income countries (38%), lower-middle-income countries (28%), and low-income countries, where only 8% of the wastewater generated is treated (Jones et al., 2021). The most interesting and challenge aspect for the researcher across the world is the most of the countries not releasing the data on the waste generation. Only ~35 countries revelling the complete data on the waste generation and ~22 countries releasing the partial data on the wastewater generation. While rest of the countries across the globe not revealing the data on waste generation or waste processing that create obstacle to clear the world waste generation scenario.

At present, around 85% of global energy demand is fulfilled by non-sustainable resources (fossil fuels) such as petroleum (34%), coal (27%), and natural gas (24%), which have adverse impacts on global climate, human health and ecosystems as well as generate a huge amount of wastewater. The rising usage of limited fossil fuel resources and their depletion triggered the attention of the world towards the energy scenario in upcoming future (Chaurasia, Shankar, et al., 2021).

Along with this, alarming and continuously rising global warming focused the world to immediate initiation of clean and renewable energies. Fourth Assessment Report (AR4) of the United Nations Intergovernmental Panel on Climate Change (IPCC) demonstrated that emission of anthropogenic greenhouse gases is totally accountable in global warming increment (Alley et al.,2007). Hosseini & Wahid, (2016) prescribed that 12, 25 and 30 trillion Watts of power can stabilize atmospheric carbon dioxide concentrations at 550, 450 and 350 ppm levels by 2050. In order to successfully achieve this outstanding goal, a revolution is demanded in production, storage, conversion and distribution of energies in clean, green and sustainable manner.

To mitigate these challenges for the treatment of the waste or waste to green energy generation for the sustainable development of society required international collaboration, research funding and advancement in the waste to energy generation technologies. The renewable energy resources such as hydrogen production from the waste and wastewater is getting strong interest as hydrogen from waste will not run out ever while other sources of energy are finite and will someday be depleted. It is also realized that the hydrogen production from the wastewater should be economical and sustainable (IEA (2020), 2020).

Hydrogen is the energy carries and fuel for the future. Despite, hydrogen being capable to be served as a clean energy, very minimal of hydrogen is generated for energy purposes. Most of the hydrogen is used as a chemical feedstock in various processing industries such as, food, metallurgical, petrochemical and electronics (Mondal et al., 2017; Shankar et al., 2017). Konieczny et al., (2008) demonstrated that around 49% of hydrogen is employed in ammonia production, 37% in petroleum refining, 8% in methanol production and 6% for other purposes. Balat, (2007) reported that hydrogen can act as a competent fuel in internal combustion engines. It is beneficial for automobiles as a fuel due to certain advantages like quick burning speed, high effectual octane number and zero toxic emission (Hosseini & Wahid, 2016).

Various categories of catalyst are being prepared to provide good hydrogen yield. With the implementation of nano-catalysts, various successful nano-catalysts are synthesized (Chaurasia & Mondal, 2021b). Despite, being capable of providing good hydrogen yield, their synthesis and exposure is not good for environment (Basheer & Ali, 2019). In order to protect environment, various authors synthesized nanoparticles with green methods (Ali et al., 2019); (Zsembik, 2006)].

Furthermore, various researchers suggested that nanotechnology can improvise renewable energy sources, such as, nano-materials can enhance the strength of rotator blades for wind energy; they can prevent corrosion in tidal energy equipment's; can help in making fatigue-resistance drilling machines for geothermal energy and many more (Hussein, 2015). The elaborated applications of nanotechnology in solar, hydrogen, wind and bioenergy are represented in this chapter. Recent advances in nanotechnology offer tremendous opportunities for wastewater treatment.

Currently, the water treatment processes which totally depend on centralized and conveyance system are not sustainable. The highly modular, efficient and multifunctional processes in combination with nanotechnology are competent to furnish affordable wastewater treatment solutions with elevated performance (Qu et al., 2013). The key goal of this chapter is to derive a strong lineage among nanotechnology and clean, green and sustainable hydrogen generation from the waste materials.

This chapter lays emphasis on the role of nanotechnology in the various hydrogen production technology, current technology advancement and their applications. World hydrogen production prospect along with future of hydrogen economy. It also collects evidences about tremendous need for renewable energy and international contributions towards the promotion of green energy. It enlightens the way how nanotechnology implementations in existing renewable technologies fosters production of pristine energy along with environmental remediation. The role of various cathodes, nano-cathode materials or cathode nano-catalyst in the bio-electrochemical system for the green energy of hydrogen generation from waste materials.

Necessity for Clean Energy

Nowadays, climate change and corresponding global environmental out-turns are at verge of crossing critical point. No doubt that these impacts will be ruinous for the future generation (Janusz Nowotny et al., 2016). The worsening of the environment is most observable in developing countries with dramatic consequences. In most part of the world, air pollution has reached far beyond the levels of safety for human health (Wasif et al., 2021). As per the Journal of the American Medical Association, most cases of lung cancer are due to exposure to air pollution. According to the European Environmental Agency (EEA), air pollution is main culprit of asthma and pulmonary diseases. World Health Organization (WHO) declared that polluted air is responsible for high probability of stroke, heart diseases and respiratory diseases (Janusz Nowotny et al., 2016). Apart from air quality, deterioration of the environment affects contamination of water as well (Sucher et al., 2012). According to United Nations, 1.1 billion people do not have access to clean drinking water and some 2 million (mostly children) die due to lack of pure drinking water (WHO, 2005); (Janusz Nowotny et al., 2016) . Statistics show that large populations of Asia and Africa lack availability of clean, safe and pure drinking water. According to World Health Organization (WHO) report in Bangladesh, only 40 million people drink safe water, while other have access to unsafe water (containing high amounts of arsenic) (Karim, 2000). The consumption of fossil fuels in order to generate energy is the major factor playing key role towards fabrication of undesired and detrimental consequences. Fossil fuel consumption via burning and utilization inflates the atmosphere with greenhouse gases like carbon dioxide (CO_2), methane (CH_4), nitrous oxide (N_2O) and many more (Zona et al., 2013; Fearnside et al., 2006); Tremblay et al., 2004). Wasif et al., (2021) reported that Asia-Pacific Economic Cooperation (APEC) countries accounts for 60% of world's energy demand and contribute almost 57% of global Gross Domestic Product (GDP). They also stated that these countries reach majority of their energy demands through fossil fuels thereby contributing more than 70% of global carbon dioxide emission. Water is used in various ways during energy extraction and production from fossil fuels. The water utilized in energy generation from non-renewable energy resources and nearby water bodies are contaminated with wide variety of sediment & chemical pollutants along with liquid or solid waste generated during extraction processes [(Rozell & Reaven, 2012); (Su et al., 2019); (Gleick, 2013)]. As signalled by the recent historic Paris Agreement, it is wise to develop extremely high determination to reduce and eliminate use of fossil fuels in energy production (Janusz Nowotny et al., 2016). More prominence should be given towards development of alternate sources of energy generation. As, energy generation is highly impactful towards climate and environment (Wuebbles & Jain, 2001). Therefore, it becomes critical to emphasize over the consequences of energy generation. More decisive emphasis towards global environment sustainability will lead to more global prosperity in the longer run (Janusz Nowotny et al., 2016). An enthusiastic perspective is demanded to tackle the detrimental concerns of environment, such as continuous melting of polar ice caps, rising sea levels, extreme weather conditions leading to increased frequency of natural disasters mainly droughts and floods along with human health effects (Wuebbles & Jain, 2001). Energy generation through unsustainable means possess deadly outcomes towards ecosystem, water resources, agriculture, sea level rise, human health and infrastructure. The Future Earth Program established by United Nations, is driving the whole world towards innovation of novel approaches to establish sustainable development (Janusz Nowotny et al., 2016). The success to thrive novel approaches is only feasible through a strong linkage between diverse disciplines. Energy is an extensive field encompassing various disciplines with diverse conceptual backgrounds and takes into consideration various key factors such as climate change, water management, pollution, etc. It is pivotal to fully understand conversion of energy and identify the utilitarian materials that are obligatory to achieve optimal performance of the corresponding system. More awareness is growing towards nano size systems as they are capable of exhibiting outstanding properties, entirely disparate to bulk phase. With desired processing and better design, nano-size materials are competent enough to develop clean, green and sustainable technology for energy generation (Janusz Nowotny et al., 2016).

Energy and Environment

Huge scale combustion of fossil deposits is aggregating in an unsustainable generation of carbon dioxide to the atmosphere, thereby contributing to an elevated greenhouse effect along with terrible climatic consequences (Wuebbles & Jain, 2001); (Tingem & Rivington, 2009); (Panwar et al., 2011). This large scale burning and utilization of fossil fuels is potentially active for initiating catastrophic changes to the environment. The human induced climate variation is the matter of danger for the upcoming generations (Panwar et al., 2011). If it keeps on going, then access to clean water, air and food will be an aspiration in the coming years [(Gleick, 2013); (Janusz Nowotny et al., 2018)]. In order to protect the environment (for our sake as well), it is totally clear and urgent that human activity induced climate change is to be reduced. As reported by Panwar et al., (2011), the largest cause of greenhouse gas emission is burning of carbon (in fossil fuels) for energy generation. Projections on world population and energy consumption are made by various specialized entities such as the one shown in Figure 1.

Figure 1. Projections on world population and energy consumption (IEA, 2019)

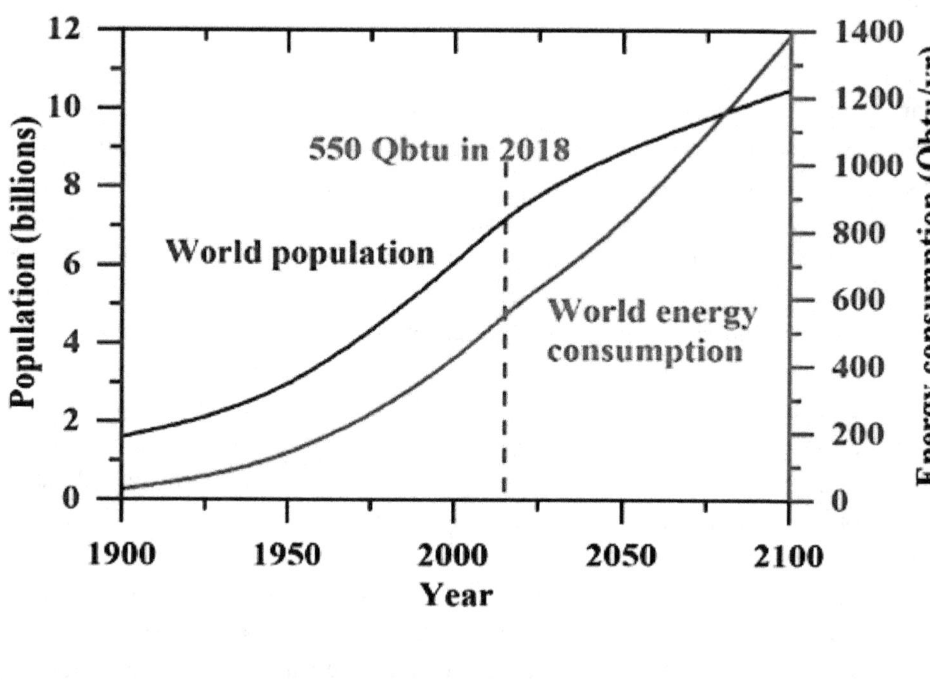

Wuebbles & Jain, (2001) reported that with increase in carbon dioxide levels, temperature will warm that will shift the geographical distribution and composition of many ecosystems. With change in earth's radiation balance change, climate changes will become more frequent. The agriculture sector will suffer too as the crop yields will decrease in some areas resulting in elevated risk of hunger and famine. As the climate will be warmer, sea level will eventually rise due to thermal expansion of ocean and snow, ice melting from polar ice caps and mountain glaciers. With the sea level rise, risk of flood will be more. Intergovernmental Panel on Climate

Change (IPCC) has stated that adverse impact of global warming would be dealing with human health issues. Frequent weather and climate changes will subsequently increase the risk of diseases and death, high temperature will initiate the formation of secondary pollutant from primary one (generated in fossil fuel combustion) and less food production will eventually cause hunger and malnutrition. Fossil fuels are predominantly utilized during electricity production in power plants, in transportation as well and many more (Gleick, 2013). With increasing awareness towards climate impact on environment seeks development of environment friendly systems. These systems should be competent enough to generate energy in a sustainable manner with less or no detrimental impact towards the environment (Janusz Nowotny et al., 2016). Large volumes of water are consumed and polluted extraction, refining and combustion of fossil fuels. The contamination occurs via regular operations as well as through accidental releases or by any other means as well. But the thing is that very minimal consideration is given to the water quality innuendo of fossil fuel production and utilization (Gleick, 2013). Despite that quality of water is degraded at each and every step. No doubt fossil fuels themselves are the significant water contaminants, but chemicals employed during process and refining of these fuels are a sincere threat to water bodies and environment. Gleick, (2013) reported that on an average 15-18 billion m^3 of freshwater resources are annually affected due to fossil fuel production. It is evident that coal, natural gas and petroleum are produced in every region around the globe. Similarly, the water gets contaminated due to fossil fuels across the world (Su et al., 2019) . This intense degree of contamination of water is terribly affecting ecosystems and other communities as well that depend on water for drinking and livelihood.

Various literatures suggest that systems like ecological footprint buildings, photovoltaic farms, hydrogen fuelled transportation, and etc. can be promising for achieving better future. Renewable energy production offers massive opportunities in this context. In fact, implementation of solar and wind based energy sources are rising dramatically from last decade, and are constituting 3.2% of overall electricity produced around the globe, and 4.2% in the USA (Janusz Nowotny et al., 2016). Despite impressive deployment of renewable energy technologies like solar and wind, there are some shortcomings too. These clean technologies demand availability of large-scale energy storage systems that can act as a buffer against the generated energy. So, the rapid development and instant deployment of photoelectrochemical and electrochemical technologies for hydrogen energy are capable of providing awesome opportunities for large-scale energy storage. At the same time, there is a tremendous need to replace the current technology of hydrogen generation, utilizing natural gas as the feedstock, with technologies established on utilizing green and renewable resources. Introduction of sustainable and entirely clean technology for the generation of green energy like hydrogen is the game changing approach to the elimination of fossil fuels.

Economy

From past few years, the use of renewable energies like wind, solar, biomass and hydro, enhanced substantially globally. As reported by Blazejczak et al., (2014), global wind capacity is seen increased to 318 GW in 2013 compared to 6.1 GW in 1996, along with increment in solar photovoltaic capacity to 13 9GW from 0.7 GW for the same time period. The expanding consumption of renewable energy sources strongly promises elevated economic activity. Bunch of employments can be created in manufacturing, operating and maintaining of renewable energy installations along with transportation of biofuels and biomass. Indirect employments can also emerge in other sectors linked with fulfilment of demands of inputs for renewable energy production such as electronics (Wei et al., 2010). Edler et al., (2009) reported in their study that indirect and direct employment for renewable energy in Germany, for the year 2013, calculated to 371400 persons. If more emphasis of a nation will be on renewable energy production, then the cost for imports of fossil fuels will also get reduced. Not only this, the renewable energy production will create enormous opportunities to export. Blazejczak et al., (2014) reported that renewable energy production in Germany will reduce the cost of imports by €33 billion by 2030 and with export of renewable energy facilities and components these exports will generate €30 billion by 2030. However, renewable energy production will altogether raise the Gross Domestic Product (GDP) of the country, directly accelerating the economy of the nation. With the expansion of the renewable energy generation, there lies a positive net effect on gross economy.

World Hydrogen Energy Prospects

Globally, it is widely recognized that hazard and risks of energy generation through unsustainable means, affecting climate and environment, can only be taken into consideration through international cooperation. It is crucial to acknowledge the dangers as consequent from energy generation technology. The first noteworthy international move towards formation of an international body to promote renewable energy was through UN Conference of New and Renewable Sources of Energy in Nairobi, Kenya (1981)(Wright et al., 2018). This program mainly focused on energy assessment, planning, research and development. A decision on "Energy for Sustainable Development" was adopted by UN Commission on Sustainable Development in April 2001 (Wright et al., 2018). In 2004, Renewable Energy Network for the 21st century was established. The need for international cooperation was emphasized in the Beijing International Renewable Energy Conference (2005). In January 2009, International Renewable Energy Agency was established to act as a driving force in encouraging a rapid transformation towards the sustainable use of renewable energy (Lima & Gupta, 2013). United Nations (UN) initiated to set emission targets for greenhouse gases and to introduce emission trading. In 2014, 164 countries adopted renewable energy policy (Adib, 2015). EU set the target to meet twenty percent of its total energy requirements through renewable energy by 2020 (Burgherr, 2011). An international agreement was adopted by 2015 UN Climate Conference in Paris to modify the fossil fuel driven economy and to minimize the speed of global warming (Janusz Nowotny et al., 2018). International initiative is set to implement hydrogen as the lead energy carrier. This decision is considered as most remarkable initiatives of the international community. The main crux of this initiative is to achieve global energy security along with preservation of environment. George W. Bush formally presented hydrogen as the fuel of the future in 2003. Within this proposal, International Partnership for hydrogen economy (IPHE) was formed by US Department of Energy and US Department of Transportation in 2003. This step aimed to reduce the usage of gasoline for transportation. Around 20% of energy is utilized by transportation and this energy is actually derived from gasoline. However, to implement hydrogen as energy source demands rapid development of hydrogen specific technologies such as hydrogen generation, storage, transportation, distribution and safety (Janusz Nowotny et al., 2018). Wasif et al., (2021) reported that Asia-Pacific Economic Cooperation (APEC) countries have initiated to multiply the renewable energy usage by 2030 via sustainable technology innovations and minimal use of fossil fuels.

CLASSIFICATION OF CLEAN, GREEN AND SUSTAINABLE ENERGY

A variety of renewable energy generation technologies are employed across the world for sustainable development of the society. Because renewable energy technologies deliver an excellent opportunity for mitigation of greenhouse gas emission and minimizing global warming through substituting conventional energy sources. The renewable are the domestic, primary and inexhaustible energy resources that are also referred to as alternative energy sources. Sustainable development demands technological advancements to measure and differentiates the various industrial or energy generation process. Nowadays, along with improved quality of life, consumption of fossil fuels is also increasing. The energy resources have been categorized mainly into three sectors viz. fossil fuels, renewable and nuclear (Panwar et al., 2011). In simple language, renewable resources are those resources which can be used again and again to generate energy that consist of solar energy, wind energy, hydropower energy, ocean energy, geothermal energy, electrochemical energy, photochemical and photo-electrochemical energy and biomass energy. Clean, green and sustainable energy systems can resolve the present crucial tasks such as improvement in energy supply, organic fuel economy, local energy, water supply, enhancing living standards, generating employments and environmental protection. There is a significant scientific and technology progress across the world in the waste to energy generation technologies.

The recent emerging renewable energy are the solar energy that can easily harvested for useful purpose but it is available in a limited fraction of the world. Solar energy straps the power of sun to generate electricity either through photovoltaic cells (direct) or concentrated solar power (indirect). In concentrated solar power technology, arrays of mirrors are employed to track the sun and simultaneously reflect the rays to a certain point to heat the working liquid, which is mostly employed to produce electricity in a conventional turbine. Concentrated solar power is effective in large areas, while photovoltaic cells are distributed on any surface exposed to sunlight (Gasparatos et al., 2017). In the past decades, solar energy generation show the potential growth world-wide. Globally, 177 GW of photovoltaic cells and 4.4 GW of concentrated solar power are installed as documented in 2014 (Gasparatos et al., 2017). Solar energy possesses negligible ecological impacts (Katzner et al., 2013). The common usage of solar energy is in solar heating. However, the solar photons have the potential to generate electricity directly or accomplish hydrogen production photochemically or photoelectrochemically. Most prominently used solar cells to generate electricity are made up of silicon. Silicon based solar cells are economical too (Janusz Nowotny et al., 2018). The challenging aspect of the solar energy is the high capital cost, not available uniformly across the globe and variation with geographical location. The second renewable energy source is the wind energy that has potential to amplify. It has been used historically, in sails, windmills and windpumps. Wind power is a popular, sustainable, renewable energy source that has a much smaller impact on the environment than burning fossil fuels. Wind energy technologies utilized the wind turbine to generate electricity. Large rotating blades are employed to power turbines for wind energy generation (Gasparatos et al., 2017). Tabassum et al., (2014) reported that largest blades, even greater than 100m in diameter, rotating 100-120m above the ground and generating 5 MW of power are employed. Over the last two decades, wind power is seen as the fastest growing energy generation technology (Peste et al., 2015). The wide energy also has the drawback of not uniformly available across the globe. High capital cost and technological advancement is also the second most important hurdle for the wider applications of the wind energy.

Hydropower is another renewable energy sources that available across the globe with certain limitation. Hydropower is produced with the help of fresh flowing water. The water is used to run the turbine, which in order generates electricity. On the basis of geographical constraints and human demand patterns, various hydropower technologies are deployed. These technologies are conventional hydropower from dams; run of river hydropower; and pumped storage hydropower (Gasparatos et al., 2017). Adib, (2015) reported that in 2012, hydropower was regarded as the largest source of renewable energy, comprising of 16.2% of global electricity production. No doubt this technology is renewable and dependable source of low carbon electricity. Hydropower is the commercial technology for the electricity generation but technologies advancement is required to enhanced the its energy efficiency and reduction in the energy losses.

Ocean energy abbreviates a broad range of technologies to extract energy from the ocean. One way employed similar to conventional dams, tidal barrages, where incoming tide is trapped and slowly released to generate electricity. Another approach is to capture the energy of ocean currents and tides with the help of devices installed under the ocean surface to generate hydrokinetic energy. Various other ways viz. utilizing the surface wind wave's energy to produce electricity through a systematic device installed at bank of ocean or the sea surface such as power from offshore wind generators or by using the temperature differential between warm surface water and cold water from ocean depths; using osmotic energy i.e. pressure differential among salt and fresh water. Among the above technologies, literatures reported tidal barrages as mature one but some countries are in doubt about the economic viability of tidal barrages (Kidd et al., 2015).

Geothermal energy refers to the heat derived from the crust of earth. This energy comprises of high temperature hydrothermal resources, hot rock resources and deep aquifer systems with low and medium temperatures. Production/ reinjection boreholes, silencers, turbines/generators, separators, connecting/delivery pipelines and cooling towers are the various components of geothermal power plants (Bayer et al., 2013). Only 6.5% of overall worldwide geothermal potential is tapped i.e. 12.8 GW (Janusz Nowotny et al., 2018).

Electrochemical energy comprises of range of processes such as fuel cells, electrolysis, photolysis and photocatalysis. Electrochemical energy such as Fuel cells allow the conversion of chemical energy of fuels into electrical energy. The driving force of the conversion is the decrease of the Gibb's free energy of the related redox reactions that take place at the anode (oxidation) and cathode (reduction). Presently, more focus is given towards the development of new materials with increased ionic conduction and charge capacity. The more electrode materials development and enhancement in the energy conversion efficiency needs to be improved. The electrochemical energy conversion technologies have promising and great potential to be commercialized in the larger scale. Presently, more focus is given towards the development of new materials with increased ionic conduction, optimized materials properties to enhanced electrochemical conversion, durability, economical cost and lower operating temperatures is the emerging research areas.

The very often stated photochemical energy conversion process harvest energy by light induced water oxidation, hydrogen production via water splitting and partial oxidation of the organic compounds. Nowotny et al., (2018) reported that whole technologies is based on corrosion resistant oxide semiconductors immersed in water. They further stated that process of total oxidation generates oxygen at anode and hydrogen at cathode. The most common strategy to develop oxide semiconductors, like TiO_2, to elevate light absorption via modified electronic structure. Newly published reports state that performance of these systems is determined by the associated functional properties and local properties of the interface layer (J. Nowotny et al., 2014). They reported some of the essential properties like concentration of surface-active sites, electric fields essential for charge separation and gradient of defect concentration within the layer. Extreme and intensive efforts are focused in last decade for the wide-scale introduction of hydrogen-fuelled vehicles to reduce CO_2 emissions.

The most important and promising renewable energy technologies is the energy generation from the waste or biomass are termed as bioenergy. Bioenergy is referred as the renewable energy generated from plant and animal based matter (Gasparatos et al., 2017). The sources of bioenergy are diverse including residues from forest sectors (wood), livestock waste residues from agricultural sector, manufacturing waste, food waste, domestic waste and municipal waste (Roberts et al., 2015). Adib (2015) reported that total energy (primary) demand from bioenergy was 58.5 EJ and bioenergy's share in the total worldwide primary energy consumption was 10%. Traditional bioenergy such as dung, charcoal and woodfuel, used mainly for cooking and heating, accounts 54-60% [(Adib, 2015); (Gasparatos et al., 2017)] . Modern uses of bioenergy generating Green Economy include bio-heating, bio-fuels and bio-power. They mainly focus on biodiversity and ecosystem impacts (UNEP, 2014). The prominent technologies to drive modern bioenergy are classified into thermochemical conversion and biochemical conversion. Thermochemical conversion technologies, such as combustion, gasification and pyrolysis, produce bio-heat and bio-power. While as, biochemical conversion technologies, like digestion and fermentation, produce liquid bio-fuel (bio-ethanol, bio-diesel) for transport, cooking and heating [(Adib, 2015); (Roberts et al., 2015)]. Most of the attention in renewable energy circle is attracted by biomass energy. It incorporates residues from forest, crops, wood, human sewage, solid waste, energy crops and food processing waste (Dogan & Inglesi-Lotz, 2017). Far from other renewable, biomass is a readily obtainable source, and its ampleness makes it a suitable alternative to replace fossil fuels (Miranville, 2020). In 2009, biomass provided 55% of total renewable energy supply. Various developing nations have met 35% of energy demand and 13% of its global consumption through biomass. It is notable that agricultural, municipal and forests wastes represent capacity of 13 billion metric tons i.e. 10 times higher than present global energy demand (Sarkodie et al., 2019). On the other hand, biomass plays a key role in enhancing environmental quality [(Adewuyi & Awodumi, 2017); (Sulaiman et al., 2020)]. Wasif et al., (2021) reported that consumption of modern biomass energy sources has less environmental impacts compared to traditional ones. So, the new trend is focused generate renewable energy through modern biomass technology like anaerobic digestion. Biomass energy is termed as important alternative energy as it can reduce foreign oil dependency along with mitigate greenhouse gases emission (Bilgili & Ozturk, 2015). On the basis of feedstock and conversion technology employed, liquid bio-fuels are categorized as first, second and third generation fuel. Bioenergy, if generated in a sustainable manner, it shows the potential to mitigate carbon dioxide emissions. Gao & Zhang, (2021) reported that traditional use of biomass is expected to decline and modern use of biomass is rising exponentially.

Hydrogen Production Technologies

Globally, hydrogen is currently generated from coal, oil, natural gas and water electrolysis. (Gupta & Rao, 2017) reported that 50% global hydrogen demand is fulfilled by steam reforming of natural gas, 30% from oil reforming, 18% from coal gasification, 3.9% from water electrolysis and 0.1% from other sources. Some literatures reported that catalytic reforming of biomass, hydrocarbon or CH_4 as the major source of hydrogen production (Tao et al., 2017). Basheer & Ali, (2019) reported that the above method of hydrogen production consists of severe drawbacks, such as, low yield, high cost and environment challenging. These hydrogen production technologies by water through photo-catalysis and photolysis are the green process and highly efficient. There are some other waters splitting methods employed for hydrogen productions like photo-electrochemical, photo-biological and thermal decomposition. Various methods for green hydrogen energy generation are totally based on biomass, waste and wastewater using thermal, biochemical, electrical, photonic, photo-thermal, electro-thermal, photo-electric, thermal-biochemical and photo-biochemical. The substrate for the above process is selected on the basis of various fundamental criteria such as availability, efficiency, cost effectiveness, system integration option and many more. Green hydrogen generation requisites the utilization of sustainable energy resources in place of fossil fuels. Biological hydrogen production is considered as promising methods for green and clean hydrogen production. It is believed that in the upcoming future, green methods of hydrogen production will be implemented to supply hydrogen, production of fertilizers and other chemicals; in oil upgrade and many more (Dincer, 2012). They have also reported about biological methods, such as dark fermentation, photo-fermentation and direct & indirect photolysis, for hydrogen generation. The author further characterized different forms of energy that are utilized for hydrogen generation from renewable sources, namely, biochemical, photonic, electrical and thermal (Chaurasia, Siwach, et al., 2021). The electrical and thermal energy can be extracted from renewable energies such as solar, tidal, wind, wave, hydro, biomass, geothermal and ocean thermal. The photonic energy is comprised in solar energy. The biochemical energy is present in organic matter and can be harvested by various microbial species that release hydrogen from various substrates. There are some hybrid energy systems for generating hydrogen like electrical + thermal, photonic + biochemical, biochemical + thermal and electrical + photonic.

Importance of Hydrogen Energy

- Efficiency: Hydrogen (H_2) has a very high energy density (122 KJ/g), three times which of petrol or diesel.
- Environmental Benefits: Hydrogen is clean sources of energy that have a zero environmental impact than conventional energy technologies
- Feasibility to harvest wind renewable energy: hydrogen can be generated from a variety of renewable sources such as biomass, wastewater and waste materials.
- Not dependence on foreign oil: Oil fuel are uniformly distributed on our earth.
- Zero GHGs emissions or no any environmental impacts. No any noise generation.

Safety Implications

Hydrogen is a flammable gas at room temperature. Hydrogen also has various properties that may need some precautions in handling or great care while using directly or indirectly.

- Hydrogen is colourless and odourless gas
- High activity with oxygen and other oxidizers
- Very low ignition energy (1:10:H2: Petrol)
- Higher flame temperature (around 2273 K)
- Small molecular size offer difficulties as may promote diffusion and/or leakage
- Easily reacts with transition metals.

Table 1. Comparisons of hydrogen production technologies (Chaurasia & Mondal, 2021a; Kadier et al., 2020)

Process	Advantages	Disadvantages	Efficiency	Cost ($/kg)
Steam Reforming	Commercialized stage	Produced CO, CO_2 Unsteady supply	75-85	2.27
Partial Oxidation	Commercialized stage	H_2 Production with value added products (coke and oil) production	61-74	1.48
Auto thermal Reforming	Commercialized stage	H_2 Production with CO_2, use of fossil fuels	60-75	1.48
Bio photolysis	Utilize CO_2, Produced O_2, Under development stage	Low yields of H_2, sunlight needed, large reactor required, O_2 sensitivity, high cost of material.	10-11	2.13
Dark Fermentation	Waste reduction with H_2 production, light andO_2 not required, Technology development required	Low efficiency and low yield Required high volume of digester	65-78	2.57
Photo Fermentation	Degrade almost any wastewater, light required, under technology development.	Low hydrogen yield, Low efficiency, required light, high volume of reactor needed, sensitive to O_2	0.1	2.83
Gasification	Waste reduction, low-cost substrate, low pollutants generation, Commercialized stage	Impurities, unstable rate of H_2 production, geographical suitability	35-45	1.77-2.05
Pyrolysis	Utilized low-cost substrate or waste materials, CO_2-neutral. Established technology	Impurities, unstable rate of H_2 production, geographical suitability	40-50	1.59-170
Thermolysis	Sustainable technology, O_2 as by-product, high cost, no pollutants generation	High capital and operation cost, corrosion issue.	25-50	7.98-8.40
Photolysis	Sustainable technology under development, O_2 as by-product, high cost, no pollutants generation	Light required, catalyst under development, low yield	0.06	8-10
Electrolysis	Commercialized and greener technology, O_2 as by-product, costly process	Stationery technology, high cost	62-78	10.30

THE ROLE OF NANO-TECHNOLOGY ENERGY/ HYDROGEN PRODUCTION

Nanotechnology is used to produced materials at nanometre scale level such as nanometres in size but possess improved physical, chemical and electro-catalytic properties. Variety of formulations of nanoparticles have been developed including nanotubes, nanofilms, nanowires, nanoparticles, colloids and quantum dots. (Chaurasia et al., 2020) reported various benefits of nanotechnology for renewable energy generation i.e., increased efficiency and electro-catalytic activity of the materials that enhanced the energy harvesting process. Any sort of material can have various properties at nanoscale such as lighter, stronger, conductive, stable, reflective, and chemically active and enhanced magnetic properties (Rani et al., 2017). Literature reported that nanotechnology can provide significant contributions to hydrogen and solar energy sectors. Due to nanotechnology the elevated photovoltaic (PV) solar cells have improved the properties and reduction in their manufacturing and electricity generation costs. The application of nanotechnology has increased hydrogen adsorption capacity of fuel cells and make it economical and efficient. Few literature suggested that nanoparticles implementation in manufacturing sector can save non-renewable fuels, distribution of nanowires in electricity sector can reduce electricity losses and nanomaterials employed in buildings can prevent thermal and cooling energy loss. Hussein, (2015) mentioned that solar energy collection and efficiency can be enhanced through more conductivity and increased surface area to volume ratio of nanomaterials. They further reported that nanoparticles can facilitate release of more number of electrons when hit by a photon thereby increasing the efficiency of solar devices. Taylor et al., (2011) reported about the advantages of nanotechnology in solar energy production. They stated that nano size materials can easily pass via pumps and plumbing without resulting any adverse effects; nanomaterials are competent to absorb energy directly; they can be optically selective in nature; they can maintain uniform temperature thereby reducing material constraints; they can enhance heat transfer by elevated convection and thermal conductivity and increased absorption efficiency can be achieved by turning nanoparticle size and shape to the required application. Quantum dots are reported to be three times more efficient than current materials employed in solar cells (Shrair, 2014). Otanicar & Golden, (2009) concluded that nanofluids based solar collector possesses lower embodied energy and higher levels of efficiency. The nanofluids can be more effective than conventional fluids as a heat transfer medium in the solar water heater. The solar cell design with a nanowire based solar cell to use economical, durable and environment friendly semiconducting photovoltaic (PV) components. Otanicar et al., (2010) reported from their experimental observations that efficiency of solar thermal collectors can increase up to five percent by utilizing nanofluids made from carbon nanotubes, graphite and silver. Nanotechnology even possesses the potential to solve the major drawbacks of fuel cells used to generate hydrogen energy, namely, high cost, operability issues and durability issues (Hussein, 2015). Abbaraju et al., (2008) reported the use of TiO_2/SnO_2 nanoparticles on Nafion membranes to improve proton exchange membrane fuel cells. Nanomaterials can be employed for clean, green and economical hydrogen generation via photocatalysis (Jang et al., 2008).

Table 2. Current scenario of hydrogen production technologies

Hydrogen Production Techniques	Driving Energy	Renewable Resources	Comment	References
Electrolysis	Electrical Energy	Water	O_2 & H_2 is formed from H_2O through electrochemical reactions by passing a direct current	(Zeng & Zhang, 2010)
Thermolysis	Thermal Energy	Water	Decomposition of H_2O thermally	(Baykara, 2004)
Thermocatalysis	Thermal Energy	Hydrogen Sulfide	Sea derived H_2S is cracked thermo-catalytically	(Cong et al., 2016)
Thermochemical	Thermal Energy	Biomass, Water	H_2 is formed via thermo-catalytic biomass conversion, water molecule splitting via chemical reactions	(Tolga Balta et al., 2009)
PV-electrolysis	Photonic Energy	Water	Electricity produced by PV panels in order to drive electrolyzer	(Khaselev et al., 2001)
Photo-catalysis	Photonic Energy	Water	Photo-catalysts employed to generate H_2 from H_2O	(Idriss, 2021)
Bio-photolysis	Photonic Energy	Water	Cyanobacterial systems generate H_2	(Briefs & Energy, n.d.)
Dark Fermenatation	Biochemical Energy	Biomass	Anaerobic fermentation without light	(De Gioannis et al., 2013)
High Temperature Electrolysis	Electrical Energy & Thermal Energy	Water	Chemical reactions for water splitting in electrolyte cells	(O'Brien et al., 2010)
Photo-electrolysis	Electrical Energy & Photonic Energy	Water	Photoelectrodes employed along with external electric source	(Bendaikha & Larbi, 2013)
Thermophillic Digestion	Biochemical Energy & Thermal Energy	Biomass	Biomass digestion with heating at low grade temperature	(Tyagi et al., 2014)
Bio-photolysis	Photonic Energy & Biochemical Energy	Biomass, Water	Bacterial species employed for photo-generation of H_2	(Meher Kotay & Das, 2008)
Photo-fermentation	Photonic Energy & Biochemical Energy	Biomass	Fermentation in presence of light	(Meher Kotay & Das, 2008)
Artificial photosynthesis	Photonic Energy & Biochemical Energy	Biomass, Water	Chemically engineered molecules and systems to copy photosynthesis	(Dincer, 2012)

Use of Nano-Technology in the Wastewater Treatment

With continuous rise in world population, safe and clean drinking water for each and every person is a biggest challenge of this century. In 2015, WHO reported that more than 1.1 billion strive for sufficient drinking water and continuous contamination of freshwater by organic and inorganic pollutants is a major challenge in the water supply chain. The treatment of wastewater and drinking water can help to overcome these concerns to some extent (Ferroudj et al., 2013). But the main problem is that the traditional methods of wastewater treatment are not competent enough to remove the emerging contaminants and meet the strict standards of water quality. Ferroudj et al., (2013) mentioned the drawbacks of existing wastewater technologies such as, generation of toxic sludge, incomplete pollutant removal and requirement of high energy. Zelmanov & Semiat, (2008) reported that biological wastewater treatment is no doubt widely implied but these treatment technologies are slow, limited due to presence of non-biodegradable contaminant and sometimes toxic to microbial cultures via toxic contaminants. While there are some physical processes of wastewater treatment such as filtration which are capable of removing contaminants by transforming one phase to another. But they also possess a serious concern i.e., production of highly concentrated toxic sludge. Various literatures have enlightened the pivotal need of development of efficient and powerful wastewater treatment technologies (Burkhard et al., 2000); (Anjum et al., 2019). Various researchers suggested that advanced nanotechnology can have potential to remediate wastewater and can act as nano-adsorbents, nano-catalyst and many more for treatment of wastewater. Yin et al., (2013) showed that combination of biological wastewater treatment process with advance nanotechnology resulted in efficient water purification system. The types and useful properties of some nanomaterials used with biological processes are presented in Table. Some of the nanotechnology combined biological processes are described below:

Table 3. Impacts of nano-technology in the energy production process

Renewable Energy	Advantages	References
Solar Energy	1. Larger effective optical path for adsorption 2. Reduced recombination losses 3. Desired energy band gap value by varying the size of nanoparticles	(Sethi. et al., 2011)
Hydrogen Energy	1. Improving performance as multifunctional materials 2. Reduce cost of components 3. Enhance mechanical strength 4. Enhance electrical conductivity 5. Amplify hydrogen storage	(Hussein, 2015)
Bioenergy	1. Reduced capital costs 2. Enhanced performance of bioreactors 3. Improves efficiency of biofuel production	(Goh et al., 2012)
Wind Energy	1. Construction of longer and strong blades 2. Increase wind turbine efficiency 3. Improve wear resistance	(Hussein, 2015)

Table 4. Some used nano-materials in the wastewater treatment (A. Zare et al., 2019)

Nanomaterial	Properties
Gold Nanomaterials	• Chemically Stable • Excellent Conductivity • Ability to scatter light
Magnetic Nanomaterials	• Excellent Conductivity • Employed to concentrate particles in assays
Quantum Dots	• Energy donors • Photo-stability • Tuneable emission spectra
Carbon Nanotubes	• Excellent electrical properties • Exhibit photoluminescence

Table 5. Some used nano-materials in the pollutant's removal

Nanomaterials	Pollutant	Reference
CuO/ZnO	Methylene blue	(Sathishkumar et al., 2011)
Polyaniline /ZnO	Methylene blue Malachite green	(Eskizeybek et al., 2012)
ZnO/Zn	Methylene blue	(Lin et al., 2014)
AgBr/ZnO	Methylene Blue	(Dai et al., 2014)
3D SnO	Methyl Orange	(Cui et al., 2015)
ZnO nanorods	Rhodamine B	(Fang et al., 2015)
TiO_2/trititanate	Rhodamine B	(Chen et al., 2015)
Carbon nanorods- TiO_2	2,4-dichlorophenol	(Ortega-Liebana et al., 2016)
Immobilized TiO_2	Propranolol Diclofenac Carbamazepine	(He et al., 2016)
Ti_2CO	Methylene blue	(Guan et al., 2016)
Y_2O_3:Eu^{3+}	Methylene blue	(Anjum et al., 2019)

Use of Nanomaterials in the Feedstock Pretreatment for the Aerobic Digestion

Conventionally, treatment of municipal wastewater with biological process is less challenging in comparison to industrial wastewater as it contains more toxic and minimal biodegradable pollutants (Rittmann and McCarty, 2001). The nanomaterials are capable of degrading organic contaminants and are reported as well for the remediation of chlorinated organic compounds from ground water (Anjum et al.,

2019). Ma & Zhang, (2008) reported that combination of ZVI nanoparticle treatment with aerobic digestion is highly promising for biodegradation of organic pollutants from wastewater. They further mentioned that some of the contaminants will be remediated by nanoparticles and the remaining ones with aerobic digestion. This process becomes more efficient and less complex. As per reported by Anjum et al., (2019), that the products formed by ZVI nanoparticles are more biodegradable for aerobic microbial species. Some researchers have also reported the treatment of azo dye and nitro-aromatic compounds through coupling of ZVI nanoparticles with biological processes [(Oh et al., 2001); (Saxe et al., 2006)]. A pilot scale study was performed by Ma & Zhang, (2008) where they revealed that the effect of ZVI nanoparticles on biological treatment resulted in 96.5% removal efficiency and 86% Biological Oxygen Demand (BOD) & Chemical Oxygen Demand (COD).

Nanomaterials with Algae Membrane Based Bioreactor

Cultivation of algae in wastewater is regarded as one of the prominent technique for wastewater treatment along with energy generation (Sahu et al., 2019). With the availability of micronutrients, such as, trace metals and vitamins (cyanocobalamin, thiamin), and macronutrients, such as, calcium, sodium and potassium, which are crucial for algal growth, many algae species can effectively grow in wastewater [(Chong et al., 2000); (Abou-Shanab et al., 2013)]. Anjum et al., (2019) mentioned that chemical salts present in wastewater act as nutrient solution along with light and carbon dioxide can create a perfect niche for algal growth. This will result in removal of nutrients from wastewater and formation of algal biomass for energy generation. Brennan & Owende, (2010) mentioned about various algal biomass harvesting techniques such as, sedimentation, air flotation and centrifugation with chemical flocculation. They also stated that these techniques demand huge capital when implemented on large scale. W. Hu et al., (2015) reported that membrane technology in which membrane bioreactors are employed for high density algae cultivation is the most significant approach towards algae cultivation and biomass harvesting among advance techniques. Among this advantage, there are other benefits of membrane technology as well, unlike other harvesting methods there is no necessity of addition of chemicals like coagulants in the membrane filtration thereby making easing the reuse of water after filtration and biomass separation (Ríos et al., 2012). Furthermore, algal biomass recovery is better without cell damage and minimal energy requirement in comparison with conventional methods (Hu et al., 2015). Polysulfone (PSF), polyethersulfone (PES) and Polyvinylidene fluoride (PVDF) membranes are mostly employed [(Anjum et al., 2019); (Sahu et al., 2019)]. Despite being chemically and physically stable the only issue is the membrane fouling due to hydrophobic mechanism between membrane materials and microbial cells

(Maximous et al., 2009). Several researchers have discussed various approaches, such as, plasma treatment, surface coating and incorporation of nanoparticles, to enhance hydrophilicity and reduce membrane fouling [(Kim et al., 2011); (Madaeni & Ghaemi, 2007); (Yin et al., 2013)]. Madaeni & Ghaemi, (2007) mentioned that coating of TiO_2 nanoparticles coating modified the polyvinyl reverse osmosis membrane along with minimal fouling through self-cleaning mechanism under UV radiation. With high surface area and hydrophilic properties TiO_2 nanoparticles can be employed for pollution control (Martínez et al., 2013). Moghimifar et al., (2014) reported that due to possession of above characteristics, nanoparticles can be incorporated with membranes for reduction of fouling and hydrophobicity.

Use of Nanotechnology in the Bio-Electrochemical System for the Hydrogen Production

Microbial Electrolysis Cells (MECs) have attracted attention of various researchers due to their benefit for wastewater treatment along with renewable energy production by using microbes as biocatalysts [(Ghasemi, Daud, et al., 2013); (Logan, 2009)]. The energy produced by MECs can possibly reduce the electricity need for a conventional treatment process while complex organic molecules like acetate are split into hydrogen and carbon dioxide by bacterial species [(Ghangrekar & Shinde, 2007); (Logan et al., 2008)]. Overall low performance and expensive components are major barriers for commercialization of this technology (Ghasemi, Ismail, et al., 2013). Di Lorenzo et al., (2010) suggested that implementation of cathode catalyst would make MECs economically feasible. Along with that Imran Jafri et al., (2009) mentioned that selection of appropriate material for performance enhancement is also pivotal. This comprehends modification of electrode surface to achieve higher catalytic activity for oxygen reduction reaction (Zhuang et al., 2009). Some authors believe that nanoparticles will be biologically more active with regard to other materials as they possess high surface area and can effectively interact with bio-systems [(Vaddiraju et al., 2010); (Uskoković, 2007)]. Yuan et al., (2010) reported the use of economical nanocomposite materials like nanostructured carbon in electrodes. They further concluded that these electrodes possess better electrochemical catalytic activity, high conductivity and large surface area. Ghasemi et al., (2011) concluded in their study that due to high surface area of carbon nanotubes, the catalytic activity of Platinum enhanced. X. Wang et al., (2006) reported that unique structural properties and high electrical conductivity of carbon nanotubes derived the above results. Ghasemi, Ismail, et al., (2013) suggested that carbon nanotube/platinum can work as cathode catalyst for commercial purposes replacing platinum. Zou et al., (2008) reported that electrodes incorporated with nanoparticles possess large surface area with more active sites for electrochemical reactions and microbial attachment. Thus, biofilm formation will take place easily necessary for electron transfer towards anode surface. Biohydrogen production in MEC process and electrode reaction performances along with hydrogen yield can be enhanced by using efficient cathode materials or cathode catalyst materials. At the beginning of the MEC technology started extensive research on the cathode materials or cathode catalyst. The cathode or cathode catalyst materials are the mainly responsible for the ion's recombination at the cathodes surface in the MEC for the hydrogen generation as well as to overcome the thermodynamic barrier of the spontaneous MEC process. Table 6 enlisted some promising nanomaterials and nano catalyst as electrode material in the bio electrochemical system such as microbial fuel cells or microbial electrolysis cells etc. These nanomaterials show higher energy recovery from the waste or wastewater along with improved the process efficiency.

Table 6. Various cathode nano-catalyst materials in the MEC for the simultaneous treatment of the wastewater and hydrogen production

Study	E_{app} (V)	Cathode	Substrate	Q (m^3/m^3d)	Y_{H2} (mol H_2/mol substrate)
Liu et al., 2005	0.45	Carbon Paper/Pt	Acetate	0.37	2.52
Cheng & Logan, 2007a	0.6	CC/Pt	Cellulose	0.11	8.20
	0.6	CC/Pt	Butyric acid	0.45	8.01
	0.6	CC/Pt	Propionic acid	0.72	6.25
	0.6	CC/Pt	Valeric acid	0.14	8.77
Lu et al., 2010	0.6	Carbon Cloth (CC)/Pt	BSA	0.14-0.28	18.0
	0.8	CC/Pt	BSA	0.34-0.38	22.6-23.8
	0.6	CC/Pt	Peptone	0.07-0.09	5.1-6.5
	0.8	CC/Pt	Peptone	0.08-0.10	3.6-5.8
Kyazze et al., 2010	0.6	CC/Pt	Acetate	0.1	0.52
	0.85	CC/Pt	Acetate	0.2	1.1
	1.0	CC/Pt	Acetate	0.3	0.92
	0.7	CC/Pt	Glucose	0.34	1.63
Selembo et al., 2009	0.9	Ni 201	Acetate	0.34-0.42	-
	0.9	Ni 400	Acetate	0.31-0.51	-
	0.9	Stainless Steel (SS) 304	Acetate	0.58-0.60	-
	0.9	SS 316	Acetate	0.27-0.43	-
	0.9	SS 420	Acetate	0.51-065	-
	0.9	Ni 625	Acetate	0.79	-
Manuel et al., 2010b	0.8	Biocathode	Acetate	0.03	

continues on following page

Table 6. Continued

Study	E_{app} (V)	Cathode	Substrate	Q (m^3/m^3d)	Y_{H2} (mol H_2/mol substrate)
Escapa et al., 2012	0.75		domestic wastewater	0.3	-
Tenca et al., 2013	0.7	SS	Industrial wastewater/ food wastewater	0.8-0.12	-
Gil-Carrera et al., 2013b.	0.2	SS	Municipal wastewater	0.3	-
Sangeetha et al., 2016(Li, 2019)[137](Li, 2019)(Li, 2019)(Li, 2019)(Li, 2019)(Li, 2019)(Li, 2019)(Li, 2019)	0.7	Carbon Paper (CP)	Acetate	0.12-0.11	-
	0.8	SS	Beer industry wastewater	0.11	-
	0.8	Ni	‘’	0.1	-
	0.8	Cu	‘’	0.13	-
Mitov et al., 2017	0.6	Ni-MO/Ni-foam	Acetate	0.14	-
	0.6	Ni-W/ Ni-foam	Acetate	0.13	-
	0.6	Ni-foam	Acetate	0.04	-
Jayabalan et al., 2021	1	Ni-foam	Sugar industry wastewater	0.12	-
Jayabalan et al., 2021	1	$NiMoO_4$	“	0.14	-
Jayabalan et al., 2021	1	Nickel molybdate	”	0.12	-
Jayabalan et al., 2020b	1	NiO.rGO/ Co_3O_4.rGO	”	0.11-0.13	-
Jayabalan et al., 2021	1	Ni_2P	”	0.29	

CONCLUSION

The present chapter provides a comprehensive understanding about the various state of art and advancement in the waste to energy generation technologies. Their various process parameter is discussed along with advantages and disadvantages. It becomes evident that nanotechnology plays a pivotal role in increasing the efficiency of fuel cell, solar cell and wind turbine. With the utilization of nanotechnology, developed countries can reduce the hazardous impacts of fossil fuel burning for energy generation and can enhance their economy as well. Green and sustainable nanotechnology is advantageous for clean energy production in every aspect such as, performance, efficiency, cost and durability. Several countries across the globe have been funding and promoting the scientific advances in the energy generation from waste materials but still it needs the further technologies advancement. Advancement of green and sustainable nanotechnology toward the direction of clean energy generation should increment in providing robust solutions for environmental remediation and fulfilment of clean energy demand. Various green and sustainable technology led clean energy production processes are about to find their niche. But in order to overcome the challenge to commercialize all these green energy processes collaboration between researchers, industry, government and other stakeholders is very much essential.

REFERENCES

Abbaraju, R. R., Dasgupta, N., & Virkar, A. V. (2008). Composite Nafion Membranes Containing Nanosize TiO[sub 2]/SnO[sub 2] for Proton Exchange Membrane Fuel Cells. *Journal of the Electrochemical Society, 155*(12), B1307. doi:10.1149/1.2994079

Abou-Shanab, R. A. I., Ji, M. K., Kim, H. C., Paeng, K. J., & Jeon, B. H. (2013). Microalgal species growing on piggery wastewater as a valuable candidate for nutrient removal and biodiesel production. *Journal of Environmental Management, 115*, 257–264. doi:10.1016/j.jenvman.2012.11.022 PMID:23270891

Adewuyi, A. O., & Awodumi, O. B. (2017). Biomass energy consumption, economic growth and carbon emissions: Fresh evidence from West Africa using a simultaneous equation model. *Energy, 119*, 453–471. doi:10.1016/j.energy.2016.12.059

Adib, R. (2015). *Renewables 2015 Global Status Report Distrbuted Renewable Energy For Energy Access.* Academic Press.

Ali, I., Peng, C., Khan, Z. M., Naz, I., Sultan, M., Ali, M., Abbasi, I. A., Islam, T., & Ye, T. (2019). Overview of microbes based fabricated biogenic nanoparticles for water and wastewater treatment. *Journal of Environmental Management, 230*(August), 128–150. doi:10.1016/j.jenvman.2018.09.073

Alley, Berntsen, Bindoff, Chen, Chidthaisong, Friedlingstein, Gregory, Hegerl, Heimann, Hewitson, Hoskins, Joos, Jouzel, Kattsov, & Lohmann. (2007). Climate Change 2007 - The physical science basis: Contribution of working group I to the fourth assessment rep. *Bulletin Fuer Angewandte Geologie, 18*(2), 5–19.

Anjum, M., Miandad, R., Waqas, M., Gehany, F., & Barakat, M. A. (2019). Remediation of wastewater using various nano-materials. *Arabian Journal of Chemistry, 12*(8), 4897–4919. doi:10.1016/j.arabjc.2016.10.004

Balat, M. (2007). Hydrogen in fueled systems and the significance of hydrogen in vehicular transportation. *Energy Sources. Part B, Economics, Planning, and Policy, 2*(1), 49–61. doi:10.1080/15567240500400911

Basheer, A. A., & Ali, I. (2019). Water photo splitting for green hydrogen energy by green nanoparticles. *International Journal of Hydrogen Energy, 44*(23), 11564–11573. doi:10.1016/j.ijhydene.2019.03.040

Bayer, P., Rybach, L., Blum, P., & Brauchler, R. (2013). Review on life cycle environmental effects of geothermal power generation. *Renewable & Sustainable Energy Reviews*, *26*, 446–463. doi:10.1016/j.rser.2013.05.039

Baykara, S. Z. (2004). Experimental solar water thermolysis. *International Journal of Hydrogen Energy*, *29*(14), 1459–1469. doi:10.1016/j.ijhydene.2004.02.011

Bendaikha, O., & Larbi, S. (2013). Hydrogen production system analysis using direct photo-electrolysis process in Algeria. *Proceedings of 2013 International Conference on Renewable Energy Research and Applications, ICRERA 2013,* 1123–1128. 10.1109/ICRERA.2013.6749921

Bilgili, F., & Ozturk, I. (2015). Biomass energy and economic growth nexus in G7 countries: Evidence from dynamic panel data. *Renewable & Sustainable Energy Reviews*, *49*, 132–138. doi:10.1016/j.rser.2015.04.098

Blazejczak, J., Braun, F. G., Edler, D., & Schill, W. P. (2014). Economic effects of renewable energy expansion: A model-based analysis for Germany. *Renewable & Sustainable Energy Reviews*, *40*, 1070–1080. doi:10.1016/j.rser.2014.07.134

Brennan, L., & Owende, P. (2010). Biofuels from microalgae-A review of technologies for production, processing, and extractions of biofuels and co-products. *Renewable & Sustainable Energy Reviews*, *14*(2), 557–577. doi:10.1016/j.rser.2009.10.009

Briefs, S., & Energy, I. N. (n.d.). *Solar Based H2 Prod Systems*. Academic Press.

Burgherr, P. (2011). Accidents and risks BT. *IPCC Special Report on Renewable Energy Sources and Climate Change Mitigation, 9.3.4.7.*

Burkhard, R., Deletic, A., & Craig, A. (2000). Techniques for water and wastewater management: A review of techniques and their integration in planning. *Urban Water*, *2*(3), 197–221. doi:10.1016/S1462-0758(00)00056-X

Chaurasia, A. K., Goyal, H., & Mondal, P. (2020). Hydrogen gas production with Ni, Ni–Co and Ni–Co–P electrodeposits as potential cathode catalyst by microbial electrolysis cells. *International Journal of Hydrogen Energy*, *45*(36), 18250–18265. doi:10.1016/j.ijhydene.2019.07.175

Chaurasia, A. K., & Mondal, P. (2021a). *Hydrogen Production From Waste and Renewable Resources*. doi:10.4018/978-1-7998-4945-2.ch002

Chaurasia, A. K., & Mondal, P. (2021b). *Simultaneous Removal of Organic Load and Hydrogen Gas Production Using Electrodeposits Cathodes in MEC*. doi:10.1007/978-981-15-6595-3_34

Chaurasia, A. K., & Mondal, P. (2022). Enhancing biohydrogen production from sugar industry wastewater using Ni, Ni–Co and Ni–Co–P electrodeposits as cathodes in microbial electrolysis cells. *Chemosphere, 286*(November). doi:10.1016/j.chemosphere.2021.131728

Chaurasia, A. K., Shankar, R., & Mondal, P. (2021). Effects of nickle, nickle-cobalt and nickle-cobalt-phosphorus nanocatalysts for enhancing biohydrogen production in microbial electrolysis cells using paper industry wastewater. *Journal of Environmental Management, 298*(July), 113542. doi:10.1016/j.jenvman.2021.113542 PMID:34426219

Chaurasia, A. K., Siwach, P., Shankar, R., & Mondal, P. (2021). Shankar, · Ravi, & Mondal, · Prasenjit. (2021). Effect of pre-treatment on mesophilic anaerobic co-digestion of fruit, food and vegetable waste. *Clean Technologies and Environmental Policy, 1*, 1–14. doi:10.100710098-021-02218-5

Chen, C. C., Lin, C. L., & Chen, L. C. (2015). Functionalized carbon nanomaterial supported palladium nano-catalysts for electrocatalytic glucose oxidation reaction. *Electrochimica Acta, 152*, 408–416. doi:10.1016/j.electacta.2014.11.116

Cheng, S., & Logan, B. E. (2007). Sustainable and efficient biohydrogen production via electrohydrogenesis. *Proceedings of the National Academy of Sciences of the United States of America, 104*(47), 18871–18873. doi:10.1073/pnas.0706379104 PMID:18000052

Chong, A. M. Y., Wong, Y. S., & Tam, N. F. Y. (2000). Performance of different microalgal species in removing nickel and zinc from industrial wastewater. *Chemosphere, 41*(1–2), 251–257. doi:10.1016/S0045-6535(99)00418-X PMID:10819208

Cong, T. Y., Raj, A., Chanaphet, J., Mohammed, S., Ibrahim, S., & Al Shoaibi, A. (2016). A detailed reaction mechanism for hydrogen production via hydrogen sulphide (H2S) thermolysis and oxidation. *International Journal of Hydrogen Energy, 41*(16), 6662–6675. doi:10.1016/j.ijhydene.2016.03.053

Cui, Y., Wang, F., Iqbal, M. Z., Wang, Z., Li, Y., & Tu, J. (2015). Synthesis of novel 3D SnO flower-like hierarchical architectures self-assembled by nano-leaves and its photocatalysis. *Materials Research Bulletin, 70*, 784–788. doi:10.1016/j.materresbull.2015.06.021

Dai, K., Lv, J., Lu, L., Liu, Q., Zhu, G., & Li, D. (2014). Synthesis of micro-nano heterostructure AgBr/ZnO composite for advanced visible light photocatalysis. *Materials Letters, 130*, 5–8. doi:10.1016/j.matlet.2014.04.190

De Gioannis, G., Muntoni, A., Polettini, A., & Pomi, R. (2013). A review of dark fermentative hydrogen production from biodegradable municipal waste fractions. *Waste Management (New York, N.Y.), 33*(6), 1345–1361. doi:10.1016/j.wasman.2013.02.019 PMID:23558084

Di Lorenzo, M., Scott, K., Curtis, T. P., & Head, I. M. (2010). Effect of increasing anode surface area on the performance of a single chamber microbial fuel cell. *Chemical Engineering Journal, 156*(1), 40–48. doi:10.1016/j.cej.2009.09.031

Dincer, I. (2012). Green methods for hydrogen production. *International Journal of Hydrogen Energy, 37*(2), 1954–1971. doi:10.1016/j.ijhydene.2011.03.173

Dogan, E., & Inglesi-Lotz, R. (2017). Analyzing the effects of real income and biomass energy consumption on carbon dioxide (CO2) emissions: Empirical evidence from the panel of biomass-consuming countries. *Energy, 138*, 721–727. doi:10.1016/j.energy.2017.07.136

Edler, D., Ottm, M., & Lehr, U. (2009). *Bruttobeschäftigung durch erneuerbare Energien in Deutschland im Jahr 2008*. Academic Press.

(n.d.).. . *Environmental Biotechnology*.

Escapa, A., Gil-Carrera, L., García, V., & Morán, A. (2012). Performance of a continuous flow microbial electrolysis cell (MEC) fed with domestic wastewater. *Bioresource Technology, 117*, 55–62. doi:10.1016/j.biortech.2012.04.060 PMID:22609714

Eskizeybek, V., Sari, F., Gülce, H., Gülce, A., & Avci, A. (2012). Preparation of the new polyaniline/ZnO nanocomposite and its photocatalytic activity for degradation of methylene blue and malachite green dyes under UV and natural sun lights irradiations. *Applied Catalysis B: Environmental, 119–120*, 197–206. doi:10.1016/j.apcatb.2012.02.034

Fang, J., Fan, H., Ma, Y., Wang, Z., & Chang, Q. (2015). Surface defects control for ZnO nanorods synthesized by quenching and their anti-recombination in photocatalysis. *Applied Surface Science, 332*, 47–54. doi:10.1016/j.apsusc.2015.01.139

Fearnside, P., Rosa, L. P., Saut, P., Guiana, F., Paris, T., & Nations, U. (2006). Methane Quashes Green Credentials of Hydropower Preprint Analysis Quantifies Scientific Plagiarism. *Nature, 444*(November), 524–525.

Ferroudj, N., Nzimoto, J., Davidson, A., Talbot, D., Briot, E., Dupuis, V., Bée, A., Medjram, M. S., & Abramson, S. (2013). Maghemite nanoparticles and maghemite/silica nanocomposite microspheres as magnetic Fenton catalysts for the removal of water pollutants. *Applied Catalysis B: Environmental, 136–137*, 9–18. doi:10.1016/j.apcatb.2013.01.046

Gao, J., & Zhang, L. (2021). Does biomass energy consumption mitigate CO2 emissions? The role of economic growth and urbanization: Evidence from developing Asia. *Journal of the Asia Pacific Economy*, *26*(1), 96–115. doi:10.1080/13547860.2020.1717902

Gasparatos, A., Doll, C. N. H., Esteban, M., Ahmed, A., & Olang, T. A. (2017). Renewable energy and biodiversity: Implications for transitioning to a Green Economy. *Renewable and Sustainable Energy Reviews, 70*(December), 161–184. doi:10.1016/j.rser.2016.08.030

Ghangrekar, M. M., & Shinde, V. B. (2007). Performance of membrane-less microbial fuel cell treating wastewater and effect of electrode distance and area on electricity production. *Bioresource Technology*, *98*(15), 2879–2885. doi:10.1016/j.biortech.2006.09.050 PMID:17107789

Ghasemi, M., Daud, W. R. W., Hassan, S. H. A., Oh, S. E., Ismail, M., Rahimnejad, M., & Jahim, J. M. (2013). Nano-structured carbon as electrode material in microbial fuel cells: A comprehensive review. *Journal of Alloys and Compounds*, *580*, 245–255. doi:10.1016/j.jallcom.2013.05.094

Ghasemi, M., Ismail, M., Kamarudin, S. K., Saeedfar, K., Daud, W. R. W., Hassan, S. H. A., Heng, L. Y., Alam, J., & Oh, S. E. (2013). Carbon nanotube as an alternative cathode support and catalyst for microbial fuel cells. *Applied Energy*, *102*, 1050–1056. doi:10.1016/j.apenergy.2012.06.003

Ghasemi, M., Shahgaldi, S., Ismail, M., Kim, B. H., Yaakob, Z., & Wan Daud, W. R. (2011). Activated carbon nanofibers as an alternative cathode catalyst to platinum in a two-chamber microbial fuel cell. *International Journal of Hydrogen Energy*, *36*(21), 13746–13752. doi:10.1016/j.ijhydene.2011.07.118

Gleick, P. H. (2013). The World's Water Volume 7. *Berkeley Planning Journal*, *26*(1).

Goh, W. J., Makam, V. S., Hu, J., Kang, L., Zheng, M., Yoong, S. L., Udalagama, C. N. B., & Pastorin, G. (2012). Iron oxide filled magnetic carbon nanotube-enzyme conjugates for recycling of amyloglucosidase: Toward useful applications in biofuel production process. *Langmuir*, *28*(49), 16864–16873. doi:10.1021/la303046m PMID:23148719

Guan, S., Hao, L., Lu, Y., Yoshida, H., Pan, F., & Asanuma, H. (2016). Fabrication of oxygen-deficient TiO2 coatings with nano-fiber morphology for visible-light photocatalysis. *Materials Science in Semiconductor Processing*, *41*, 358–363. doi:10.1016/j.mssp.2015.09.018

Gupta, U., & Rao, C. N. R. (2017). Hydrogen generation by water splitting using MoS2 and other transition metal dichalcogenides. *Nano Energy*, *41*(July), 49–65. doi:10.1016/j.nanoen.2017.08.021

He, Y., Sutton, N. B., Rijnaarts, H. H. H., & Langenhoff, A. A. M. (2016). Degradation of pharmaceuticals in wastewater using immobilized TiO2 photocatalysis under simulated solar irradiation. *Applied Catalysis B: Environmental*, *182*, 132–141. doi:10.1016/j.apcatb.2015.09.015

Hosseini, S. E., & Wahid, M. A. (2016). Hydrogen production from renewable and sustainable energy resources: Promising green energy carrier for clean development. *Renewable & Sustainable Energy Reviews*, *57*, 850–866. Advance online publication. doi:10.1016/j.rser.2015.12.112

Hu, W., Yin, J., Deng, B., & Hu, Z. (2015). Application of nano TiO2 modified hollow fiber membranes in algal membrane bioreactors for high-density algae cultivation and wastewater polishing. *Bioresource Technology*, *193*, 135–141. doi:10.1016/j.biortech.2015.06.070 PMID:26125613

Hussein, A. K. (2015). Applications of nanotechnology in renewable energies - A comprehensive overview and understanding. *Renewable & Sustainable Energy Reviews*, *42*, 460–476. doi:10.1016/j.rser.2014.10.027

Idriss, H. (2021). Toward Large-Scale Hydrogen Production from Water: What Have We Learned and What Are the Main Research Hurdles to Cross for Commercialization? *Energy Technology (Weinheim)*, *9*(2), 1–9. doi:10.1002/ente.202000843

IEA. (2019). *The Future of Hydrogen – Analysis*. IEA.

IEA. (2020). *(2020). World Energy Outlook 2020*. IEA.

Imran Jafri, R., Sujatha, N., Rajalakshmi, N., & Ramaprabhu, S. (2009). Au-MnO2/MWNT and Au-ZnO/MWNT as oxygen reduction reaction electrocatalyst for polymer electrolyte membrane fuel cell. *International Journal of Hydrogen Energy*, *34*(15), 6371–6376. doi:10.1016/j.ijhydene.2009.05.084

Jang, J. S., Kim, H. G., Joshi, U. A., Jang, J. W., & Lee, J. S. (2008). Fabrication of CdS nanowires decorated with TiO2 nanoparticles for photocatalytic hydrogen production under visible light irradiation. *International Journal of Hydrogen Energy*, *33*(21), 5975–5980. doi:10.1016/j.ijhydene.2008.07.105

Jayabalan, T., Matheswaran, M., Preethi, V., & Naina Mohamed, S. (2020). Enhancing biohydrogen production from sugar industry wastewater using metal oxide/graphene nanocomposite catalysts in microbial electrolysis cell. *International Journal of Hydrogen Energy*, *45*(13), 7647–7655. doi:10.1016/j.ijhydene.2019.09.068

Jayabalan, T., Matheswaran, M., Radhakrishnan, T. K., & Naina Mohamed, S. (2021). Influence of Nickel molybdate nanocatalyst for enhancing biohydrogen production in microbial electrolysis cell utilizing sugar industrial effluent. *Bioresource Technology*, *320*, 124284. doi:10.1016/j.biortech.2020.124284 PMID:33137640

Jones, E. R., van Vliet, M. T. H., Qadir, M., & Bierkens, M. F. P. (2021). Country-level and gridded estimates of wastewater production, collection, treatment and reuse. *Earth System Science Data*, *13*(2), 237–254. doi:10.5194/essd-13-237-2021

K., S. V., M., P., & P., S. (2011). Use of Nanotechnology in Solar PV Cell. *International Journal of Chemical Engineering and Applications, 2*(2), 77–80.

Kadier, A., Chaurasia, A. K., Sapuan, S. M., Ilyas, R. A., Ma, P. C., Alabbosh, K. F. S., Rai, P. K., Logroño, W., Hamid, A. A., & Hasan, H. A. (2020). Essential Factors for Performance Improvement and the Implementation of Microbial Electrolysis Cells (MECs). In Bioelectrochemical Systems (pp. 139–168). Springer Singapore. doi:10.1007/978-981-15-6872-5_7

Karim, M. M. (2000). Arsenic in groundwater and health problems in Bangladesh. *Water Research*, *34*(1), 304–310. doi:10.1016/S0043-1354(99)00128-1

Katzner, T., Johnson, J. A., Evans, D. M., Garner, T. W. J., Gompper, M. E., Altwegg, R., Branch, T. A., Gordon, I. J., & Pettorelli, N. (2013). Challenges and opportunities for animal conservation from renewable energy development. *Animal Conservation*, *16*(4), 367–369. doi:10.1111/acv.12067

Khaselev, O., Bansal, A., & Turner, J. A. (2001). High-efficiency integrated multijunction photovoltaic/electrolysis systems for hydrogen production. *International Journal of Hydrogen Energy*, *26*(2), 127–132. doi:10.1016/S0360-3199(00)00039-2

Kidd, I. M., Fischer, A., Chai, S., & Davis, J. A. (2015). A scenario-based approach to evaluating potential environmental impacts following a tidal barrage installation. *Ocean and Coastal Management*, *116*, 9–19. doi:10.1016/j.ocecoaman.2015.06.016

Kim, E. S., Yu, Q., & Deng, B. (2011). Plasma surface modification of nanofiltration (NF) thin-film composite (TFC) membranes to improve anti organic fouling. *Applied Surface Science*, *257*(23), 9863–9871. doi:10.1016/j.apsusc.2011.06.059

Konieczny, A., Mondal, K., Wiltowski, T., & Dydo, P. (2008). Catalyst development for thermocatalytic decomposition of methane to hydrogen. *International Journal of Hydrogen Energy*, *33*(1), 264–272. doi:10.1016/j.ijhydene.2007.07.054

Kyazze, G., Popov, A., Dinsdale, R., Esteves, S., Hawkes, F., Premier, G., & Guwy, A. (2010). Influence of catholyte pH and temperature on hydrogen production from acetate using a two chamber concentric tubular microbial electrolysis cell. *International Journal of Hydrogen Energy*, *35*(15), 7716–7722. doi:10.1016/j.ijhydene.2010.05.036

Li, J. (2019). Investigation of hydrogen evolution on electrodeposited Ni/P coated carbon paper electrode in microbial fuel cell. *International Journal of Electrochemical Science*, *14*, 7582–7593. doi:10.20964/2019.08.04

Lima, M. G. B., & Gupta, J. (2013). Studying Global Environmental Meetings. *Global Environmental Politics*, *13*(August), 46–64. doi:10.1162/GLEP_a_00166

Lin, S. T., Thirumavalavan, M., Jiang, T. Y., & Lee, J. F. (2014). Synthesis of ZnO/Zn nano photocatalyst using modified polysaccharides for photodegradation of dyes. *Carbohydrate Polymers*, *105*(1), 1–9. doi:10.1016/j.carbpol.2014.01.017 PMID:24708945

Liu, H., Grot, S., & Logan, B. E. (2005). Electrochemically assisted microbial production of hydrogen from acetate. *Environmental Science & Technology*, *39*(11), 4317–4320. doi:10.1021/es050244p PMID:15984815

Logan, B. E. (2009). Fuel Cells Fuel cells. *Fuel Cells: Technologies for Fuel Processing*, 1–32.

Logan, B. E., Call, D., Cheng, S., Hamelers, H. V. M., Sleutels, T. H. J. A., Jeremiasse, A. W., & Rozendal, R. A. (2008). Microbial electrolysis cells for high yield hydrogen gas production from organic matter. In Environmental Science and Technology (Vol. 42, Issue 23, pp. 8630–8640). doi:10.1021/es801553z

Lu, L., Xing, D., Xie, T., Ren, N., & Logan, B. E. (2010). Hydrogen production from proteins via electrohydrogenesis in microbial electrolysis cells. *Biosensors & Bioelectronics*, *25*(12), 2690–2695. doi:10.1016/j.bios.2010.05.003 PMID:20537524

Ma, L., & Zhang, W. X. (2008). Enhanced biological treatment of industrial wastewater with bimetallic zero-valent iron. *Environmental Science & Technology*, *42*(15), 5384–5389. doi:10.1021/es801743s PMID:18754450

Madaeni, S. S., & Ghaemi, N. (2007). Characterization of self-cleaning RO membranes coated with TiO2 particles under UV irradiation. *Journal of Membrane Science*, *303*(1–2), 221–233. doi:10.1016/j.memsci.2007.07.017

Martínez, F., López-Muñoz, M. J., Aguado, J., Melero, J. A., Arsuaga, J., Sotto, A., Molina, R., Segura, Y., Pariente, M. I., Revilla, A., Cerro, L., & Carenas, G. (2013). Coupling membrane separation and photocatalytic oxidation processes for the degradation of pharmaceutical pollutants. *Water Research*, *47*(15), 5647–5658. doi:10.1016/j.watres.2013.06.045 PMID:23863375

Maximous, N., Nakhla, G., & Wan, W. (2009). Comparative assessment of hydrophobic and hydrophilic membrane fouling in wastewater applications. *Journal of Membrane Science*, *339*(1–2), 93–99. doi:10.1016/j.memsci.2009.04.034

Meher Kotay, S., & Das, D. (2008). Biohydrogen as a renewable energy resource-Prospects and potentials. *International Journal of Hydrogen Energy*, *33*(1), 258–263. doi:10.1016/j.ijhydene.2007.07.031

Miranville, A. (2020). Annual report 2019. *AIMS Mathematics*, *5*(1), i–v. doi:10.3934/math.2020i

Moghimifar, V., Raisi, A., & Aroujalian, A. (2014). Surface modification of polyethersulfone ultrafiltration membranes by corona plasma-assisted coating TiO2 nanoparticles. *Journal of Membrane Science*, *461*, 69–80. doi:10.1016/j.memsci.2014.02.012

Mondal, P., Kumari, P., Singh, J., Verma, S., Chaurasia, A. K., & Singh, R. P. (2017). Oil from algae. *Sustainable Utilization of Natural Resources*, 214–253. doi:10.1201/9781315153292

More power from water. (2018). *Nature Nanotechnology*, *13*(12), 1087. doi:10.1038/s41565-018-0340-7

Nowotny, J., Atanacio, A. J., Bak, T., Belova, I. V., Fiechter, S., Ikuma, Y., Ionescu, M., Kennedy, B. J., Majewski, P., Murch, G. E., & Wachsman, E. D. (2014). Photosensitive oxide semiconductors for solar hydrogen fuel and water disinfection. *International Materials Reviews*, *59*(8), 449–478. doi:10.1179/1743280414Y.0000000039

Nowotny, J., Dodson, J., Fiechter, S., Gür, T. M., Kennedy, B., Macyk, W., Bak, T., Sigmund, W., Yamawaki, M., & Rahman, K. A. (2017, February). Towards global sustainability: Education on environmentally clean energy technologies. *Renewable & Sustainable Energy Reviews*, *81*, 2541–2551. doi:10.1016/j.rser.2017.06.060

Nowotny, J., Hoshino, T., Dodson, J., Atanacio, A. J., Ionescu, M., Peterson, V., Prince, K. E., Yamawaki, M., Bak, T., Sigmund, W., Veziroglu, T. N., & Alim, M. A. (2016). Towards sustainable energy. Generation of hydrogen fuel using nuclear energy. *International Journal of Hydrogen Energy*, *41*(30), 12812–12825. doi:10.1016/j.ijhydene.2016.05.054

O'Brien, J. E., Stoots, C. M., Herring, J. S., McKellar, M. G., Sohal, M. S., & Condie, K. G. (2010). *High Temperature Electrolysis for Hydrogen Production from Nuclear Energy - Technology Summary*. Academic Press.

Oh, B. T., Just, C. L., & Alvarez, P. J. J. (2001). Hexahydro-1,3,5-trinitro-1,3,5-triazine mineralization by zerovalent iron and mixed anaerobic cultures. *Environmental Science & Technology*, *35*(21), 4341–4346. doi:10.1021/es010852e PMID:11718353

Ortega-Liebana, M. C., Hueso, J. L., Ferdousi, S., Yeung, K. L., & Santamaria, J. (2016). Nitrogen-doped luminescent carbon nanodots for optimal photo-generation of hydroxyl radicals and visible-light expanded photo-catalysis. *Diamond and Related Materials*, *65*, 176–182. doi:10.1016/j.diamond.2016.03.021

Otanicar, T. P., Carlson, J. D., Golden, J. S., Kaloush, K. E., & Phelan, P. E. (2010). Impact of the Urban Heat Island on light duty vehicle emissions for the phoenix, AZ Area. *International Journal of Sustainable Transportation*, *4*(1), 1–13. doi:10.1080/15568310802428552

Otanicar, T. P., & Golden, J. S. (2009). Comparative environmental and economic analysis of conventional and nanofluid solar hot water technologies. *Environmental Science & Technology*, *43*(15), 6082–6087. doi:10.1021/es900031j PMID:19731722

Panwar, N. L., Kaushik, S. C., & Kothari, S. (2011). Role of renewable energy sources in environmental protection: A review. *Renewable & Sustainable Energy Reviews*, *15*(3), 1513–1524. doi:10.1016/j.rser.2010.11.037

Peste, F., Paula, A., da Silva, L. P., Bernardino, J., Pereira, P., Mascarenhas, M., Costa, H., Vieira, J., Bastos, C., Fonseca, C., & Pereira, M. J. R. (2015). How to mitigate impacts of wind farms on bats? A review of potential conservation measures in the European context. *Environmental Impact Assessment Review*, *51*, 10–22. doi:10.1016/j.eiar.2014.11.001

Qu, X., Brame, J., Li, Q., & Alvarez, P. J. J. (2013). Nanotechnology for a safe and sustainable water supply: Enabling integrated water treatment and reuse. *Accounts of Chemical Research*, *46*(3), 834–843. doi:10.1021/ar300029v PMID:22738389

Rani, M., Shanker, U., & Chaurasia, A. K. (2017). Catalytic potential of laccase immobilized on transition metal oxides nanomaterials: Degradation of alizarin red S dye. *Journal of Environmental Chemical Engineering*, *5*(3), 2730–2739. Advance online publication. doi:10.1016/j.jece.2017.05.026

Ríos, S. D., Salvadó, J., Farriol, X., & Torras, C. (2012). Antifouling microfiltration strategies to harvest microalgae for biofuel. *Bioresource Technology*, *119*, 406–418. doi:10.1016/j.biortech.2012.05.044 PMID:22739578

Roberts, J. J., Cassula, A. M., Osvaldo Prado, P., Dias, R. A., & Balestieri, J. A. P. (2015). Assessment of dry residual biomass potential for use as alternative energy source in the party of General Pueyrredón, Argentina. *Renewable & Sustainable Energy Reviews*, *41*, 568–583. doi:10.1016/j.rser.2014.08.066

Rozell, D. J., & Reaven, S. J. (2012). Water Pollution Risk Associated with Natural Gas Extraction from the Marcellus Shale. *Risk Analysis*, *32*(8), 1382–1393. doi:10.1111/j.1539-6924.2011.01757.x PMID:22211399

Sahu, J. N., Zabed, H., Karri, R. R., Shams, S., & Qi, X. (2019). Applications of nano-biotechnology for sustainable water purification. In *Industrial Applications of Nanomaterials*. Elsevier Inc. doi:10.1016/B978-0-12-815749-7.00011-6

Sarkodie, S. A., Strezov, V., Weldekidan, H., Asamoah, E. F., Owusu, P. A., & Doyi, I. N. Y. (2019). Environmental sustainability assessment using dynamic Autoregressive-Distributed Lag simulations—Nexus between greenhouse gas emissions, biomass energy, food and economic growth. *The Science of the Total Environment*, *668*, 318–332. doi:10.1016/j.scitotenv.2019.02.432 PMID:30852209

Sathishkumar, P., Sweena, R., Wu, J. J., & Anandan, S. (2011). Synthesis of CuO-ZnO nanophotocatalyst for visible light assisted degradation of a textile dye in aqueous solution. *Chemical Engineering Journal*, *171*(1), 136–140. doi:10.1016/j.cej.2011.03.074

Saxe, J. P., Lubenow, B. L., Chiu, P. C., Huang, C.-P., & Cha, D. K. (2006). Enhanced Biodegradation of Azo Dyes Using an Integrated Elemental Iron-Activated Sludge System: II. Effects of Physical-Chemical Parameters. *Water Environment Research*, *78*(1), 26–30. doi:10.2175/106143005X84486 PMID:16553163

Selembo, P. A., Merrill, M. D., & Logan, B. E. (2009). The use of stainless steel and nickel alloys as low-cost cathodes in microbial electrolysis cells. *Journal of Power Sources*, *190*(2), 271–278. doi:10.1016/j.jpowsour.2008.12.144

Shankar, R., Pathak, N., Chaurasia, A. K., Mondal, P., & Chand, S. (2017). Energy production through microbial fuel cells. *Sustainable Utilization of Natural Resources*, 354–380. doi:10.1201/9781315153292

Shrair, J. S. A. (2014). *Advances in Nanotechnology can Provide Clean Energy Resources and Sustainable Development*. Academic Press.

Su, Y., Liang, Y., Chai, L., Han, Z., Ma, S., Lyu, J., Li, Z., & Yang, L. (2019). Water degradation by China's fossil fuels production: A life cycle assessment based on an input-output model. *Sustainability (Switzerland)*, *11*(15), 4130. Advance online publication. doi:10.3390u11154130

Sucher, N. J., Carles, M. C., Nowotny, J., & Bak, T. (2012). Photocatalytic water disinfection on oxide semiconductors: Part 2 - Structure, functional properties and reactivity of microbial agents. *Advances in Applied Ceramics*, *111*(1–2), 16–33. doi:10.1179/1743676111Y.0000000039

Sulaiman, C., Abdul-Rahim, A. S., & Ofozor, C. A. (2020). Does wood biomass energy use reduce CO2 emissions in European Union member countries? Evidence from 27 members. *Journal of Cleaner Production*, *253*, 119996. doi:10.1016/j.jclepro.2020.119996

Tabassum, A., Premalatha, M., Abbasi, T., & Abbasi, S. A. (2014). Wind energy: Increasing deployment, rising environmental concerns. *Renewable & Sustainable Energy Reviews*, *31*, 270–288. doi:10.1016/j.rser.2013.11.019

Tao, K., Arano, H., Zhang, P., Ai, P., Han, L., & Tsubaki, N. (2017). Enhanced Hydrogen Production from Steam Reforming of Vegetable Oil over Bimodal ZrO2-SiO2 Supported Ni Catalyst. *ChemistrySelect*, *2*(1), 527–532. doi:10.1002lct.201601395

Taylor, R. A., Phelan, P. E., Otanicar, T. P., Walker, C. A., Nguyen, M., Trimble, S., & Prasher, R. (2011). Applicability of nanofluids in high flux solar collectors. *Journal of Renewable and Sustainable Energy*, *3*(2), 023104. Advance online publication. doi:10.1063/1.3571565

Tenca, A., Cusick, R. D., Schievano, A., Oberti, R., & Logan, B. E. (2013). Evaluation of low cost cathode materials for treatment of industrial and food processing wastewater using microbial electrolysis cells. *International Journal of Hydrogen Energy*, *38*(4), 1859–1865. doi:10.1016/j.ijhydene.2012.11.103

Tingem, M., & Rivington, M. (2009). Adaptation for crop agriculture to climate change in Cameroon: Turning on the heat. *Mitigation and Adaptation Strategies for Global Change*, *14*(2), 153–168. doi:10.100711027-008-9156-3

Tolga Balta, M., Dincer, I., & Hepbasli, A. (2009). Thermodynamic assessment of geothermal energy use in hydrogen production. *International Journal of Hydrogen Energy*, *34*(7), 2925–2939. doi:10.1016/j.ijhydene.2009.01.087

Tremblay, A., Lambert, M., & Gagnon, L. (2004). Do hydroelectric reservoirs emit greenhouse gases? *Environmental Management*, *33*(S1), 509–517. doi:10.100700267-003-9158-6

Tyagi, V. K., Angériz Campoy, R., Álvarez-Gallego, C. J., & Romero García, L. I. (2014). Enhancement in hydrogen production by thermophilic anaerobic co-digestion of organic fraction of municipal solid waste and sewage sludge - Optimization of treatment conditions. *Bioresource Technology*, *164*, 408–415. doi:10.1016/j.biortech.2014.05.013 PMID:24880931

UNEP. (2014). *A Guidance Manual for.* UNEP.

Uskoković, V. (2007). Nanotechnologies: What we do not know. *Technology in Society*, *29*(1), 43–61. doi:10.1016/j.techsoc.2006.10.005

Vaddiraju, S., Tomazos, I., Burgess, D. J., Jain, F. C., & Papadimitrakopoulos, F. (2010). Emerging synergy between nanotechnology and implantable biosensors: A review. *Biosensors & Bioelectronics*, *25*(7), 1553–1565. doi:10.1016/j.bios.2009.12.001 PMID:20042326

Wang, X., Li, W., Chen, Z., Waje, M., & Yan, Y. (2006). Durability investigation of carbon nanotube as catalyst support for proton exchange membrane fuel cell. *Journal of Power Sources*, *158*(1), 154–159. doi:10.1016/j.jpowsour.2005.09.039

Wasif, M., Sinha, A., Ahmed, Z., & Qin, Q. (2021). Effects of biomass energy consumption on environmental quality : The role of education and technology in Asia-Pacific Economic Cooperation countries. *Renewable & Sustainable Energy Reviews*, *142*(March), 110868. doi:10.1016/j.rser.2021.110868

Wei, M., Patadia, S., & Kammen, D. M. (2010). Putting renewables and energy efficiency to work: How many jobs can the clean energy industry generate in the US? *Energy Policy*, *38*(2), 919–931. doi:10.1016/j.enpol.2009.10.044

WHO. (2005). The World Health Report 2005: Make every mother and child count. *World Health Report.*

Wright, G., Renewable, S., Law, E., Review, P., & Wright, G. (2018). *The International Renewable Energy Agency : A Global Voice for the Renewable Energy The International Renewable Energ A Global Voice for the Renewable*. Academic Press.

Wuebbles, D. J., & Jain, A. K. (2001). Concerns about climate change and the role of fossil fuel use. *Fuel Processing Technology*, *71*(1–3), 99–119. doi:10.1016/S0378-3820(01)00139-4

Yin, J., Zhu, G., & Deng, B. (2013). Multi-walled carbon nanotubes (MWNTs)/polysulfone (PSU) mixed matrix hollow fiber membranes for enhanced water treatment. *Journal of Membrane Science*, *437*, 237–248. doi:10.1016/j.memsci.2013.03.021

Yuan, Y., Zhou, S., & Zhuang, L. (2010). Polypyrrole/carbon black composite as a novel oxygen reduction catalyst for microbial fuel cells. *Journal of Power Sources*, *195*(11), 3490–3493. doi:10.1016/j.jpowsour.2009.12.026

Zare, Khanipour, Hojjat, Sarverstani, Iman, Kakavandi, Ehsan, Shokroo, Farniaei, Baghbani, & Shokroo. (2019). Hydrogen and carbon dioxide recovery from the petrochemical flare gas to methanol production using adsorption and absorption combined high-efficient method. *Applied Petrochemical Research, 9*(2), 127–145. doi:10.1007/s13203-019-0232-2

Zelmanov, G., & Semiat, R. (2008). Phenol oxidation kinetics in water solution using iron(3)-oxide-based nano-catalysts. *Water Research*, *42*(14), 3848–3856. doi:10.1016/j.watres.2008.05.009 PMID:18657285

Zeng, K., & Zhang, D. (2010). Recent progress in alkaline water electrolysis for hydrogen production and applications. In Progress in Energy and Combustion Science (Vol. 36, Issue 3, pp. 307–326). doi:10.1016/j.pecs.2009.11.002

Zhuang, L., Zhou, S., Wang, Y., Liu, C., & Geng, S. (2009). Membrane-less cloth cathode assembly (CCA) for scalable microbial fuel cells. *Biosensors & Bioelectronics*, *24*(12), 3652–3656. doi:10.1016/j.bios.2009.05.032 PMID:19556120

Zona, D., Janssens, I. A., Aubinet, M., Gioli, B., Vicca, S., Fichot, R., & Ceulemans, R. (2013). Fluxes of the greenhouse gases (CO2, CH4 and N2O) above a short-rotation poplar plantation after conversion from agricultural land. *Agricultural and Forest Meteorology*, *169*, 100–110. doi:10.1016/j.agrformet.2012.10.008

Zou, Y., Xiang, C., Yang, L., Sun, L. X., Xu, F., & Cao, Z. (2008). A mediatorless microbial fuel cell using polypyrrole coated carbon nanotubes composite as anode material. *International Journal of Hydrogen Energy, 33*(18), 4856–4862. doi:10.1016/j.ijhydene.2008.06.061

Zsembik, B. A. (2006). Health issues in latino families and households. Handbook of Families and Health: Interdisciplinary Perspectives, 6(20), 40–61. doi:10.4135/9781452231631.n3

Chapter 8
Analysis and Comparison of Business Models of Leading Enterprises in the Chinese Hydrogen Energy Industry

Poshan Yu
https://orcid.org/0000-0003-1069-3675
Soochow University, China & Krirk University, Thailand

Xinyi Liu
Independent Researcher, China

Ramya Mahendran
https://orcid.org/0000-0001-9585-9077
Independent Researcher, India

Shengyuan Lu
https://orcid.org/0000-0002-9847-4604
Independent Researcher, China

ABSTRACT

In order to achieve the goals of peak global greenhouse gas (GHG) emissions by 2030 and carbon neutralization by 2050, it is necessary to promote the transformation of China's energy sector dominated by fossil energy. Hydrogen energy is considered as the "ultimate energy" in the 21st century. It is redefining, restructuring, and transforming the China's energy sector. Therefore, Chinese enterprises need to reshape their value proposition through business model innovation. This chapter selects 10 leading enterprises in China's hydrogen energy industry. Through the business model canvas analysis of their business models, along with policies and development trends of China's hydrogen energy industry, this chapter provides the basic knowledge of popular business models in China's hydrogen industry and puts forward suggestions and research directions for the long-term development of its enterprises.

DOI: 10.4018/978-1-7998-9810-8.ch008

INTRODUCTION

Energy is the foundation of national economic and social development, and energy security is directly related to national and social security. At present, China is highly dependent on overseas crude oil resources, which poses some hidden dangers in domestic energy security. For a long time now, China's economic growth has been considerably restricted by the relationship between energy supply and demand. Starting 1993, China became a net importer (A country that imports more than it exports) of oil. In 1996, China's crude oil import and export deficit appeared. After the Iraq war, the international oil price hit a record high. Inevitably, China's total energy consumption and energy dependence also continued to soar, which had deeply affected the development of China's national economy. Subsequently, carbon emissions from using fossil fuels, imposed pressure on environment protection and preservation. Strategic restructuring, transformation, and upgradation of the national energy sector has become the Trend of The Times. Also low-carbon emissions and clean energy consumption has entered mainstream development (Tan & Yu, 2020). As a clean and carbon free secondary energy, hydrogen energy is taking its place in the energy strategies of various countries. With the in-depth development of technology, hydrogen energy will play a vital role in achieving the goals of "emission peak" and "carbon neutrality" defined by Chin[1]a.

Hydrogen is the simplest and the most abundant element in existence. While the gases in the universe are composed with 90% hydrogen in plasma or atomic state (in the stars, interstellar matters, other planets in solar systems etc.,) the earth's atmosphere is composed of only 14% hydrogen. Hydrogen is not available in natural gaseous form on Earth because it cannot be held by Earth's gravity, it exists in abundance in combined stable forms like water and hydrocarbons (found in organic materials) and can be produced from these sources (Ratnakar et al., 2021). Hydrogen energy is also a clean energy that can promote the transformation from traditional non-renewable energy to green energy. Its energy density is 140 megajoules per kilogram, which is 3 times that of oil and 4.5 times that of coal. It is regarded as the subversive technical direction of the future energy revolution (Liu et al., 2021).

Since 2011, China has issued "Energy Technology Revolution and Innovation Action Plan (2016-2030)", "Made in China 2025" and other top-level plans. "Promoting hydrogen economy infrastructure construction" has also been written into the government work report to encourage the development of hydrogen energy and fuel cell technology and thereby the industry at large, which has been listed as strategic emerging industries and plays an important part of China's energy strategy. The hydrogen energy industry is thus embracing favorable development opportunities.

Peter F. Drucker (2006), the inventor of modern management, once said: "the competition between enterprises is not the competition of products, but the competition of business models". At present, China is promoting the sustainable development strategy and is focusing on establishing a resource-saving and environment-friendly society. In order to accelerate the realization of this strategy, the Chinese government and enterprises are working together to promote the innovation of the hydrogen energy industry in business models, which is mainly oriented on sustainability, industrialization, and marketization.

A study conducted by the World Business Council for Sustainable Development (WBCSD) explored sustainable innovation in large enterprises across multiple business areas and found that while 50% of the enterprises studied were actively involved to process innovation and 42% to product innovation, only 18% were actively focusing on more systematic innovative business model (as cited in Bocken & Short, 2021, p. 2). Ritala et al. (2018) based on a content analysis also found that understanding and application of enterprises to sustainable business models is limited. Although there are pieces of evidence to suggest that different types of sustainable business models are becoming increasingly prominent, S&P 500 enterprises mainly follow broad and profitable societal and environmental trends rather than adopt a proactive profile toward business model innovation. Elkington (2020) holds that enterprises need to go further than current initiatives such as net-zero and circular economy and try to convey meaningful net-positive environmental and social value. Moreover, it's necessary to introduce the concept of "sustainable business models" as a pursuing goal, in order to rectify the environmental (and societal) damage done over the past decades.

The article focuses on the business models implemented by China's hydrogen energy industry to achieve sustainable development. The authors have chosen the tool of Business Model Canvas - a common language for describing, visualizing, evaluating business models proposed by Alexander Osterwalder and Yves Pigneur (2010) in Business Model Generation. Through analyzing the business models of ten leading enterprises in China's hydrogen energy industry, this article will help readers form a preliminary understanding of:

a. What are the most widely used and pursued business models of leading hydrogen energy enterprises in China?
b. Summary of the development status and tendency of hydrogen energy policies of the Chinese government.
c. Several development suggestions and research directions are put forward for the Chinese enterprises and researchers to further promote the industrialization, marketization, and socialization of China's hydrogen energy business models.

INDUSTRIAL CHAIN STRUCTURE OF CHINA'S HYDROGEN ENERGY INDUSTRY

The upstream of the hydrogen energy industry chain is hydrogen production, hydrogen purification and hydrogen liquefaction. The midstream of the hydrogen energy industry chain is the storage and transportation, which involves three main ways of gas, liquid and solid, and using liquid organic hydrides is also a common method. The downstream of hydrogen energy industry chain is the construction of hydrogen filling stations and comprehensive applications, and the details are shown in Figure 1 (Anonymous, 2021a). At present, the application of hydrogen fuel cell electric vehicles (FCEV) has become the main direction for the future. For example, in June 2021, Toyota Motor Corporation and SinoHytec Corporation jointly funded the establishment of Toyota SinoHytex Fuel Cell Co., Ltd., with a total investment of about 8 billion yen, with 50% shares held by both sides, so as to popularize hydrogen fuel cell vehicles through the production and sales of fuel cell systems. In 2022, according to the data released by the Beijing Winter Olympics Organizing Committee, this winter Olympics demonstration will run more than 1000 hydrogen energy vehicles, equipped with more than 30 hydrogen refueling stations, which is the world's largest fuel cell vehicle demonstration (Du, 2022).

Present Situations of Chinese Hydrogen Energy Industry Development

The Figure 2 shows the distribution map of enterprises in China's hydrogen energy industry chain. In terms of regional distribution, China's hydrogen energy production enterprises are mainly distributed in Shandong Province, Zhejiang Province and Guangdong Province. Among them, Shandong Province, the province which already has competitive advantages in chemical industry produces hydrogen energy such as industrial by-product hydrogen. Therefore, there are many enterprises involved in hydrogen production. From the distribution of representative enterprises, representative enterprises in China's hydrogen energy industry are mainly distributed in North China dominated by Beijing, East China dominated by Jiangsu, Zhejiang, Shanghai and coastal areas, and Sichuan and Chongqing. It can be seen that most representative hydrogen energy enterprises in North China are upstream hydrogen production enterprises that focus on hydrogen production, purification and liquefaction, of which Beijing has layout in all links. East China and coastal areas are covered by the whole hydrogen energy industry, among which there are many enterprises in the application of downstream hydrogenation stations and hydrogen FCEV.

Figure 1. Industrial chain structure of China's hydrogen energy industry
Source: Forward Industry Research Institute; https://www.sohu.com/a/474687654_473133

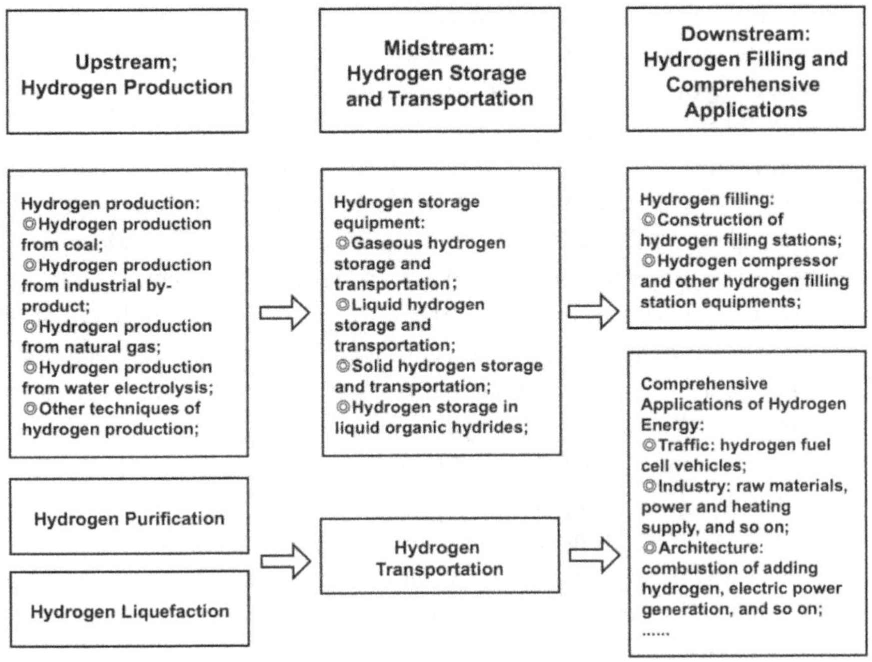

Figure 2. Distribution map of enterprises in China's hydrogen energy industry chain
Source: Forward Industry Research Institute; https://www.sohu.com/a/474687654_473133

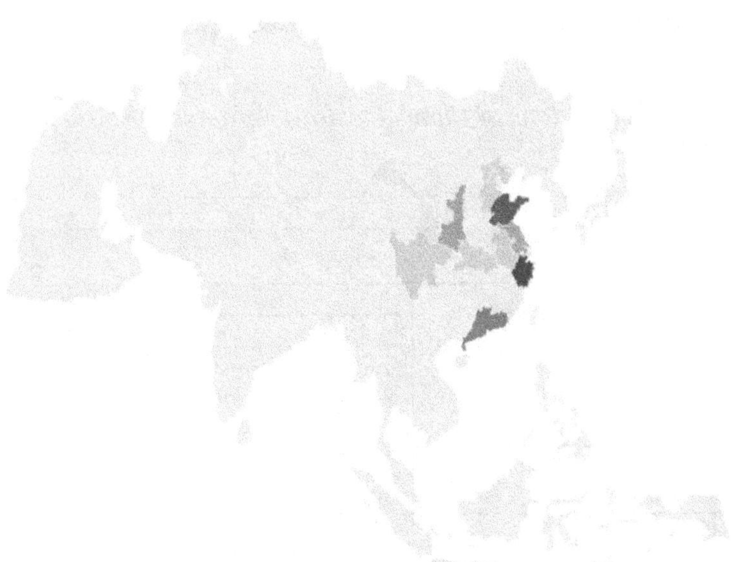

Moreover, China is a major producer of hydrogen. According to industry statistics, the annual output of industrial hydrogen production is about 30 million tons, most of which are used as industrial raw materials. The industrial hydrogen by-products that can be used to meet the new demand for hydrogen is about 6 million tons per year. At present, hydrogen production in China is basically by fossil energy, of which coal accounts for the highest proportion, which is shown in Figure 3. And hydrogen from coal is the most mature technology. However, to achieve the goal of carbon neutrality, non-fossil energy consumption should account for 70% of China's primary energy consumption by 2050, but stands at 16.4% by the end of 2020, according to research by relevant institutions. That means a significant reduction in non-fossil energy consumption in the future. So far, due to the immature and high cost of carbon capture and storage technology, it is difficult to sustain hydrogen production from fossil energy. Therefore, hydrogen industry in China should head in the direction of "green hydrogen" and carry out hydrogen production from renewable energy according to local conditions (Zhang, 2021).

In recent years, China's hydrogen energy industry has developed rapidly. By the end of 2020, China will have more than 7,000 fuel cell vehicles commercially available and over 100 hydrogen filling stations, which makes China the world's largest producer of fuel cell electric vehicle. The distribution of local hydrogen industry has been accelerated. So far, more than 20 provinces (municipalities) have issued hydrogen energy plans and guidelines. More than 30 hydrogen energy industrial parks have been built (or planned), among which the Yangtze River Delta, Pearl River Delta and Beijing-Tianjin-Heibei Urban Agglomeration have begun to take shape, attracting many hydrogen energy enterprises and research and development (R&D) institutions, which shows a trend of cluster development.

Figure 3. Present conditions of China's hydrogen production structure
Source: Zhang, X. Q. (2021)

Hydrogen Production		
Methods	Raw materials	Percentage
Hydrogen production from fossil energy	Coal-to-hydrogen	62%
	Natural gas reforming to hydrogen	19%
	Hydrogen production from petroleum, coke oven gas, Chlor-Alkali Tail gas, etc.	18%
Hydrogen production by the electrolysis of water	-	1%
Other methods	Other methods	Other methods

At present, the industrialization of hydrogen energy and fuel cells in China presents the following four remarkable features:

1. The Chinese government continues to introduce policies that support the development of hydrogen energy industry, central enterprises actively develop hydrogen energy industry, and large-scale energy and manufacturing enterprises accelerate the layout. Unlike the active involvement of foreign industrial giants in hydrogen energy and fuel cell field, the hydrogen energy and fuel cell industry in China is dominated by small and medium-sized enterprises and private enterprises in the initial stage of development, and the involvement of large enterprises in energy and manufacturing industry is limited (Wan, 2019). With the establishment of China's hydrogen energy and fuel cell industry innovation strategic alliance, large-scale enterprises accelerate the deployment of hydrogen energy industry. By 2018, there were about 309 domestic hydrogen energy and fuel cell industry chain enterprises with scale and nearly 20% of them were large-scale enterprises in energy and manufacturing. Relevant research shows that: firstly, most enterprises choose to invest in hydrogen fuel cells industry; Secondly, enterprises rich in fossil resources choose to produce hydrogen and build hydrogen filling stations to build an integrated system in the early part of the industrial chain. Thirdly, equipment manufacturers choose to develop hydrogen fuel cell vehicles and fuel cells to build an integrated platform for the latter stage. In addition to clear development strategies, it also takes supporting measures such as establishing industry-university-research-application system, share ownership encouragement and industrial investment fund which is established by local governments (Liu, et al., 2020). In the purview of hydrogen research and development, industry-university-research-application system is aimed at encouraging enterprises to actively cooperate with universities and research institutions with human and intellectual resources and promoting all stakeholders to form a user centric and market-orientated view. As for share ownership encouragement, many hydrogen energy technology companies in China have achieved a balance between equity and management relationship and provide equity-based incentives to the founding employees of the company. According to the Interim Measures for Equity and Dividend Incentives of State-Owned Science and Technology Enterprises, the total amount of equity incentive will not exceed 30% of the total share capital, and the incentive equity obtained by a single incentive object will not exceed 3% of the total share capital of the enterprise.
2. Infrastructure is weak and needs to be improved. Most enterprises are mainly distributed in the field of fuel cell and application, and the development

of hydrogen energy storage, transportation and hydrogen filling station infrastructure is weak.

3. The local governments function as an important new driving force to promote the development of hydrogen industry development. And the whole industrial chain layout has the characteristics of significant regional industrial agglomeration effect. China has initially formed hydrogen energy industry clusters and demonstration applications in the Yangtze River Delta, Pearl River Delta, Beijing-Tianjin-Heibei Urban Agglomeration. However, due to the lack of objective analysis on the comparative advantages of local hydrogen energy industry, the phenomenon of homogenous development and disordered competition has emerged. Due to the Covid-19 pandemic, the whole industry is fragile, and it will take some time for large-scale commercial use. Local governments should also be aware of the risks of overcapacity caused by excessive subsidies, local protectionism, and single GDP performance view that local governments only focus on meeting the national targets, regardless of the real development of hydrogen industry, which leads to oversupply of hydrogen fuel cell vehicles and power battery.
4. The capital market of hydrogen energy industry is active. At present, the hydrogen energy industry is in its infancy, and there is still a long way to go before large-scale industrialization. However, based on the expectation of rapid growth in the future, listed enterprises have invested in the hydrogen energy industry. In 2019, more than 70 A-share listed enterprises and hundreds of related enterprises participated in the hydrogen industry through direct investment or indirect holding. According to the investment field, it can be roughly divided into seven categories: storage, transportation and hydrogenation station, battery, core materials, industrial park, parts and components, whole industry chain and fund. At the present stage, capital is more inclined to distribute in storage, transportation and hydrogenation stations and fuel cells. But relatively few participants are distributed in the high technical threshold of key components and core materials (Liu, et al., 2020).

ANALYSIS OF MAIN POLICIES OF CHINESE HYDROGEN ENERGY INDUSTRY

In recent years, the hydrogen energy industry has received unprecedented attention in China. From the early promotion stage when it was first mentioned in the government report in 2019, and then later appeared in the "14th Five-Year Plan", the policy development of the hydrogen energy industry has realized the rapid changes from scratch. The current development trend of China's hydrogen energy industry is to implement the demonstration project of cross-border industrial integration, create more application scenarios of hydrogen energy technology, and accelerate the formation of industry (Anonymous, 2021d). At the same time, China's energy development in the new era mentioned the requirement of accelerating the development of hydrogen energy industry chain technology and equipment such as green hydrogen production, storage, transportation, and application, and promote the development of hydrogen fuel cell technology chain and hydrogen fuel cell automobile industry chain. Figure 4 shows the development history of China's hydrogen energy policy as of May 2021.

Figure 4. Development history of China's hydrogen energy policy as of May 2021
Source: Forward Industry Research Institute; https://www.sohu.com/a/473497939_121119270

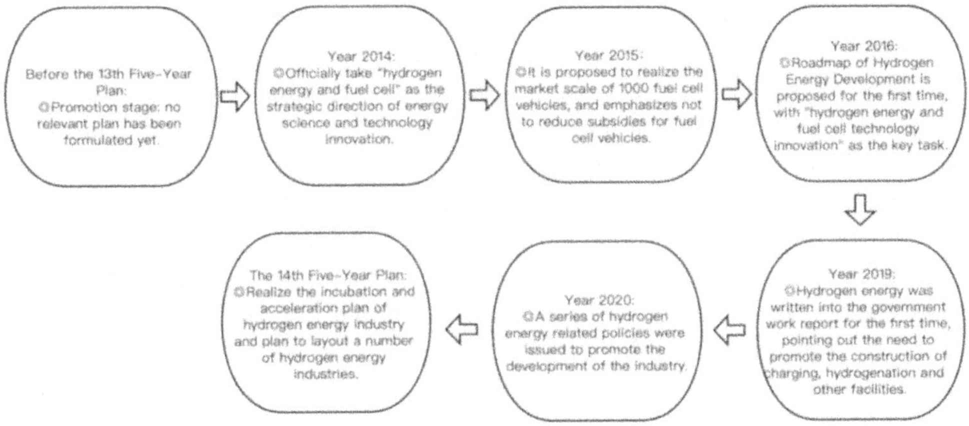

In terms of subsidy policies of hydrogen energy, the promotion of hydrogen energy related industries is an important measure for domestic hydrogen energy development, including policy support such as fuel vehicle promotion subsidy and demonstration application subsidy. China's tax policies at the consumer end focus on vehicle purchase subsidies, with less actual financial support for fuel cell core technology R&D and upstream of the industrial chain. In all, the supporting policies have not yet formed a system. China's fuel cell vehicle subsidy policy has been continuously adjusted and improved, mainly showing the following three characteristics: (1) Raise the subsidy technical threshold. Before 2017, the subsidy amount is mostly determined according to the driving miles, but after 2017, the subsidy amount and coefficient are determined by the comprehensive consideration of vehicle energy consumption, driving miles, battery performance, safety requirements and so on; (2) Replace the inclusive subsidies with targeted subsidies, and transfer the subsidies for fuel cell light passenger and freight vehicles to fuel cell large and medium-sized passenger and freight vehicles with high miles; (3) With the progress of technology, the subsidy for fuel cell passenger vehicles shows a downward trend (You et al., 2019).

In terms of policies in the hydrogen energy consumption, the rapid growth of new energy needs to balance speed and quality, and the consumption of new energy is always a key point. According to expert analysis, to achieve efficient consumption of new energy, we should not only use technology-driven promotion, but also bring about policy guidance and market mechanism cooperation. Recently, driven by the strategic goal of "emission peak and carbon neutralization", a new round of investment in China's energy sectors have been opened. A new round of investment in the new energy industry has accepted the help of industrial funds (Anonymous, 2021b). In terms of hydrogen energy consumption, since the 13th Five-Year Plan, relevant departments have also successively issued relevant policies to promote the consumption of clean energy such as hydrogen energy. For example, the guidance on establishing and improving a long-term mechanism for clean energy consumption (Exposure Draft) issued by the National Energy Administration in May 2020 mentioned that in areas rich in clean energy, the local government should encourage the promotion of hydrogen production by electricity, expand local consumption space and promote the local consumption of clean energy in multiple ways.

Moreover, the policies of important industrial clusters (such as, Guangdong-Hong Kong-Macao Bay area, the Yangtze River Delta and the Beijing-Tianjin-Heibei Urban Agglomeration) are important for foreign investors to evaluate financing prospects and the feasibility of building strategic alliances with Chinese companies. As one of the four Bay areas in the world, the construction of Guangdong-Hong Kong-Macao Bay area has attracted much attention. In recent years, Guangzhou has continuously expanded and improved the layout of hydrogen energy industry. Guangzhou Huangpu District and Guangzhou Development Area are committed to building 'a world hydrogen valley'. On July 30, 2020, Guangzhou Development and Reform Commission issued the development plan of hydrogen energy industry in Guangzhou (2019-2030), as shown in Figure 5, which clearly established Guangzhou as a hydrogen energy R & D center in the Greater Bay Area (Li, 2021).

In 2019, the Yangtze River Delta Hydrogen Corridor Construction and Development Plan, which is the first inter provincial hydrogenation infrastructure network construction plan in China, was officially released. The plan shows that in the future, the Yangtze River Delta expressway network will be used as a link to strengthen regional linkage, form a regional application environment, and create a world leading hydrogen energy and fuel cell vehicle industry economic belt (Sun, 2019). The construction and development of hydrogen corridor will be divided into three stages (Weng et al., 2021), which is shown in the Figure 6. In order to fight the "Blue Sky Defense War", according to the Beijing-Tianjin-Heibei Urban Agglomeration Energy Coordinated Development Plan (2016-2025), in addition to "reducing coal and increasing natural gas", promoting the development and consumption of renewable energy should be put on agenda. Hydrogen energy can play an important role, which is embodied in actively exploring the purification and utilization of industrial by-product hydrogen and hydrogen production from renewable energy power generation, and improving the infrastructure construction of hydrogen "production, storage, transportation and distribution".

Figure 5. Development plan of hydrogen energy industry in Guangzhou (by 2025)
Source: Guangzhou Development and Reform Commission issued the development plan of hydrogen energy industry in Guangzhou (2019-2030)

Objective	
◎Hydrogen Production Scale	
◎Hydrogen Fuel Cell	
◎Hydrogen Filling Stations	
Key technology development	
◎PEM electrolysis of water to produce hydrogen, solar photolysis of water to produce hydrogen and other hydrogen sources are prepared with low cost and high efficiency.	
◎Low and high temperature fuel cell stacks, key materials, components and system integration.	
◎Metal plate hydrogen fuel cell stack, high pressure storage tank, low pressure solid storage tank, low temperature liquid hydrogen system,etc.	
Key industrial layout	**Key layout cities**
◎Hydrogen fuel cell industrial parks	Foshan(Yunfu) industrial-transfer industrial park, Foshan Nanhai and Gaoming District, Guangzhou Development Area
◎R&D and manufacturing base of high temperature fuel cell and system	Guangzhou, Shenzhen
◎Hydrogen energy high-end equipment industry	Guangzhou, Foshan, Dongguan, Yunfu
◎Hydrogen energy production, storage and transportation industry	Huizhou, Maoming, Dongguan, Zhanjiang

SUSTAINABLE BUSINESS MODELS

Lüdeke-Freund (2010) defined the sustainable business model as "a business model that creates competitive advantage through superior customer value and contributes to the sustainable development of the company and society". Sustainable business models (SBMs) play an important role in solving systemic societal and environmental issues in a business context (Stubbs & Cocklin, 2008). SBMs allow enterprises to pursue sustainability through the alignment of their business interests with contributions to sustainable development of natural environment and society (as cited in Lüdeke-Freund et al., 2018, p.147). Due to the improving awareness of the impacts of climate change, even some biggest oil and gas producers have announced to transform their business models to achieve the goal of carbon neutrality by 2050. And Sustainable business opportunities can be classified as the following three types (He & Ortiz, 2021):

a. Energy efficiency - Reducing energy consumption per unit by improving efficiency.
b. Renewable Energy - Generating electricity or heat from sun, wind, biomass, geothermal or hydrogen resources.
c. Cleaner production - Minimize waste and emissions in industrial processes and maximize product output.

At present, China mainly realizes the above three points through the business model of the combination of circular economy and strategic alliance, coupled with policy support, and government subsidy in new energy industry.

Figure 6. The Yangtze River Delta Hydrogen Corridor Construction and Development Plan
Source: Xinhuanet-qwfb.chinareports.org.cn/qcfc/52963.html

Stage	Number of fuel cell vehicles	Number of hydrogen filling stations	Number of high-speed connection stations	Number of highways covered	Characteristics
2019-2021	>5000	>40	8	>4	Hydrogen corridor realizes the belt linkage of leading cities;
2022-2025	>50000	>200	36	>10	Hydrogen corridor realizes network linkage of key cities;
2026-2030	>200000	>500	48	>20	The hydrogen corridor realizes full coverage linkage;

Business Model Canvas

The concept of a business model first emerged in the 1950s but did not become popular until the advent of the Internet in the 1990s. Although the concept has been in common use since then, there exists a lack of consensus in its definition. Drucker (2012) argued that a business model must answer the four basic questions: "Who are the customers? What is valuable for them? How does the company make money from the business? How can value be provided to customers at an appropriate cost level?". In simple terms, business model refers to the logic of how an enterprise conducts its business (Yang et al., 2017). The business model canvas (BMC) developed by Osterwalder and Pigneur (2010) is a very effective tool in helping users understand an organization's business model. This BM framework is defined by three main elements, i.e., value proposition, value creation and value capture (Hamwi et al., 2021), which is convenient for users to visualize the elements of a business model and their potential interconnections and impact on value creation. As a visual tool, BMC facilitates discussion, debate, and exploration of potential business model innovations (Joyce & Paquin, 2016). Users can take a more systematic view of the organization and emphasize its impact on creating value. BMC is used since it is the most popular business model assessment framework (as shown in Figure 7), focusing on the business key activities required to create value and revenue, in turn, it can encourage strategic relationships. Because of its comprehensiveness, it provides a tool ready to help reshape the business (Reis et al., 2021). Therefore, this paper will use BMC, as shown in Figure 8, to analyze the business models of ten well-known leading Hydrogen energy enterprises in China.

Figure 7. Three dimensions of the business model as analytical framework
Source: Hamwi et al. (2021)

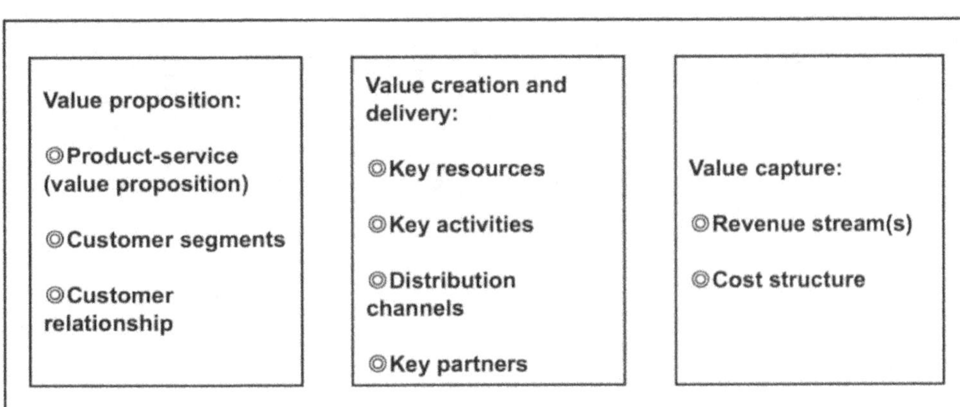

CASE STUDY AND RESULT ANALYSIS AND SOLUTIONS

Research Objects

Based on categories of hydrogen energy industry chain (downstream, middle-stream and upstream) showed in the literature review, the authors select the following ten enterprises as the research objects, and which are chosen from the relevant reports in 2021 published by China's three famous news websites Tencent News, Sohu News and Netease News. They are Baofeng Energy, Zhangjiagang Furui Special Equipment Co., Ltd, Oriental Energy Company Limited, Houpu Clean Energy Co., Ltd., Beijing Sinohytec Co., Ltd., Weichai Power, Fujian Snowman Co., Ltd., Sinoma Science & Technology Co., Ltd., China Petroleum and Chemical Corporation, Shanxi Meijin Energy Co., Ltd. Basic information of which are shown in Table 1, including enterprise type, major position in the supply chain and main business advantages.

Figure 8. Business model canvas

Source: Business Model Canvas originally templated by A Osterwalder and Y Pigneur website: creatively (https://creately.com/blog/diagrams/business-model-canvas-explained/)

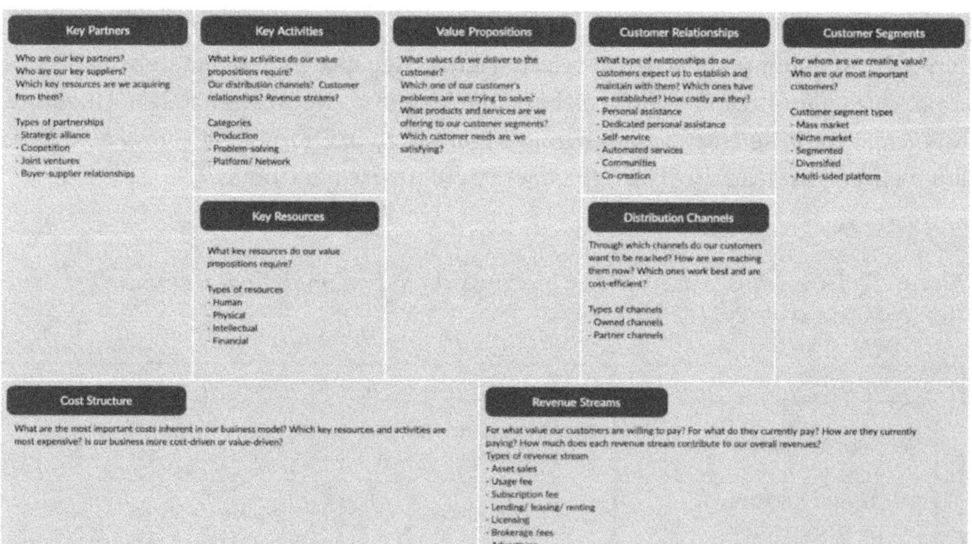

Business Model Canvas Application and Result Analysis

The authors use a widely cited framework which is shown in Figures 7 & 8 as the theoretical basis to construct the scheme for analyzing the characteristics of the business models of these ten China's leading hydrogen energy enterprises.

Build a Customer-Centered Business Model

According to the Table 1, almost every leading enterprise prioritize the idea of "the principle of the customer supremacy". This kind of business model features a user-centered approach, where businesses need to meet the basic diversified energy needs of its consumers and users and take users' convenience as the core of the business model. Chen et al. (2015) pointed out that the customer-oriented principle could induce users to change their consumption habits and provide innovative goods and services, so as to create new value for the whole energy system. This value can be delivered to the user side, or have an impact on the system operator, service provider, or equipment manufacturer, which conduces to speed up the marketization and communization of hydrogen energy industry in China. Many leading enterprises take "customer-oriented principle" as their tenet, and their customer commitment is not only limited to product life cycle management, but also comprises of later improvement services and expert guidance. For example, Beijing Sinohytec enterprise allows customers to participate in relevant decisions, promises to implement lifelong service for products and even helps to train operators for customers (Table 3). At the same time, it promises to organize and establish a service expert team. Once a fault occurs, the expert team will respond to the customer's needs within one hour, and even can rush to the site to deal with the fault according to the customer's requirements. Also, if the customer needs training or technical guidance, the expert team will provide help to the customer free of charge.

Build a Business Model Targeted at Marketization

According to the above analysis, every enterprise is willing to substitute traditional energy with hydrogen energy and other clean energy to offer a more sustainable value proposition. This offers clean energy more avenues to enter into different product and service segments and into all aspects of human life, like household electricity, household cars and so on. Moreover, in order to adapt to the changes of China's energy supply and demand situation, promoting the revolution of energy production and consumption has become a long-term strategy in China. In a more relaxed and competitive market environment, energy production, transportation, transformation, storage, consumption, and other links will be opened up, and its commodity attributes will be restored. The oligopoly of energy sales business will be broken, the investment and operation of energy system will allow social third-party capital to enter, and the energy transaction will break the regulation and return to the essence of market allocation of resources. For example, Fujian Snowman, Weichai Power and other leading enterprises have carried out M&A and established branches overseas. Therefore, taking the reform as an opportunity, a batch of new business models will emerge, and new growth points will be found through obtaining reform dividends. For instance, Huangpu district of Guangzhou tried to replace the demonstration operation project purely relying on financial subsidies through "government-led bidding and market-oriented operation of enterprises". In February 2021, this district issued a public bidding announcement for the procurement and leasing of 500 fuel cell dump garbage trucks. Hydrogen fuel garbage trucks will be purchased by state-owned enterprises in the district and then handed over to enterprises with market operation experience. The method of hydrogen car procurement by state-owned enterprises solves the problems of huge capital investment in the early stage of the enterprises, and the operation by a third-party enterprise forms a market-oriented business model exploration (Li & Peng, 2021).

Build a Business Model featuring Research and Development Pattern

The R&D pattern in strategic emerging industries is mainly guided by government investment, led by enterprises, combined with universities and research institutes. According to the above analysis, most leading enterprises choose to cooperate with governments and build strategic alliance with renowned research institutes and companies at home and abroad. From the distribution properties of listed hydrogen energy enterprises in China, one third of the 24 leading enterprises in the hydrogen production link of the industrial chain are central state-owned enterprises or local state-owned enterprises. Among the seven leading enterprises in the hydrogen storage and transportation link of the industrial chain, there are five state-owned enterprises, and among the 17 leading enterprises in the hydrogenation link of the industrial chain, eight are state-owned enterprises (Anonymous, 2021c). On July 16, 2021, at the press conference of the State Council Information Office, the Chinese spokesman Peng Huagang said that at present, more than one third of the central enterprises had been laying out the whole hydrogen energy industrial chain and had achieved a number of achievements in technology R&D and demonstration application, and central state-owned enterprises would constantly enter this market (Chen, 2021). At present, the development motivation of China's hydrogen energy industry can be roughly divided into two categories: one is dominated by large traditional energy supply enterprises, and another is dominated by other types of enterprises which are energy enterprises, energy service enterprises and science and technology enterprises (He & Lu, 2019):

Table 1. Basic information of ten leading enterprises in China's hydrogen energy industry

Enterprise Name	Enterprise Type	Position	Business Advantages
Baofeng Energy	Joint venture and listing among Taiwan, Hong Kong, Macao, and Chinese mainland	Upstream	Actively expanding solar hydrogen production and hydrogen production by water electrolysis and having a comprehensive green hydrogen demonstration project with an annual output of 160 million m^3
Zhangjiagang Furui Special Equipment Co., Ltd.	A private technology enterprise which has been listed	Midstream	As early as 2017, the company had built the production line of phase I high-pressure hydrogen storage cylinder project, which was disturbed locally and exported overseas
Oriental Energy Co., Ltd.	A Sino-foreign joint venture which has been listed	Midstream	Having Built the first commercial operation hydrogen filling station in the field of hydrogen fuel cell vehicles in Jiangsu
Houpu Clean Energy Co., Ltd.	A listed enterprise	Midstream	"Low-pressure solid hydrogen storage project" and construction of hydrogen filling stations
Sinomascience & Technology Co., Ltd.	A state-owned enterprise which has been listed	Midstream	Having absolute advantages in production of hydrogen storage tubes, such as one of the newest R&D achievements – domestic maximum volume 320L fuel cell hydrogen storage tube
Fujian Snowman Co., Ltd.	A listed enterprise	Downstream	The four main business segments of the hydrogen energy chain are liquid hydrogen equipment, fuel cell integration system, fuel cell core parts, hydrogen filling stations
Beijing Sinohytec Co., Ltd.	A listed enterprise	Downstream	Mastering the core technologies in the hydrogen fuel cell engine, which is an energy conversion device that generates electric energy by electrochemical reaction between hydrogen and oxygen
Weichai Power	Joint venture and listing among Taiwan, Hong Kong, Macao and Chinese mainland	Downstream	Having advantages in the field of hydrogen energy commercial vehicles, and having successfully developed an integration advantage in a new energy power system of "battery + motor + electronic control"
Shanxi Meijin Energy Co., Ltd.	Other joint stock limited enterprise which has been listed	The whole supply chain	Having the largest investment in domestic hydrogen industry chain and the most perfect industrial layout. In the business field of coking, coke oven is rich in more than 50% hydrogen gas in the coking process, which can produce hydrogen at low cost
China Petroleum and Chemical Corporation	A state-owned enterprise which has been listed	The whole supply chain	Having great advantages in the whole chain layout and the integrated development of "manufacturing, storage, transportation and processing"

Source: Related company's official website

Table 2. Business Model Canvas analysis in the value proposition level

<table>
<tr><th colspan="4">Value proposition</th></tr>
<tr><th>Enterprise name</th><th>Product-service (value proposition)</th><th>Customer segments</th><th>Customer relationship</th></tr>
<tr><td>Baofeng Energy</td><td>• Substitute new energy for fossil energy
• The principle of the customer supremacy</td><td>• Buyers of olefin and coking products, steam power system, high-end coal-based new materials</td><td rowspan="4">• Direct interactions
• Sign strategic cooperation agreements or contracts</td></tr>
<tr><td>Zhangjiagang Furui Special Equipment Co., Ltd.</td><td>• Promote the development of clean energy and remanufacturing industries in China, including LNG (Liquefied Natural Gas) and hydrogen</td><td>• Small and medium-sized LNG liquefaction plants at home and abroad
• Buyers of LNG industrial chain solution and hydrogen relevant products</td></tr>
<tr><td>Oriental Energy Co., Ltd.</td><td>• Promote R&D of clean energy and new materials industry
• A comprehensive operator of alkane resources
• The principle of the customer supremacy</td><td>• Polypropylene new material market
• Hydrogen energy supply market</td></tr>
<tr><td>Houpu Clean Energy Co., Ltd.</td><td>• Supplier of overall solutions for clean energy equipment
• Participate in the drafting of national and local standards and norms in hydrogen energy industry
• The principle of the customer supremacy</td><td>• Governments
• Clean energy markets such as natural gas / hydrogen energy
• Purchasers of intelligent Internet of Things information integration and supervision platform</td></tr>
<tr><td>Sinomascience & Technology Co., Ltd.</td><td>• Promote R&D of new materials industry
• Principle of Customer First</td><td>• Governments
• Application field of high-pressure hydrogen storage tank, high-pressure gas cylinders, etc.</td><td>• Direct interactions
• Sign strategic cooperation agreements or contracts
• Stable directional cooperation relationship</td></tr>
<tr><td>Fujian Snowman Co., Ltd.</td><td>• Take ice making equipment and compressor equipment as the core industry, extend to the fields of new energy, etc.</td><td>• Hydrogen energy market
• Commercial refrigeration, industrial refrigeration, vaccine cold chain, cold chain logistics, etc.</td><td>• Direct interactions
• Sign strategic cooperation agreement or contracts
• Merger and Acquisition (M&A)</td></tr>
<tr><td>Beijing Sinohytec Co., Ltd.</td><td>• Technological innovation of hydrogen fuel cell engine system
• Lead the industrialization of China's hydrogen fuel cell vehicle industry
• The principle of the customer supremacy</td><td>• Hydrogen fuel cell engine market
• Fuel cell commercial vehicle market</td><td>• Direct interactions
• Sign strategic cooperation agreements or contracts
• Undertake major national special projects and UN demonstration projects
• Government-enterprise cooperation</td></tr>
</table>

continues on following page

Table 2. Continued

Value proposition			
Enterprise name	**Product-service (value proposition)**	**Customer segments**	**Customer relationship**
Weichai Power	• The principle of the customer supremacy • Take the vehicle and machine as the leading and power system as the core technology support	• Domestic market & Export trade: subsidiaries, offices, and authorized service stations all over the world • Local manufacturing: the asset-light localization cooperation mode characterized by "technology + brand" output	• Direct interactions • Sign strategic cooperation agreements
China Petroleum and Chemical Corporation	• The construction of industrial structure and production mode conducive to resource conservation • The principle of the customer supremacy	• Foreign energy companies • Domestic market: governments, energy companies, etc.	• Direct interactions • Sign strategic cooperation agreements • As a contractor of international and regional major projects
Shanxi Meijin Energy Co., Ltd.	• Overall layout of "industrial chain + region + integrated energy station network" • Development concept of combination traditional energy and new energy • A comprehensive energy supplier	• Governments • Main business market, focusing on the whole industrial chain of hydrogen energy and its application	• Direct interactions • Sign strategic cooperation agreements

Source: Related company's official website

a. Based on large-scale traditional energy supply enterprises: this model can be divided into two forms: independent promotion by only large-scale energy supply enterprises and joint promotion by enterprises and local governments. The latter one is generally initiated by the provincial branches of large enterprises in conjunction with the local government. With the help of the existing marketing channels and customer resources of large energy supply enterprises and the policy support of the local government, it can quickly integrate resources and carry out comprehensive energy services locally. This new business entity usually has the characteristics of abundant funds and various forms of investment. The representative enterprises of this model are China Petroleum and Chemical Corporation, Sinomascience & Technology Co., Ltd., etc (Table 4).

b. Mainly guided by other types of enterprises: this model occupies a considerable share in China's current hydrogen energy market. The survey found that there are roughly three types of enterprises participating in the comprehensive energy service market, including enterprises focusing on the production and R&D of a certain link of the hydrogen energy industry chain, enterprises committed to providing overall solutions for the hydrogen energy industry chain and science and technology enterprises. Although these three types of enterprises are different and operate in specialized fields, their advantage lies in their combined professional expertise, project experience and investment decision-making. This allows them to quickly dominate the fields they are good at and cooperate with each other. For example, many leading enterprises choose to cooperate with computer companies or artificial intelligence companies to jointly establish big data sharing and intelligent factories in the hydrogen energy industry. However, due to its limitations in mastering energy resources, this type of hydrogen energy enterprises are relatively limited in business, the "comprehensiveness" of the services provided is weaker than the previous model, and the enterprise investment risk is relatively high, but the development prospect is broad. The representative enterprises of this model are Beijing Sinohytec Co., Ltd., Weichai Power enterprise, etc.

The breakthrough of key technologies is an important "catalyst" for the birth of a new business model, and the business model itself will also play a vital role in promoting the development and maturity of technology (Chen et al., 2015). Therefore, with support from the government and formation of strategic alliances, these companies have enhanced their independent R&D strength. By 2020, the number of hydrogen energy related patent applications in China has reached 269 (Figure 9). Based on the technical standards of independent intellectual property rights, enterprises form several industrial alliances to attract domestic and foreign capital and private capital in various ways, which conduces to form industrialization in the hydrogen energy.

Build a Circular Economy Business Model

According to the Table 2, it can be found that Baofeng energy enterprise has applied the circular economy industry chain into the actual production, which is a prevalent trend in China. The National Development and Reform Commission has issued the "14th Five-Year plan" for the development of circular economy. Geissdoerfer et al. (2020) defined circular economy as an economic system whose aim is to minimize waste of resources, emissions and energy losses through recycling, expansion, intensification and dematerialization of materials and energy circuits. Recently, there are more and more literature on circular economy in the business model of strategic management, which is called circular business model (CBM, sometimes named Circular Economy Business Model or CEBM). CBM is the element of how an organization circulates its business model, that is, being: proposition, creation, delivery, and value capture in closed cycles of materials (as cited in Machado & Morioka, 2021, p. 2). Fehrer & Wieland (2021) found that CBMs are more effective than traditional business models in terms of cost savings and supply chain challenges. As an important way to improve resource utilization efficiency and reduce greenhouse gas emissions, the development of circular economy can realize the closed cycle of resources and effectively promote the economical and intensive recycling of resources, which is of great significance to support the realization of Chinese goal of "emission peak" and "carbon neutralization".

Table 3. Business Model Canvas analysis in the value creation and delivery level & value capture level

<table>
<tr><th colspan="5">Value creation and delivery</th></tr>
<tr><th>Enterprise name</th><th>Key resources</th><th>Key activities</th><th>Distribution channels</th><th>Key partners</th></tr>
<tr><td>Baofeng Energy</td><td rowspan="4">• Physical conditions
• Research resources
• Financial, intellectual and human resources</td><td>• Circular economy industrial cluster integrating “coal, coke, gas, new energy, etc.,”</td><td rowspan="2">• On-the-spot investigation
• Online contact, such as telephone, email, customer careline
• Exhibitions</td><td>• Strategic cooperation between non-competitors, such as Industrial and Commercial Bank of China, Bank of Communications, etc.</td></tr>
<tr><td>Zhangjiagang Furui Special Equipment Co., Ltd.</td><td>• Problem solving in clean energy overall solution provider</td><td>• Strategic partnership with energy suppliers
• Establish several wholly-owned and holding subsidiaries for new business
• Carry out industry-university-research cooperation</td></tr>
<tr><td>Oriental Energy Co., Ltd.</td><td>• Propane Dehydrogenation Technology
• Hydrogen fuel cell industry</td><td>• On-the-spot investigation
• Online contact, such as telephone, email, customer careline
• Polyentang e-commerce platform
• Exhibitions</td><td>• Strategic cooperation with energy enterprises
• Sign strategic cooperation agreements with local governments</td></tr>
<tr><td>Houpu Clean Energy Co., Ltd.</td><td>• R&D and production of hydrogen energy filling equipment, etc.
• Engineering, procurement, and construction (EPC) contracts of hydrogen energy and other related projects
• R&D of Intelligent Internet of Things informatization integrated supervision platform
• After-sales service covering the whole industrial chain</td><td rowspan="2">• On-the-spot investigation
• Online contact, such as telephone, email, customer careline
• Exhibitions</td><td>• Sign strategic cooperation agreements on energy projects with governments and companies
• Establish the Houpu fund partnership</td></tr>
<tr><td>Sinomascience & Technology Co., Ltd.</td><td>• Rich R&D financial support
• One of the A shares with most patents</td><td>• The application fields of new energy, aerospace, energy conservation, etc.</td><td>• Strategic cooperation among energy companies and local governments
• Scientific Research Institute
• Holding subsidiaries and branches</td></tr>
</table>

continues on following page

Table 3. Continued

Value creation and delivery				
Enterprise name	**Key resources**	**Key activities**	**Distribution channels**	**Key partners**
Fujian Snowman Co., Ltd.	• Core technology and advance production equipment and instruments • Financial support	• Take compressor engine as the core industry • Integrating R&D, design, manufacturing, sales, engineering installation and after-sales service of new energy, industrial and commercial refrigeration, etc.	• On-the-spot investigation • Online contact, such as telephone, email, customer careline • Exhibitions • Marketing centers distributed throughout the country	1.Universities and research institutions 2.Global M&A 3.Strategic cooperation among enterprises and companies 5.Subsidiaries
Beijing Sinohytec Co., Ltd.	• Independently established science and technology R&D team • Numerous proprietary intellectual property rights and patented technologies	• Vertical integration product and service system with autonomous hydrogen fuel cell engine as the core, including fuel cell laboratory full set of solutions, etc.	• On-the-spot investigation • Online contact, such as telephone, email, customer careline • Exhibitions • Service centers in eight cities	• National Major Special Projects and UN demonstration projects • Strategic cooperation between companies and governments
Weichai Power	• R&D institutions and personnel all over the world	• Power system, automobile business, intelligent logistics, etc.	• On-the-spot investigation • Online contact, such as telephone, email, customer careline • Exhibitions • A global collaborative R&D platform	• M&A and strategic restructure of overseas groups • Strategic cooperation among enterprises • Country: establish production bases • Subsidiaries all over the world
China Petroleum and Chemical Corporation	• Physical conditions • Research resources • intellectual and human resources • Government financial support	• New energy, geothermal and other energy products • Self-support and agency for import & export of relevent commodities and technologies	• On-the-spot investigation • Long-term contracts • Online contact, such as telephone, email, customer careline • Exhibitions	• Strategic cooperation among enterprises and governments • Wholly owned enterprises, holding enterprises and joint-stock enterprises
Shanxi Meijin Energy Co., Ltd.		• Coal-coke-gas-chemical integrated industrial chain • Invest in the hydrogen energy industry chain		• Strategic cooperation between enterprises and governments • Cooperation with universities and research institutes

Source: Related company's official website

Table 4. Business Model analysis in the value capture

Value capture		
Enterprise name	**Revenue stream(s)**	**Cost structure**
Baofeng Energy	• Sales of products	• Cost involved in the whole process of hydrogen production • Value-driven costs: Investment in "modern smart factory" and "Industrial Internet"
Zhangjiagang Furui Special Equipment Co., Ltd.	• Sales of products • Revenues generated from providing various technical services and technology transfer in the field of hydrogen energy • Revenues generated from providing clean energy overall solution	• Cost of producing hydrogen energy equipment • Value-driven costs: Investment in the wide application of clean energy in various fields of energy
Oriental Energy Co., Ltd.	• Sales of products	• Cost involved in the whole process of hydrogen production and its technology innovation • Value-driven costs: Investment in deep processing of alkane resources and promoting hydrogen university-research-application system
Houpu Clean Energy Co., Ltd.	• Sales of products • Revenues generated from providing overall solutions for clean energy injection equipment • Revenues generated from providing supervision platform and after-sale services	• Cost involved in the whole production supply chain • Value-driven costs: Investment in combining Internet of Things with hydrogen industry
Sinomascience & Technology Co., Ltd.	• Sales of products • Government subsidies	• Cost of producing fuel cell power system • Cost of R&Dsystem, product promotions and branding, etc.
Fujian Snowman Co., Ltd.	• Sales of products	• Cost of producing hydrogen fuel cell air compressor series • Cost of R&D system and global M&A
Beijing Sinohytec Co., Ltd.	• Sales of products • Revenues generated from hydrogen energy solutions	• Cost of producing fuel cell power system • Cost of R&D system, carrying out important special projects and demonstration projects
Weichai Power	• Sales of products • Government subsidies	• Cost of producing hydrogen fueled engine • Cost of R&D system, product promotions and branding, etc. • Cost of global M&A
China Petroleum and Chemical Corporation	• Sales of products • Revenues generated from R&D, application and consulting services of alternative energy products • Revenues generated from import and export of related products and technologies	• Cost involved in the whole hydrogen energy supply chain • Cost of R&D system • Cost generated by technology innovation, industrial transformation, etc.
Shanxi Meijin Energy Co., Ltd.	• Sales of products • Subsidies	• Cost involved in the whole hydrogen energy supply cha, especially in hydrogen fueled vehicles and hydrogenation stations • Cost of promoting the company's transformation in the whole hydrogen supply chain

Source: Related company's official website

Taking Baofeng energy group as an example, it has already invested 72.7 billion yuan in the core area of national Ningdong Energy Chemical Industry Base to build a "Baofeng energy circular economy industrial base" covering an area of 14000 mu (1 Mu corresponding to 1/15 ha1, about ⅔ × 1000 (or 666.7) m^2), creating a national large-scale circular economy industrial cluster integrating "coal, coke, gas, methanol, olefin, polyethylene, polypropylene, fine chemical industry and new energy". The planned total capacity includes 120,000 tons of benzene hydrogenation, 240 million Nm^3 of green hydrogen and 120 million Nm^3 of green oxygen, which effectively fills the import gap of China's chemical products and ensures the safety of the country energy.

Build a Business Model centered on Energy Internet

The above analysis shows that the leading enterprises in China's hydrogen energy industry are actively carrying out a number of actions to integrate their business with the internet. The business model under the Internet thinking is based on extensive interconnection, with information as the link, to gather a large number of scattered entities in the information system; Secondly, it needs to be user-centered and create value for users. Finally, the information contained in the data needs to be valued and turned into value (Chen et al., 2015). Internet Plus Smart Energy is a new form of energy industry with deep fusion of the Internet and energy production, transmission, storage, consumption, and market. Its main characteristics include equipment being smart, synergy of multiple energy types, information being symmetric, supply and demand being distributed, system being flat, and transactions being open. This new form of energy industry is commonly called the "Energy Internet" in the energy and information industries in China (Wang, 2016). Energy Internet is the product of the deep integration of Internet technology, spirit, and energy system, giving birth to a new way of thinking and business operation of the internet. And it will give full play to the optimization and integration of the internet in the allocation of energy production factors, deeply integrate the innovation results of the internet into the energy system and enhance the innovation and productivity of the traditional energy industry (Liu et al., 2015).

The specific details of application are as follows:

1. Intelligent transformation of energy infrastructure, such as Intelligent Operation Cloud-Platform based on Internet, to realize remote control optimization and improve operation efficiency and benefits (Wang, 2016). For example, Zhangjiagang Furui Special Equipment Co., Ltd. and IBM set up the innovation center of LNG Industry Internet of things to open the era of clean energy big data

2. Promote deep fusion of energy and information infrastructure (Wang, 2016). For example, BAOFENG ENERGY made every effort to build a modern "smart factory"
3. National Development and Reform commission (2016) proposed to cultivate an open and shared market system and cultivate new market players such as e-sellers and comprehensive energy operators. Develop B2B (Business to Business), B2C (Business to Customer), C2B (Customer to Business), C2C (Customer to Customer), O2O (Online to Online) and other business models based on the Internet (Yu et al., 2022b)
4. Develop energy big data application services and value-added services for energy production, circulation and consumption (Wang, 2016)
5. Break through the key core technology. For example, the above enterprises are trying to speed up technological innovation and cooperate with many well-known enterprises at home and abroad to establish energy Internet industry innovation alliance and technological innovation platform.

In July 2015, The State Council issued the Guiding Opinions on Actively Promoting the "Internet +" Action, which listed "Internet+ Smart Energy" as one of the key action areas. On July 4, 2016, the National Development and Reform Commission and the National Energy Administration issued the Implementation opinions on promoting the construction of Multi-Function Complementary Integrated Optimization Demonstration Project, emphasizing the innovation of management system and business model. Therefore, "Internet+ Smart Energy" is the general trend of the development of China's hydrogen energy industry.

FUTURE RESEARCH DIRECTIONS

Explore Triple Layer Business Model Canvas to carry out Sustainable Business Model Innovation

Geissdoerfer et al. (2018) argued that building a Circular Ecosystem (CE) requires not only CBMs, but also Circular Supply Chains (CSCs), circular customer behaviors, circular value chains, and CE infrastructures in terms of relationships and communities. Some scholars also stressed the need for considering CE infrastructure in CE research, such as sustainable development, sustainable supply chain and sustainable value chain (as cited in A. Asgari & R. Asgari, 2021). Joyce and L. Paquin (2016) proposes the Triple Layer Business Model Canvas (TLBMC) as a practical tool for coherently integrating economic, environmental, and social concerns into a holistic view of an organization's business model. TLBMC (Figure

10) builds on Osterwalder and Pigneur's (2010) Original Business Model Canvas, a popular and widely adopted tool that explicitly integrates environmental and social impacts to support business model innovation through additional business model layers aligned directly with the original economical-oriented canvas. TLBMC follows a triple bottom line approach to organizational sustainability, explicitly addressing and integrating economic, environmental, and social value creation as the core of an organization's business model. In particular, it uses life-cycle analysis and stakeholder management perspectives within newly created environmental and social canvases to conceptualize and connect multiple types of value creation within a business model perspective (Joyce & L. Paquin, 2016). TLBMC can help users overcome barriers to sustainability-oriented change within their organizations by creatively reconceptualizing their current business model and communicating potential innovations (Schaltegger et al. 2012).

Figure 9. Changes in patent applications related to hydrogen energy in China from 2010 to 2020
Source: SooPAT Forward Industry Research Institute

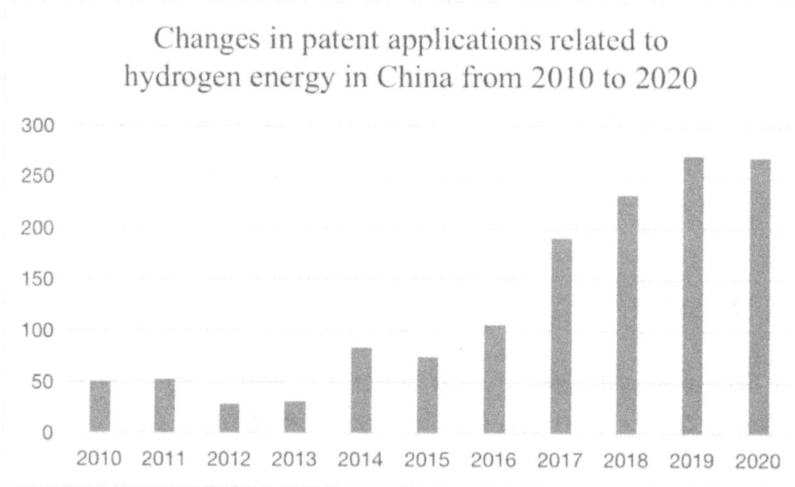

The core of TLBMC is likely to support the transition from incremental and isolated innovations to more integrated, systematic and sustainable innovations that may be better suited to the current global crisis on energy and material constraints (Joyce & L. Paquin, 2016).

Therefore, whether the traditional large petroleum and petrochemical enterprises like Sinopec choose to build the top hydrogen energy company in China, or the enterprises specializing in hydrogen energy innovate their business models of recycling and sustainable development, TLBMC helps them to quickly visualize and communicate existing business models, identify explicit data and information gaps, and creatively explore potential business model innovations which were more sustainability oriented.

Figure 10. Triple Layer Business Model Canvas
Source: Joyce & Paquin (2016)

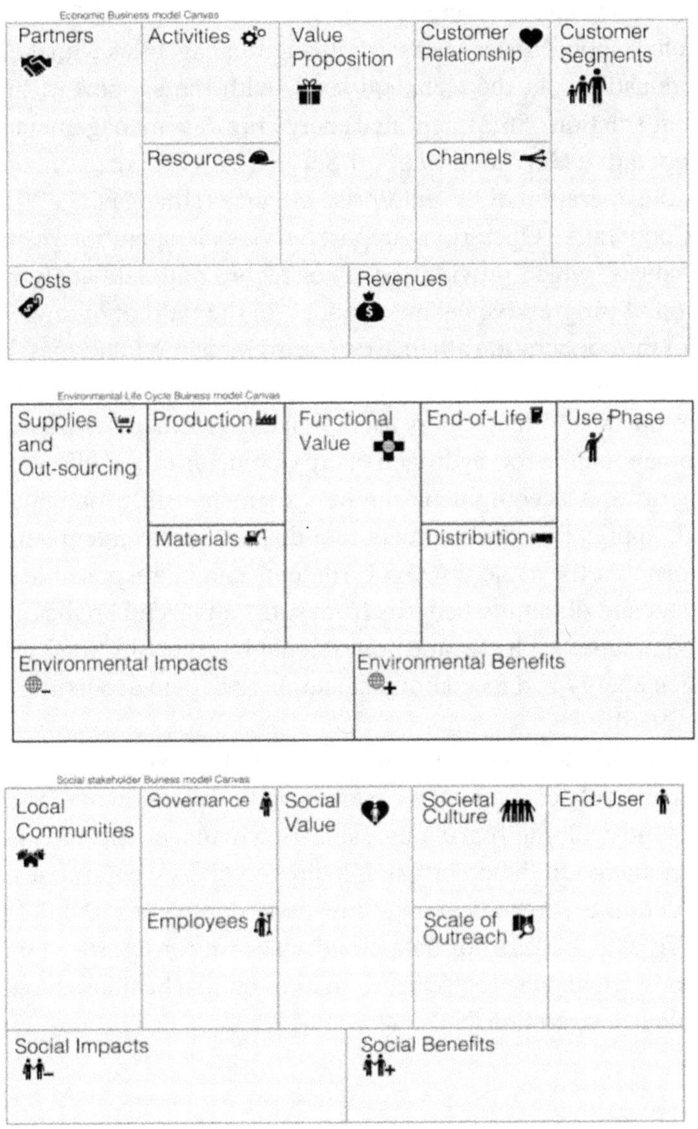

Explore the Business Model of "Shared Platform + Alliance Development"

In 2016, the National Development and Reform Commission issued the "Guiding Opinions on Promoting the Development of "Internet +" smart energy. It suggested that local governments should strive to promote the diversification and scale development of the Energy Internet in 2019-2025 and establish a more comprehensive Energy Internet market mechanism.

The Energy Internet follows the principles of openness, interaction, and mutual benefit, which is difficult to be dominated by one enterprise. Therefore, the operational mode of the future should foster enterprise alliance management with the participation of multiple industries. At the technical level, with the support of historical data records and information of the integrated energy big data management platform, the Alliance Cooperative Operation Support Service provides data, model and record reference for the operation of the enterprise alliance (Huang et al., 2015). The core of Alliance Cooperative Operation Support Services is operation management and intelligent analysis, which provides behavior record and data analysis support for the exploration of progressive business model. In the field of hydrogen energy, the participants of the cooperation alliance can be divided into four types: energy users, hydrogen energy and its various equipment suppliers, government departments and third-party companies. When energy users sign an agreement with hydrogen energy and its equipment suppliers, hydrogen energy companies are allowed to use data. Hydrogen energy and its equipment suppliers are responsible for managing the data that is mined, application development and display of the integrated energy data service platform. At the same time, governments can be responsible for providing some basic data and obtaining benefits from some advanced application functions. Third party companies such as some professional Internet of Things companies can assist hydrogen energy and its equipment suppliers in data acquisition and analysis (Chen et al., 2015).

On the one hand, driven by information technology and data core, enterprises are required to actively carry out the coverage and application of digital tools in the whole product life cycle, and carry out association management and maintenance of product related data; On the other hand, hydrogen energy enterprises are required to cooperate with the government and third-party companies to build a large-scale data center (Yu et al., 2022a) or data cloud platform for hydrogen energy and its equipment, and realize one-stop services so as to build a business model of "shared platform + alliance operation."

Explore the Business Model of "Multi-Station Integration"

"Multi-station integration" is the expansion of the integration of substation, charging and replacement (energy storage) station and data center station. In addition to the traditional three stations, "multi-station" also includes information communication, energy and environment related infrastructure and system platforms such as 5G communication base station, Beidou foundation enhancement station, distributed new energy power station, etc.,"Integration" refers to the integration of energy, information, communication, government affairs and other fields in business through the construction of "multiple stations", and the coordination of electric power enterprises, new energy enterprises, communication operators and relevant government commissions and bureaus in service subjects (Zhang et al., 2019). To sum up, "multi-station integration" aims to improve resource utilization efficiency and promote cross-border business integration, and the ability of open sharing and deep collaboration of resources and data services.

In the context of energy transformation, the integration of hydrogen energy and power system is becoming a trend. In July 2021, "the Sixth Global Offshore Wind Power Conference" was successfully held in Nanning. Hydrogen production from offshore wind power was a highlight of the meeting. In the view of experts, the development of offshore wind power and hydrogen energy will provide a new path for comprehensive energy development. Many well-known enterprise representatives in the industry declared that they would continue to innovate offshore hydrogen production and energy storage will be developed to promote the in-depth research of diversified integration test (CWEA, 2021).

Explore the Business Model of "Multi-Energy Integration Strategic Technology"

'Multi-Energy Integration' refers to the comprehensive consideration of the energy and material attributes of energy resources, and the complementary integration of traditional fossil energy and non-fossil energy through new technologies and processes, and the integrated coupling of energy flow, material flow and information flow, in order to realize the multi-objective optimization of energy efficiency, environmental benefits, economic costs and social benefits in the comprehensive energy system (Liu, 2019).

According to the division of China's existing energy system, Multi-Energy Integration technology can be classified into three main lines: The first main line is clean and efficient development and utilization of fossil energy and coupling alternatives. The second main line is the multi-energy complementarity and large-scale application of renewable energy. The third main line is low-carbon and intelligent multi-energy integration. These three lines require enterprises to give full play to large-scale advanced energy storage platforms and the multi-energy carrier role of hydrogen / low-carbon alcohol through technology and business model innovation (Zhu et al., 2021).

CONCLUSION

In order to steadily promote the realization of the goals of "emission peak" and "carbon neutralization", the Chinese government and enterprises should

1. Make full use of the characteristics of high marketization of hydrogen energy industry and active capital market, to lay the foundation for entering high-tech and high value-added links by means of equity participation or M&A (Liu et al., 2020)
2. Adhere to the combination of independent R&D and foreign cooperation and carry out demonstration project construction in due time. Establish a hydrogen energy industry incubation platform and a hydrogen energy industry fund to meet the capital needs of industrial development and reduce the risks of enterprises. Also carry out the reform of mixed ownership and equity to mobilize the innovation enthusiasm of scientific researchers (Zhang et al, 2020)
3. Learn from the successful experience of Japan, the European Union, and other countries in the hydrogen energy industry to help the reform and innovation of domestic hydrogen energy business model. For example, the research project contract "French Hydrogen Initiative" (IFHy) established by France, which is conducive to developing and coordinating public and private partnerships to ensure continuity between scientific research and product commercialization (Cuevas et al., 2021). In 2020, the European Union Commission adopted a "package for the future", which will add an additional 2 billion euros to cultivate international partnerships (Cuevas et al., 2021); Japan's Ministry of Environment has launched a project to create a hydrogen energy society by building a regional low-carbon hydrogen supply chain (Li & Nishimiya, 2021), etc. Therefore, the Chinese government and enterprises can learn from former successful cases and foreign initiatives for hydrogen energy business model innovation to promote the process of localization and marketization of the domestic hydrogen energy business model.

ACKNOWLEDGMENT

The authors extend sincere gratitude to:

- Our colleagues from Soochow University, the Australian Studies Centre of Shanghai University and Krirk University as well as the independent research colleagues who provided insight and expertise that greatly assisted the research, although they may not agree with all of the interpretations/conclusions of this chapter.
- China Knowledge for supporting our research.

REFERENCES

Anonymous. (2021a). *Carding the whole hydrogen energy industry chain and regional development hot spots map*. https://baijiahao.baidu.com/s?id=1702880835003475539&wfr=spider&for=pc

Anonymous. (2021b). *Blockbuster! Policy Summary and Interpretation of Hydrogen Energy Industry in China and 31 Provinces and Cities in 2021 (full)*. https://www.sohu.com/a/473410075_473133

Anonymous. (2021c). *The new round of investment in the new energy industry is expected to reach 12 trillion yuan, with consumption still a key issue*. https://baijiahao.baidu.com/s?id=1689778958293283188&wfr=spider&for=pc

Anonymous. (2021d). *Analysis on market status and competition pattern of China's hydrogen energy industry in 2021*. https://baijiahao.baidu.com/s?id=1706510978416834977&wfr=spider&for=pc

Anonymous. (2021e). *Summary and interpretation of hydrogen energy industry policies in China and 31 provinces and cities in 2021*. https://baijiahao.baidu.com/s?id=1702142255001709111&wfr=spider&for=pc

Asgari, A., & Asgari, R. (2021). How circular economy transforms business models in a transition towards circular ecosystem: The barriers and incentives. *Sustainable Production and Consumption*, *28*, 566–579. doi:10.1016/j.spc.2021.06.020

Bocken, N. M. P., & Short, S. W. (2021). Unsustainable business models - Recognising and resolving institutionalised social and environmental harm. *Journal of Cleaner Production*, *312*, 127828. doi:10.1016/j.jclepro.2021.127828

Chen, Q. X., Liu, D. N., Lin, J., He, J. J., & Wang, Y. (2015). Business Models and Market Mechanisms of Energy Internet (1). *Power System Technology*, *39*(11), 3050–3056.

Chen, S. T. (2021). *SASAC: More than one third of central enterprises are formulating the layout of the whole hydrogen energy industry chain.* https://baijiahao.baidu.com/s?id=1705435500101935720&wfr=spider&for=pc

Cuevas, F., Zhang, J. X., & Latroche, M. (2021). The Vision of France, Germany, and the European Union on Future Hydrogen Energy Research and Innovation. *Engineering*, *7*(6), 715–718. doi:10.1016/j.eng.2021.04.010

CWEA. (2021). *The 6th global offshore wind power conference was held in Nanning, Guangxi.* https://news.bjx.com.cn/html/20210712/1163357.shtml

Drucker, P. F. (2012). *The practice of management.* Lixin Accounting Publishing House. doi:10.4324/9780080942360

Drucker, P. F., & Joseph, A. M. (2006). The Daily Drucker 366 Days of Insight and Motivation for Getting the Right Things Done. Shanghai Translation Publishing House.

Du, Q. M. (2022). *More than 1000 demonstration vehicles help the Winter Olympics: hydrogen fuel cell vehicles meet the business inflection point.* https://baijiahao.baidu.com/s?id=1724184264204144351&wfr=spider&for=pc

Elkington, J. (2020). *Green Swans: The coming boom in regenerative capitalism.* Fast Company Pr.

Fehrer, J. A., & Wieland, H. (2021). A systemic logic for circular business models. *Journal of Business Research*, *125*, 609–620. doi:10.1016/j.jbusres.2020.02.010

Geissdoerfer, M., Morioka, S. N., de Carvalho, M. M., & Evans, S. (2018). Business models and supply chains for the circular economy. *Journal of Cleaner Production*, *190*, 712–721. doi:10.1016/j.jclepro.2018.04.159

Geissdoerfer, M., Pieroni, M. P. P., Pigosso, D. C. A., & Soufani, K. (2020). Circular business models: A review. *Journal of Cleaner Production*, *277*, 123741. doi:10.1016/j.jclepro.2020.123741

Hamwi, M., Lizarralde, I., & Legardeur, J. (2021). Demand response business model canvas: A tool for flexibility creation in the electricity markets. *Journal of Cleaner Production*, *282*, 124539. doi:10.1016/j.jclepro.2020.124539

He, J. T., & Ortiz, J. (2021). Sustainable business modeling: The need for innovative design thinking. *Journal of Cleaner Production*, *298*, 126751. doi:10.1016/j.jclepro.2021.126751

He, Y. H., & Lu, S. T. (2019). *Integrated energy service provider: User centered, one-stop solution.* https://www.sohu.com/a/335720319_269757

Huang, R. L., Pu, T. J., Liu, K. W., Yang, Z. Y., & Chen, N. S. (2015). Functional system and application scheme design of urban energy Internet. *Automation of Electric Power Systems, 9*, 26-33+40.

Joyce, A., & Paquin, R. (2016). The triple layered business model canvas: A tool to design more sustainable business models. *Journal of Cleaner Production*, *135*, 1474–1486. doi:10.1016/j.jclepro.2016.06.067

Li, H. W., & Nishimiya, N. (2021). Insight from Japan's Hydrogen Strategy and Activities. *Engineering*, *7*(6), 722–725. doi:10.1016/j.eng.2020.12.020

Li, Q. S., & Zhang, K. (2021). Research on the path of green energy development and utilization in China. *Engineering*, *1*, 101–111.

Li, S. (2021). *The "hydrogen wind" of the energy transformation of the Guangdong-Hong Kong-Macao Greater Bay Area has arrived.* https://www.hbzhan.com/news/detail/143094.html

Li, Z., & Peng, M. J. (2021). *The Greater Bay Area (GBA) cooperates to solve problems in the whole hydrogen energy industry chain, and introduces business model innovation into the market.* https://baijiahao.baidu.com/s?id=1707844202191188126&wfr=spider&for=pc

Lieder, M., & Rashid, A. (2016). Towards circular economy implementation: A comprehensive review in context of manufacturing industry. *Journal of Cleaner Production, 115*, 36–51. doi:10.1016/j.jclepro.2015.12.042

Liu, D. N., Zeng, M., Huang, R. L., Ji, L. H., Chen, Q. X., Duan, J. H., & Li, Y. F. (2015). Business Models and Market Mechanisms of Energy Internet (2). *Power System Technology*, *39*(11), 3057–3063.

Liu, Q., Zhang, H. L., Guan, S. F., & Li, Y. R. (2020). *Research on the development of hydrogen energy industry.* https://baijiahao.baidu.com/s?id=1683829736362232563&wfr=spider&for=pc

Liu, Q., Zhang, H. L., Guan, S. F., & Li, Y. R. (2020). *Development status and trend of hydrogen energy industry in China.* https://www.163.com/dy/article/FS3FSOA005509YYU.html

Liu, Y. D., Guo, H. X., & Ouyang, X. P. (2021). Development status and future prospect of hydrogen fuel cell technology. *Strategic Study of CAE, 23*(04), 162–171. doi:10.15302/J-SSCAE-2021.04.019

Liu, Z. M. (2019). *Build the 'Multi-Energy Integration' system.* https://www.zhonghongwang.com/show-278-151120-1.html

Lu, J., & Chen, G. (2005). A time-varying complex dynamical network model and its controlled synchronization criteria. *IEEE Transactions on Automatic Control, 50*(6), 841–846. doi:10.1109/TAC.2005.849233

Lüdeke-Freund, F. (2010). Towards a conceptual framework of business models for sustainability. *Environmental Management, 49*, 1–28.

Lüdeke-Freund, F., Carroux, S., Joyce, A., Massa, L., & Breuer, H. (2018). The sustainable business model pattern taxonomy-45 patterns to support sustainability -oriented business model innovation. *Sustainable Production and Consumption, 15*, 145–162. doi:10.1016/j.spc.2018.06.004

Lüdeke-Freund, F., Gold, S., & Bocken, N. (2019). A Review and Typology of Circular Economy Business Model Patterns. *Journal of Industrial Ecology, 23*(1), 36–61. doi:10.1111/jiec.12763

Machado, N., & Morioka, S. N. (2021). Contributions of modularity to the circular economy: A systematic review of literature. *Journal of Building Engineering, 44*, 103322. doi:10.1016/j.jobe.2021.103322

National Development and Reform Commission. (2016). *Guidelines on promoting the development of "Internet Plus" Smart Energy.* http://www.nea.gov.cn/2016-02/29/c_135141026.htm

Osterwalder, A., & Pigneur, Y. (2010). *Business Model Generation: A handbook for visionaries, game changers, and challengers*. John Wiley and Sons.

Ratnakar, R. R., Gupta, N., Zhang, K., Doorne, C. V., Fesmire, J., Dindoruk, B., & Balakotaiah, V. (2021). Hydrogen supply chain and challenges in largescale LH2 storage and transportation. *International Journal of Hydrogen Energy, 46*(47), 24149–24168. doi:10.1016/j.ijhydene.2021.05.025

Reis, I. F. G., Gonçalves, I., Lopes, M. A. R., & Autunes, C. H. (2021). Business models for energy communities: A review of key issues and trends. *Renewable & Sustainable Energy Reviews, 144*, 111013. doi:10.1016/j.rser.2021.111013

Ritala, P., Huotari, P., Bocken, N., Albareda, L., & Puumalainen, K. (2018). Sustainable business model adoption among S&P 500 firms: A longitudinal content analysis study. *Journal of Cleaner Production, 170*, 216–226. doi:10.1016/j.jclepro.2017.09.159

Schaltegger, S., Freund, F. L., & Hansen, E. G. (2012). Business cases for sustainability: The role of business model innovation for corporate sustainability. *International Journal of Innovation and Sustainable Development, 6*(2), 95–119. doi:10.1504/IJISD.2012.046944

Song, H. F., Wan, J. B., & Ren Z. B. (2010). Research on the connotation characteristics, industrial selection and development policy of China's strategic emerging industries. *Science & Technology for Development, 9*, 7-14.

Stubbs, W., & Cocklin, C. (2008). Conceptualizing a "sustainability business model". *Organization & Environment, 21*(2), 103–107. doi:10.1177/1086026608318042

Sun, C. L. (2019). *Building a hydrogen energy industry economic belt in the Yangtze River Delta.* http://qwfb.chinareports.org.cn/qcfc/52963.html

Tan, X. G., & Yu, Z. P. (2020). Development status and prospect of fuel cell commercial vehicle industry. *Strategic Study of CAE, 22*(5), 152–158. doi:10.15302/J-SSCAE-2020.05.019

Wan, Y. M. (2019). *Analysis of the coordinated path of Beijing-Tianjin-Heibei Urban Agglomeration in hydrogen energy development.* http://www.china-nengyuan.com/news/144437.html

Wang, H. N., Wang, L. H., Zhou, Z. C., Wang, K. S., & Cui, J. (2020). Study on The Development Strategy of Emerging Industries (2035). *Engineering, 22*(2), 1–8.

Wang, R. F. (2016). *Background Material for the Action Plan of Internet Plus Smart Energy*. http://fangtan.china.com.cn/2016-06/24/content_38738456_2.htm

Weng, S. L., Huang, Z., Yu, L. J., Xie, X. M., You, T., & Zhang, T. T. (2021). Research on the construction strategy of modern energy system in the Yangtze River Delta. *Engineering, 23*(01), 42–51.

Xu, W. W., & Shen, H. (2021). *The "Guangzhou Consensus" on hydrogen energy industry was released!* https://xw.qq.com/cmsid/20210928A0FPIE00

Yang, M., Evans, S., Vladimirova, D., & Rana, P. (2017). Value uncaptured perspective for sustainable business model innovation. *Journal of Cleaner Production, 140*, 1794–1804. doi:10.1016/j.jclepro.2016.07.102

You, S. J., Zhang, Z., Zhou, Y., Du, G. M., Wan, H., Xu, P., & Bao, L. Q. (2019). Industrial policy of hydrogen energy pioneer countries and its enlightenment. *Oil Forum, 5*, 57-66.

Yu, P., Gu, H., Zhao, Y., & Ahuja, A. (2022). Digital Transformation Driven by Internet Data Center: Case Studies on China. In D. Piaggesi, H. Landazuri, & B. Jia (Eds.), *Cases on Applying Knowledge Economy Principles for Economic Growth in Developing Nations* (pp. 203–230). IGI Global. doi:10.4018/978-1-7998-8417-0.ch011

Yu, P., Liu, Z., Hanes, E., & Mumtaz, J. (2022). Integration of IoT and Blockchain for Smart and Secured Supply Chain Management: Case Studies of China. In S. Goyal, N. Pradeep, P. Shukla, M. Ghonge, & R. Ravi (Eds.), *Utilizing Blockchain Technologies in Manufacturing and Logistics Management* (pp. 179–207). IGI Global. doi:10.4018/978-1-7998-8697-6.ch010

Zhang, X. Q. (2021). The development trend of and suggestions for China's hydrogen energy industry. *Engineering, 7*(6), 719–721. doi:10.1016/j.eng.2021.04.012

Zhang, Y., Wang, B. Y., Li, R., & Mi, X. R. (2019). Research on business model and development path of multi-station integration. *Distribution & Utilization, 6*, 62–66.

Zhu, H. X., Geng, X. Y., Xiao, Y., Cai, R., & Liu, X. M. (2021). Research on the strategic path of promoting energy revolution in Northwest China in the new era. *Engineer, 23*(01), 92–100.

KEY TERMS AND DEFINITIONS

Beidou Foundation Enhancement System: It is also known as the Beidou continuously operating reference station network, and is built on the ground surface. This system consists of reference stations, communications networks, multilevel data center, and terminals for mobile Internet and users. These terminals can process Beidou signals. The Beidou foundation enhancement system essentially belongs to the information-physical fusion system (Lu & Chen, 2005).

Blue Sky Defense War: The three-year action plan for cleaner air, issued by the State Council in June 2018, is a comprehensive strategy to improve air quality through actions across all key sectors. Key objective of the action plan is to reduce emissions of major air pollutants and greenhouse gases and decreasing the number of days with high air pollution.

Business Model: A Business model provides information about an organization's target market, their market's need, and the role that the business's products or services will play in meeting those needs.

Business Model Innovation: A business model innovation describes the process in which an organization adjusts its business model. Often times these innovations fundamentally change how a company delivers value to its customers, whether that's through the development of new revenue streams or distribution channels.

Carbon Neutrality: Carbon neutrality is a state of net-zero carbon dioxide emissions. This can be achieved by balancing emissions of carbon dioxide with its removal (often through carbon offsetting) or by eliminating emissions from society (the transition to the "post-carbon economy"). The term is used in the context of carbon dioxide-releasing processes associated with transportation, energy production, agriculture, and industry.

Circular Economy: A circular economy (also referred to as "circularity" and "CE") is "a model of production and consumption, which involves sharing, leasing, reusing, repairing, refurbishing and recycling existing materials and products as long as possible". A CE aims to tackle global challenges like climate change, biodiversity loss, waste, and pollution by emphasizing design-based implementation of the three base principles of the model. The three principles required for the transformation to a circular economy are: eliminating waste and pollution, circulating products and materials, and the regeneration of nature. CE is defined in contradistinction to the traditional linear economy.

Emission Peak: Emission peak is the point at which carbon dioxide emissions stop rising to the peak and then gradually fall back.

Equity Incentive: Compensation means all equity-based compensation awards (including stock options, restricted stock, stock appreciation rights and cash restricted units) granted under the Company's stock and incentive plans, as in effect from time to time.

Value Proposition: A value proposition refers to the value a company promises to deliver to customers should they choose to buy their product or avail their services. A value proposition is part of a company's overall marketing strategy and business model.

WBCSD: The World Business Council for Sustainable Development is a CEO-led organization of over 200 international companies. The Council is also connected to 60 national and regional business councils and partner organizations.

ENDNOTE

1 By 2030, China's carbon dioxide emissions will peak, stabilize and then decline, and by 2060, China will be carbon neutral and have fully established a green, low-carbon and circular economy.

Chapter 9
Would Crowdfunding Really Help Startups in China?

Poshan Yu
https://orcid.org/0000-0003-1069-3675
Soochow University, China & Krirk University, Thailand

Aiqi Xu
Independent Researcher, China

Emanuela Hanes
Independent Researcher, Austria

Shengyuan Lu
https://orcid.org/0000-0002-9847-4604
Independent Researcher, China

ABSTRACT

Difficulty in financing has always been a major problem faced by small companies or start-ups around the world, and China is no exception. This chapter aims to examine how crowdfunding can help startups in China to solve the problem of financing difficulties. By analyzing the Chinese startup marketplace as well as the sources of funding available to small businesses or startups, this chapter elaborates that it is reasonable to solve these funding problems and help startups survive through crowdfunding and draw forth regulations concerning it. Additionally, use cases are presented to help readers understand the development status of crowdfunding in China.

DOI: 10.4018/978-1-7998-9810-8.ch009

INTRODUCTION

Since the mid-2000s, startups have increasingly become the driving force for new jobs and a growth driver in economically advanced countries. Emerging countries have been striving to make startups viable through active government support policies. Government support refers to the extent the local government provides various forms of assistance or incentives to companies (Oliva et al., 2021). The importance of crowdfunding platforms within crowdfunding ecosystems has been identified (Berné-Martínez, Ortigosa-Blanch, Planells-Artigot, 2021). The use of crowd-based online technology (CBOT) for raising funds from a large number of people is viewed as a disruptive innovation in entrepreneurial financing as well as other forms of fundraising activities. Crowdfunding (CF), the most popular fundraising application of CBOT, is considered to be a truly global movement with significant financial and economic benefits (Kshetri, 2015).

Small enterprises or Startups play an important role in China, but their survival is difficult if compared to some developed countries like the United States, Britain and Japan because of historic financial repression policy. Before 2019, the year of the outbreak of COVID-19, Chinese startups were optimistic, expecting since 2018 that some policy documents aimed at them would be issued soon. But the current fundraising environment is challenging, and startups are worried about large companies acquiring and occupying the market. Under COVID-19, as the only country to achieve positive economic growth, China's economy has been recovering rapidly, giving small enterprises or startups chances to bounce back. The recovery of small businesses or startups during the COVID-19 situation and the innovative behavior in the financial market indicate that FinTech has followed the trend and startups can access new ways of financing.

Major sources of funding available to small businesses or startups in China include venture capital (VC), private equity, angel/ micro VC/ individual investors, corporate investors or IPO. Venture capital always ranked first, accounting for about one-third of the total funding. Some enterprises are lucky to get venture capital funds, but in the later stage of venture capital, some promising small and medium-sized enterprises will still face poor capital turnover. At this moment, they need a fund to help them recover from difficult phases. Equity crowdfunding is a good solution. Except venture capital, all the other funding methods can be realized through crowdfunding platforms (CFP). Compared to VC, the risk of crowdfunding is lower and it will not decentralize the management of the enterprise. One trend that is beginning to take shape, and that we expect to see become more prevalent in the future, is the rise of niche or specialty crowdfunding platforms. The rise of niche crowdfunding platforms can help connect interested entrepreneurs and investors alike to fund market-specific projects (Chandler, Short, & Wolfe, 2021). One successful crowdfunding case is Qingdao Raytheon Technology, a company using the method of lead and follow-up investment, which is a mature and popular way.

This chapter aims to examine how crowdfunding can help startups in China to solve the problem of financing difficulties. By analyzing the Chinese startup marketplace as well as the sources of funding available to small businesses or startups, this paper elaborates that it is reasonable to solve these funding problems and help startups survive through crowdfunding and draw forth regulations concerning it. Additionally, use cases are presented to help readers understand the development status of crowdfunding in China.

THE CHINESE STARTUP MARKETPLACE

Small businesses or startups play a very important role in China. According to the data from iResearch in 2021, small and micro enterprises account for 96.5% of China's market players, contributing 60% of GDP, 70% of the national technological innovation achievements and 80% of labor employment (Xia Shuo official account, 2021). In addition, the tax contribution accounts for half of the national tax. In the long run, with most non-religious people and an increasing number of female founders in the market, Chinese small businesses or startups are sustainable and they will remain important. This is shown in a qualitative comparative analysis of the causal patterns reveals that high GDP in combination with either (1) high shares of female founders of startups or (2) high shares of non-religious people in the population induce entrepreneurial ecosystems with relatively high levels of sustainability enterprises. (Tiba, van Rijnsoever, & Hekkert, 2021).

But the survival of small and micro enterprises is difficult. The profit growth rate of small businesses or startups has declined year after year, and the rate of profit growth decline in 2020 exceeded 30%.

Policy orientation and regulatory analysis shows that the State Council frequently promulgated policies to promote the development of small, medium-sized and micro enterprises in the hope that they "survive and live well"; The supervision of financing business of small and micro enterprises focuses on supervision during and after the financing event to encourage business innovation and development.

Analyzes of small and micro loan scales show that the loan balance of China's small and micro enterprises has increased from 4.35 trillion dollar in 2016 to 6.78 trillion dollar in 2020, with a compound annual growth rate of 12.2%. Measured by economic contribution, the loan space is expected to be 1.5-2.3 times the current scale; According to the calculation of single household financing demand, the loan space is expected to be 2.7 times the current scale (Xia Shuo official account, 2021).

Figure 1. Ease of doing business in China 2013-2019
Source: Doing Business 2020, page 4

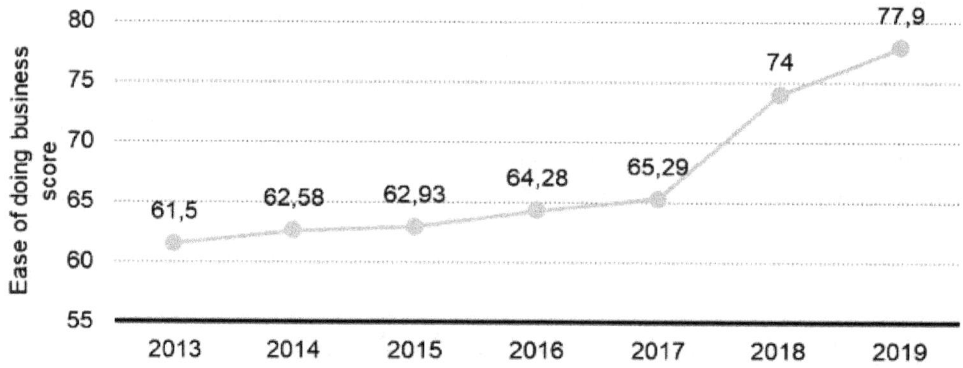

Figure 2. Business exits and exit reasons in China 2019
Source: Global entrepreneurship monitor 2019/2020, page 206-207

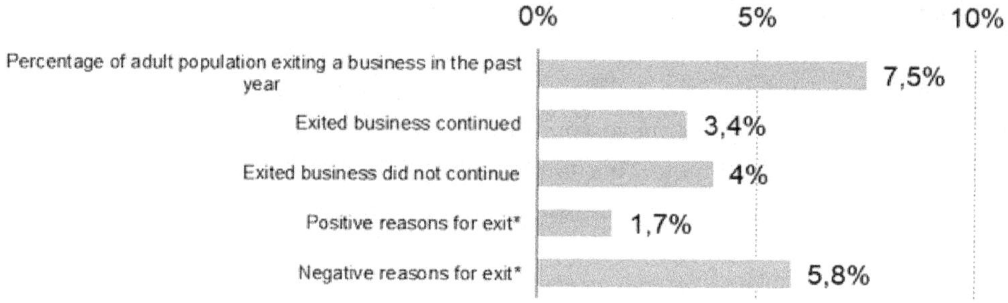

From the year of implementing the policy of reform and opening up to 2000, compared to Western startups, Chinese startups faced a challenging fundraising environment, but in the 2000s, the government's support for start-ups is gradually increasing. Especially before 2019, the year of the outbreak of COVID-19, Chinese startups were optimistic about 2018 and according to the Silicon Valley Bank Startup Outlook Survey, China has given greater support to small businesses or startups in the present situation. According to the data of doing business, we can also see that the ease of doing business in China from 2013 to 2019 is increasing year by year. Especially from 2017 to 2018, the growth rate reached the peak (Figure 1).

At the same time, start-ups are worried about large companies acquiring and occupying the market. What"s more, artificial intelligence and life science/healthcare were the top two promising areas according to Chinese startup entrepreneurs. It shows the strong support of the Chinese government for emerging technologies and the current higher demand for health.

The COVID-19 health crisis exerts a significant and negative impact on the stock returns of startups. From 2019 to 2020, many startups exited the market (accounted for 7.5%) and most of them exited due to negative reasons (Figure 2). This impact is weaker among startups with lower leverage ratios, CEO duality regime, and larger board (to some extent), as well as in countries having higher GDP growth. The negative effect on startups' stock performance is significant in the Industrial, Materials and Technology sectors, while no effect is found in the Communications, Consumer, and Health care sectors (Hoang, Nguyen, & Nguyen, 2022). As the only country to achieve positive economic growth under the COVID-19, China's economy is recovering rapidly, giving small businesses or startups chances to bounce back.

After comparing and analyzing the financing situation of small businesses or startups in major economies by using the survey data of OECD financing for small businesses or startups 2020, we found that the financing situation and financing conditions of small businesses or startups in China have reached a better level relative to the rest of the world: The proportion of loans of small businesses or startups (64.96%) is 24.55 percentage points higher than the median in the sample, and the rejection rate of credit financing (3.69%) is also the lowest in the sample, Moreover, China also has the world's largest online financing market, accounting for 62.5% of the global online financing transactions, providing an effective channel for the financing of small businesses or startups. However, in terms of financing structure, the growth rate of venture capital in China (about 10%) is lower than the average growth rate of OECD countries (20.86%), and the direct financing market of small businesses or startups is still in the cultivation stage (People"s Bank of China, 2021).

Recently, startups and new ventures are experiencing exponential growth due to the country"s recent mass entrepreneurship and innovation movement (Men, 2021). In the first quarter of 2021, the SME development index (SMEDI) was 87.5, an increase of 0.5 points over the fourth quarter of last year. It maintained an upward trend for four consecutive quarters, the highest since the first quarter of last year. Among them, the sub industry index increased comprehensively, and the sub index increased by 6 and decreased by 2. Among the small businesses or startups surveyed in this period, the social service industry with the best return to work rate is 97.50% (China Association of Small and Medium Enterprises, 2021).

At present, the world economic recovery is unstable and unbalanced, there are many uncertainties in the epidemic situation and the external environment, and the foundation for China's sustained and stable economic recovery is not solid. Some enterprises still have many production and operation difficulties and insufficient market demand. Small, medium-sized and micro enterprises still face difficulties such as rising costs, tight funds, declining benefits, arrears of accounts, insufficient investment power, weak labor supply and so on. Although the SME development index in the first quarter reached the highest value since the same period last year, it was still 5.2 points lower than that in the fourth quarter of 2019 and 5.5 points lower than that in the same period in 2019. It has not fully recovered to the pre-pandemic level (China Association of Small and Medium Enterprises, 2021).

The recovery of small and medium enterprises, COVID-19's environment and the innovative behavior in the financial market indicate that FinTech is following new trends and that startups have access to new ways of financing.

In order to reflect the innovation and entrepreneurship performance of various regions under the background of the "innovation and entrepreneurship strategy" in China, Table 1 summarizes the findings of the Peking University Center for Enterprise Research. This demonstrates the long-term innovation and entrepreneurship potential of various regions, so as to help readers to better evaluate the development trends and regional patterns of innovation and entrepreneurship in China.

Table 1. Provinces and cities with total innovation index greater than 80 in 2019

Year	Province / City	Total index	Score of new enterprises	Score of attracting foreign investment	Score of attracting venture capital	Score of patent authorization quantity	Score of trademark registration quantity
2020	Guangdong Province	100	100	99.27	98.13	99.97	99.9
2020	Jiangsu Province	99.48	99.17	99.9	98.75	99	98.44
2020	Zhejiang Province	99.38	98.54	99.58	98.23	99.48	99.69
2020	Shanghai	98.75	97.71	98.65	98.86	97.42	98.54
2020	Shandong Province	98.65	99.48	99.06	96.57	97.39	98.13
2020	Beijing	97.92	91.16	98.86	98.54	98.24	99.06
2020	Sichuan Province	96.46	97.5	96.67	95.84	94.58	96.15
2020	Fujian Province	96.25	95.42	95.63	93.96	95.48	97.71
2020	Anhui Province	96.05	96.88	96.36	93.03	96.23	95.01
2020	Henan Province	95.94	98.02	95.42	92.82	92.7	96.98
2020	Hebei Province	94.17	97.09	95.73	86.58	91.89	95.11
2020	Hubei province	93.96	93.86	92.92	92.3	94.05	93.96
2020	Jiangxi Province	92.72	92.61	96.05	91.78	90.47	91.05
2020	Hunan Province	92.61	92.82	93.96	89.49	91.91	93.24
2020	Shaanxi Province	91.26	93.03	91.36	89.18	87.37	92.09
2020	Tianjin	91.16	81.79	94.9	94.69	89.56	88.03
2020	Chongqing City	90.01	89.18	91.16	86.68	90.43	93.13

continues on following page

Table 1. Continued

Year	Province / City	Total index	Score of new enterprises	Score of attracting foreign investment	Score of attracting venture capital	Score of patent authorization quantity	Score of trademark registration quantity
2020	Liaoning Province	87.41	90.01	89.28	78.67	86.94	89.59
2020	Guangxi Zhuang Autonomous Region	86.89	89.91	91.47	84.6	81.33	88.35
2020	Guizhou Province	86.78	87.41	91.78	86.89	81.75	87.72
2020	Yunnan Province	86.58	86.58	90.43	85.74	80.1	90.84
2020	Shanxi Province	86.06	90.11	89.07	83.87	79.23	85.74
2020	Hainan	85.95	87.83	96.77	90.22	68.25	80.44
2020	Jilin Province	81.79	81.89	82.21	76.8	78.3	83.25
2020	Heilongjiang Province	80.02	80.02	82.41	70.97	76.35	83.98
2020	Xinjiang Uygur Autonomous Region	79.19	71.7	88.14	78.77	69.71	81.89
2020	Inner Mongolia Autonomous Region	77.11	78.67	83.45	57.96	76.45	83.87
2020	Gansu Province	74.71	75.75	81.06	62.43	72.87	78.36
2020	Ningxia Hui Autonomous Region	67.43	54.32	72.94	65.04	69.26	69.3
2020	Qinghai Province	58.17	44.02	59.63	55.78	59.69	63.16
2020	Tibet Autonomous Region	56.5	36.52	61.71	62.75	55.72	62.64

Source: www.cer.pku.edu.cn

It is worth adding that small and medium-sized enterprises have developed more rapidly in some provinces and cities in China, and China also tries to drive the development of enterprises in surrounding areas through these provinces and cities, so the innovation index will be used to illustrate the support of a province or city to small and medium-sized enterprises, and help readers understand where small and medium-sized enterprises are easier to develop. To begin with, the total innovation index consists of five indicators that contain score of new enterprises, score of attracting foreign investment, score of attracting venture capital, score of patent authorization quantity and score of trademark registration quantity. According to the List of Cities and Areas with a total innovation index above 80, carried out by the Center for Enterprise Research, Institute of Social Science Survey of Peking, we can find that most of the areas with active innovation in 2020 are coastal areas, such as the Yangtze River Delta, the Pearl River Delta and economic metropolis, such as Shanghai and China's capital - Beijing. As the table above shows, most of China's innovation and technology are mainly in the hands of private enterprises, while most private enterprises are small businesses or startups. Additionally, it can be seen that there are more and active small businesses or startups in these areas, more innovation opportunities and greater regional policy support. What"s more, the overall innovation index of Guangdong Province has reached 100 (Table1).

THE SOURCES OF FUNDING AVAILABLE TO SMALL BUSINESSES OR STARTUPS IN CHINA

It is difficult for startups to establish and develop without the support of informal financing. Informal financing occurs without a formal financial intermediary between suppliers and borrowers. In empirical studies, informal financing often includes, but is not restricted to, trade credit, interpersonal borrowing (money from friends or family), private money houses, pawnshops, and community cooperatives (Xu, Zhang, & Chen, 2021). Because of the high threshold of equity financing, after exhausting informal financing, small businesses or startups will consider bank credits as a next step in most cases. However, due to the large number of start-ups, small business scale, poor repayment ability, low reputation and unclear accounts, banks are more reluctant to lend to small businesses or startups than large enterprises that are stable and continuously profitable.

Excluding bank financing for small and medium-sized enterprises, as is shown in the bar chart below, major sources of funding available to technology and healthcare small businesses or startups in China are venture capital (VC), private equity, angel/ micro VC/ individual investor, corporate investor or IPO. Venture capital always ranked first and accounted for about one-third of total capital influx. Though the proportion of private equity and IPO was small, it was on the rise as a whole (Figure 3). Therefore, venture capital was especially looked-after and it played an important role with many unfavorable factors having been overcome like the negative effect of geographic distance being significantly lessened when China"s high-speed railway became available after the startup–VC partnership formation (Zhang & Gu, 2021). It needs to be pointed out that, because most enterprises in statistics in 2018 were "innovation and high-tech driven" new businesses, which were normally more likely to receive venture capital funding in China (Men, 2021), startups in other areas might not have such high expectations for VC. Nevertheless, VC is still the main way to raise funds.

However, VCs might have undisclosed conflict of interests and might not be able to offer an unbiased picture of the industry trends (Gloor, et.al.2020). When a method occupies a dominant position in a field, both investors and entrepreneurs are prone to rely too much on traditional solutions and are unwilling to try new ways. This results in improvements not being taken seriously. While VC participation significantly reduces bankruptcy hazard, it does not decrease the risk of failures (Li & Hasan, 2021).

Figure 3. Sources of funding experiences by Chinese technology and healthcare startups from 2016 to 2018
Source: svb.com/startup-outlook-report

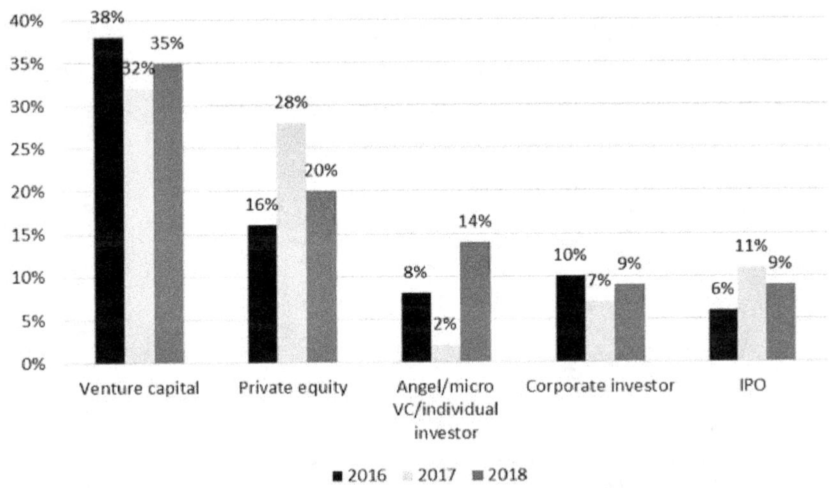

Except venture capital, the second and third method can be realized through crowdfunding platforms (CFP). China was one of leading ten countries worldwide in terms of crowdfunding volume in 2020 (Figure 4). What"s more, from 2013 to 2018, China's investment into FinTech enterprises showed an overall upward trend with slight fluctuations (Figure 5), indicating that these enterprises face a better investment environment and crowdfunding should perform better in this environment. Compared to VC, the risk of crowdfunding is lower and it will not decentralize the management of the enterprise. Moreover, even if crowdfunding is not successful in the end, startups can still have a clearer understanding of their customer groups, publicize their technologies and products in disguise, and make changes and improvements for the next financing plan, which is difficult for VC.

Figure 4. Leading ten countries worldwide in terms of crowdfunding volume in 2020 (in million U.S. dollars)
Source: The 2nd Global Alternative Finance Market Benchmarking Report, page 191; 1 yuan ≈ Approximately equal; 1 yuan ≈ 0.1569 dollar (forex.hexun.com 2021-12-20 09:28)

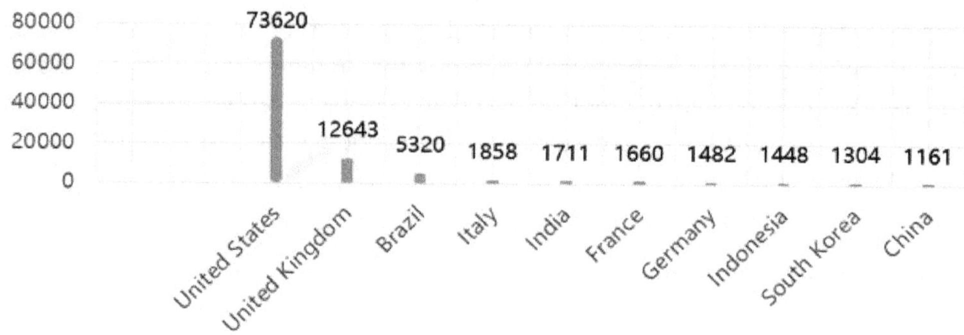

Figure 5. Value of investments into China's FinTech enterprises 2013-2020
Source: bg.qianzhan.com; 1 yuan ≈ Approximately equal; 1 yuan ≈ 0.1569 dollar (forex.hexun.com 2021-12-20 09:28)

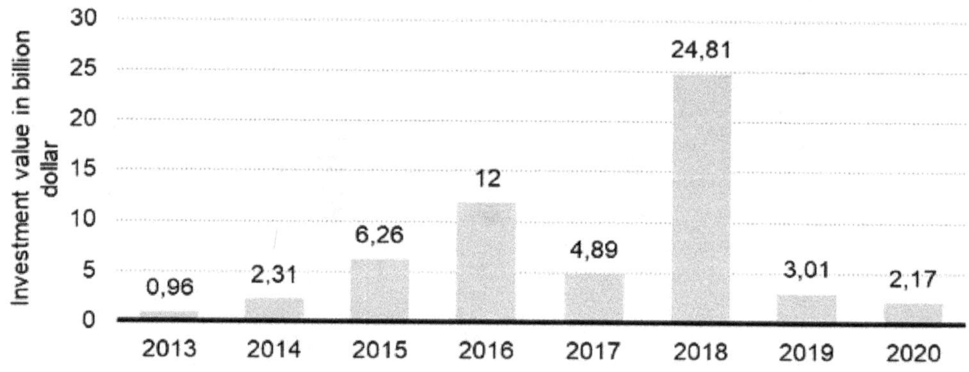

It can be seen that the traditional financing channels of small and medium-sized enterprises are too few and the cost is relatively high. The equity crowdfunding financing threshold in the developing society is relatively low and the channels are relatively wide. In the past, small and medium-sized enterprises were faced with high interest loans. The original investment and financial management method of equity crowdfunding is more suitable for them to grow and become stronger, or at least reduce the pressure on debt repayment, which is also conducive to the independent innovation and operation of small and medium-sized enterprises.

Direct financing is naturally adaptable for supporting the financing of high-tech small, medium and micro enterprises, but the overall support of equity financing in China is weak, and the stratification of the bond market is still insufficient. In terms of equity financing, the ratio of the financing amount of gem and new third board to the loan increment of small and micro enterprises in the same period in 2019 was 1:16, far lower than that of Japan in 2018, which was 1:7.4. The overall scale and individual investment amount of venture capital and risk management capital are also far below the United States. In terms of bond financing, by the end of 2019, China's various bond balances 3 for small, medium-sized and micro enterprises accounted for only 1.3% of all bond balances, which was 3.41% of the loan balance of small and micro enterprises in the same period. Moreover, more than 70% of bond financing was mainly realized by issuing bonds by banks. The real bond market for small, medium-sized and micro enterprises needs to be further cultivated and developed (People"s bank of China, 2021).

In terms of financial technology empowerment, the digital operation ability of commercial banks needs to be improved, and the application of data elements needs to be standardized. There are two major difficulties in the overall construction and market-oriented development of infrastructure such as data sharing. At present, Taizhou, Suzhou and other places have built data sharing platforms to effectively solve the problems of data collection, sharing and application for government departments. However, in most areas, government departments lack awareness of data openness and lack data openness systems, rules and standards. This results in data being closed within government departments. On the other hand, most of the sharing platforms operate in form of public utilities, with strong data collection ability, weak data analysis and application ability, a lack of market-oriented thinking, and the role of the platform cannot be reflected. The overall digital operation ability of commercial banks is not strong. Digitization currently remains at the operational level, and there is no effective breakthrough in building ecology, setting up scenes and expanding user numbers. For small and medium-sized banks, FinTech empowerment creates greater challenges. This is due to the fact that these banks are state-owned, not private and the policy orientation is to support FinTech and improve bank management. Many small and medium-sized banks are weak, but it is neither realistic nor necessary to significantly increase their science and technology investment. It is necessary to explore effective ways to cooperate with large commercial banks or FinTech companies (People"s bank of China, 2021).

WHY AND HOW CROWDFUNDING WOULD HELP STARTUPS

To recap briefly, crowdfunding is the process of funding a project by raising small monetary contributions from large numbers of individuals (Benazzouz, Malkinson, & Toth, 2021). There are four different types of crowdfunding projects: (1) crowdlending,

(2) crowdequity, (3) reward-based crowdfunding and (4) donation-based crowdfunding (Kshetri, 2015). Put donation-based crowdfunding aside, crowdlending indicates that investors invest in the project or company, obtain a certain proportion of their creditor's rights, obtain interest income and recover the principal in the future. Crowdequity refers to a mechanism where the company transfers a certain proportion of shares to ordinary investors, who can obtain future income by investing in the company. Crowdequity usually takes measures such as combination of investment. Reward-based crowdfunding is mainly used in market research and testing market potential via giving reward to investors. In crowdfunding, especially reward-based crowdfunding, entrepreneurs abandon their project if they do not raise enough money to cover expenses (Bollaert, Lopez-de-Silanes, & Schwienbacher, 2021).

As the pie chart -Share of crowdfunding platforms in China in June 2019 shows below, equity-based crowdfunding ranked first, which accounted for 37%, followed by rewards-based crowdfunding (31%) (Figure 6). Positively, from the second chart - Figure 7, there was a fluctuating increase in number of successful crowdfunding campaigns in China from 1st half 2016 to 1st half 2018, indicating crowdfunding to greatly help the market and increase attention. Crowdfunding is in fact only rarely employed as a last resort, with entrepreneurs using it more often as a tool to develop, market and validate their ideas (Junge, Laursen, & Nielsen, 2021).

Figure 6. Share of crowdfunding platforms in China in June 2019, by type
Source: bg.qianzhan.com

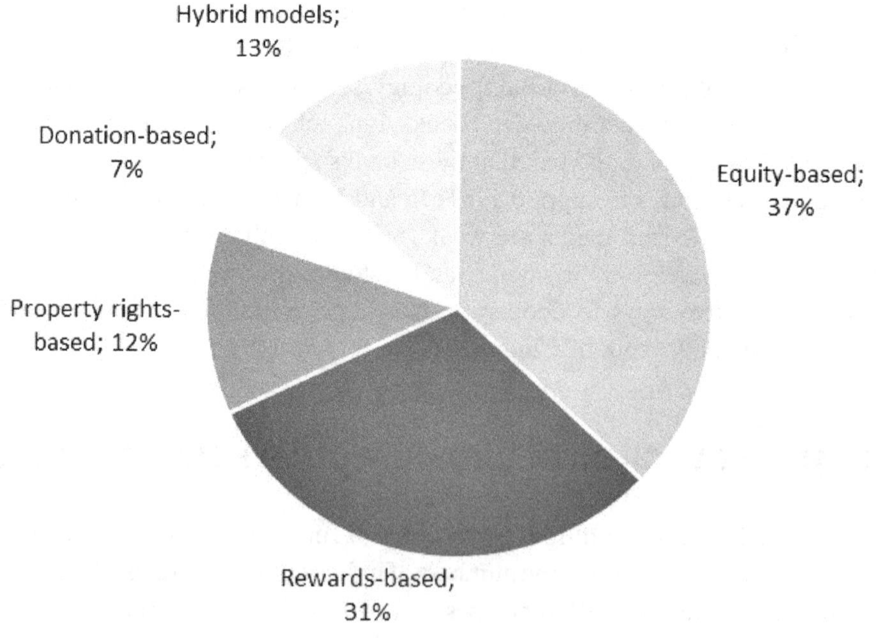

Figure 7. Number of successful crowdfunding campaigns in China from 1st half 2016 to 1st half 2018
Source: bg.qianzhan.com

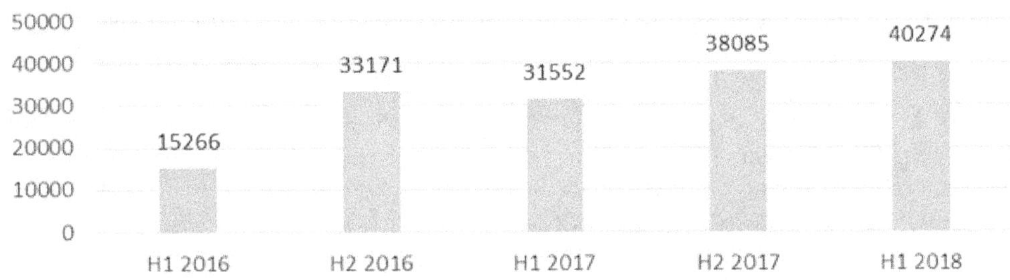

Crowdequity mainly faces investors who want to obtain capital appreciation returns through investment. Its purpose is to help seed start-ups raise initial funds and help small and medium-sized enterprises overcome the difficulty of capital turnover in the rising period. Currently, the most popular and mature method in the equity crowdfunding market is lead investment and follow-up investment. Leading investors have the largest capital contribution and are most responsible for various affairs in the early investment process. Most of their backgrounds are insiders familiar with the project or investors with rich financial expertise. Follow up investors have less capital contribution, less responsibility, lower risk, but also low income. The capital contribution of follow-up investors is not necessarily the same. When facing major projects, there are many fund choices for follow-up investment.

Reward based crowdfunding is mainly for buyers who want to get a return on goods by investing money. Its main purpose is to help entrepreneurs raise funds to produce products and conduct market research. Even if it does not raise enough funds within the specified time, the enterprise can judge whether the commodity is suitable for entering the market through crowdfunding. But at the same time, because reward based crowdfunding is not well known in China, the market survey data summarized according to it may not be accurate. If the entrepreneur's company is not well-known and the publicity effect is not in place, potential consumers may not know about the crowdfunding opportunity.

In general, FinTech initiatives currently fund small projects and start-ups (Bollaert, Lopez-de-Silanes, & Schwienbacher, 2021), because in the gray area of law, the problem of financing challenges and high interest rate of small businesses or startups has been solved to a certain extent. What"s more, crowdfunding precedes high tech and high growth entrepreneurship (Yu & Fleming, 2021). The market size of crowdfunding worldwide in 2020 is 12.27 billion U.S. dollars and with a forecast for 2027, the market size will boom to 25.8 billion (Figure 8). However, the peer-to-peer (P2P) lending industry in China has experienced the most drastic developments and destructive busts (Li & Hasan, 2021 and Yu et al., 2021), because some startups run off without repaying the money to investors. This urged the government to improve the laws and regulations on financial technology. As a result, many investors can now turn to a relatively mature crowdfunding framework without high risk.

Figure 8. Market size of crowdfunding worldwide in 2020 with a forecast for 2027 (in billion U.S. dollars)
Source: marketwatch.com

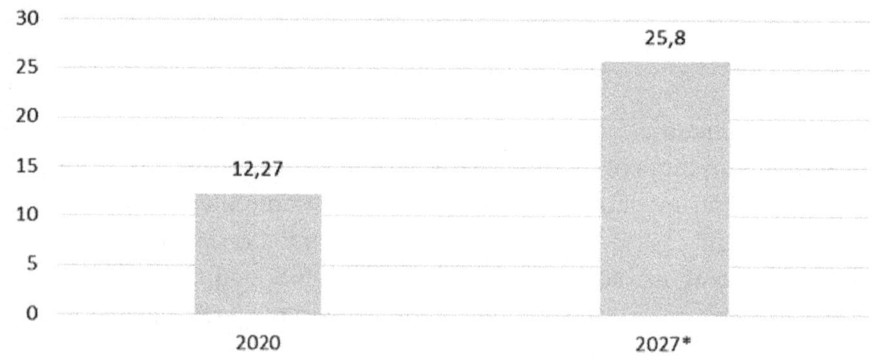

HOW GOVERNMENT REGULATES CROWDFUNDING FINANCE AND HOW THESE POLICIES IMPACT STARTUP BUSINESS STRATEGIES / PERFORMANCE

Before the outbreak of COVID-19, the most important public policy issues were corporate taxes and access to talent. Both of them were connected fundamentally to products or services and accounted for around 50% of each. The former is vital for technology research and development and the latter involves the overall development of the company (Figure 9).

Under China's national conditions, policy support has promoted the prosperity of the industry. The restriction and promulgation of laws and regulations have initially dealt a heavy blow to small businesses or startups in this field. The government was able to subsidize the small and medium but innovative firms to better exploit the knowledge spillovers effects (Yao & Hu, 2020).

However, the impact of policies has improved significantly in recent years and the government has relaxed policies for small businesses or startups and actively issued documents to help small businesses or startups tide over the difficulties. Compared with only one policy document and heavy policies on SMEs in 2016 and 2017, there were three in 2018 and 2019. After the outbreak of the epidemic, small businesses or

Figure 9. Public policy issues affecting tech & healthcare startups in China 2017-2018
Source: China Startup Outlook 2018, page 12

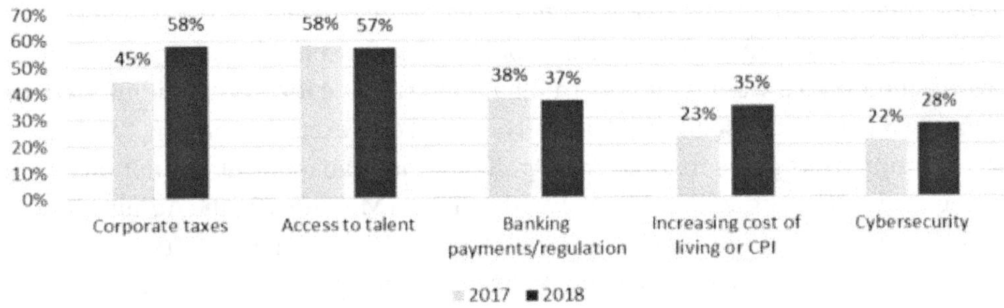

startups have suffered another major blow and faced problems in returning to work and production, and there has been a significant increase in policy documents for them. Letting startups go bankrupt potentially jeopardizes a state"s future innovativeness. Therefore, mid-, or long-term policy measures targeted future innovativeness. While unlikely to be the first responses to such a crisis, they nonetheless are essential (Kuckertz et al., 2020). The notice issued by the China Banking Regulatory Commission on January 26th on strengthening the banking industry's insurance sector and improving the prevention and control of pneumonia in New Coronavirus infection is issued. The purpose of the notice is to relax the loan regulations for small businesses or startups, such as lowering the loan interest rate appropriately and improving the policy arrangement of continued loans. It also included increasing credit loans, medium and long-term loans to help small businesses or startups initially ease the financial pressure and reduce enterprises' withdrawal from the market.

In addition, the State Administration of Taxation also extended the tax time and extended the tax declaration period in February to 24th of February. On February 2nd, the Ministry of Finance issued the notice on supporting financial strengthening services for New Coronavirus infection prevention and control. The notice said that the focus is to strengthen the financial support for small businesses or startups from the aspects of giving financial discount support to the loans of key enterprises active in epidemic prevention and control, giving discount support to the entrepreneurial guarantee loans of individuals and enterprises affected by the epidemic, optimizing the financing guarantee services for enterprises affected by the COVID-19. The above policies also explain why the non-performing loan ratio will rise in 2021.

According to the Table 2, we can define the years from 2012 to 2014 as the embryonic period, including mainly the introduction of equity crowdfunding and discussion on China's existing laws and markets. From 2014 to 2015, researchers mainly conducted research on foreign forms, and a small number of enterprises began to comply with the policy to try and establish companies. With the emergence of a large number of high-quality equity crowdfunding companies, there were also a series of cases of illegal fund-raising. Equity crowdfunding is defined as non-public online offering, that is, the online form of private placement.

However, non-public offering is contrary to the popular characteristics of crowdfunding, so it was urgent to formulate a bill suitable for crowdfunding, and the National Securities Association gradually carried out research work. In 2016, cases were summarized and sorted out and policies and management regulations were formulated to standardize the market, like the Interim Measures for the Administration of the Business Activities of Online Lending Information Intermediary Institutions and Implementation Plan for special rectification of equity crowdfunding risk.

In the regulatory policy, minimizing asymmetric information is emphasized. Understanding the decision-making process in crowdfunding requires an understanding of how information is communicated between founders and funders who have access to different information. In particular, when societal challenges are addressed with projects that are not easy to grasp from a technical perspective, information asymmetry between the founder and funders becomes substantial (Pabst, Wayand, & Mohnen, 2021). The period from 2016 to 2018 is mainly a period of enterprise screening and a stabilization of the equity crowdfunding market. The mode of equity crowdfunding gradually matured, and the leading enterprises in the market gradually emerged. With many successful cases, crowdfunding increasingly pointed the way for financing small businesses or startups, and the government departments strengthened their supervision. From 2018 to 2019, the China Securities Association added the measures for administering equity crowdfunding pilots in its legislative work plan.

However, due to the epidemic, the shutdown of small businesses or startups and the inactivity of the financial market, it has not been promulgated so far. At the same time, Beijing established a "regulatory sandbox" in 2019, a special zone specially built for FinTech companies to promote the prosperity of financial technology, trying to improve or solve the problems encountered in traditional finance through innovation, and further improve the market environment of small businesses or startups. In 2020, the outline of the 14th Five-Year Plan for National Economic and Social Development of the People's Republic of China was issued, which incorporated Internet finance, blockchain, cloud computing and the digital economy into the outline of the plan. This further promoted the innovation of financial technology, and gave policy support to Internet finance industries such as equity crowdfunding. In 2021, the State Council further strengthened policy supervision on equity crowdfunding, standardized the market structure and avoided false publicity by some enterprises for interests. Although many of the current financial rules and policies were established before the Internet era, companies offering new financial services have to ensure that their business model, operations, and products do not contravene legal requirements (Leong, Tan, Xiao, Tan, & Sun, 2017).

Table 2. Some important policies related to equity crowdfunding

Title	Issuing Authority	Date Issued	Main Implications
Measures for the administration of private equity crowdfunding (for Trial Implementation) (Draft for comments)	Securities Association of China	12/18/2014	The administrative measures clearly stipulates that equity crowdfunding shall adopt the form of non-public offering, and adopt a series of self-discipline management requirements to meet the relevant provisions on non-public offering in Article 10 of the Securities Law: first, investors must be specific objects, that is, real name registered users verified by the equity crowdfunding platform that meet the conditions specified in the administrative measures; Second, the cumulative number of investors shall not exceed 200; Third, the equity crowdfunding platform can only recommend project information to real name registered users, and neither the equity crowdfunding platform nor financiers shall conduct public publicity, promotion or persuasion.
Guiding Opinions on Promoting the Sound Development of Internet Finance (from now on referred to as the "Guiding Opinions")	the People"s Bank of China (PBOC), the Ministry of Industry and Information Technology, the Ministry of Public Security, et al	7/19/2015	1. Equity crowdfunding financing must be conducted through the equity crowdfunding financing intermediary platform (Internet website or other similar electronic media) 2. On the premise of complying with laws and regulations, equity crowdfunding financing intermediaries can innovate and explore the business model, give play to the role of equity crowdfunding financing as an organic part of multi-level capital market, and better serve innovative and start-up enterprises 3. The equity crowdfunding financier shall be a small and micro enterprise, and shall truthfully disclose the business model, operation and management, finance, fund use and other key information of the enterprise to the investors through the equity crowdfunding financing intermediary, and shall not mislead or cheat the investors 4. Investors shall fully understand the risks of equity crowdfunding activities, have corresponding risk tolerance and make small investment. The equity crowdfunding financing business shall be supervised by the CSRC.
Implementation plan for special rectification of Internet financial risks	the General Office of the State Council	4/12/2016	The equity crowdfunding platform shall not publish false targets, raise funds by itself, raise "public shares and real debts" or raise funds in disguised form. It shall strengthen the information disclosure obligations and shareholders' rights and interests protection requirements for financiers and equity crowdfunding platforms, and shall not make false statements and misleading publicity.
Interim Measures for the Administration of the Business Activities of Online Lending Information Intermediary Institutions (the "interim measures")	the China Banking Regulatory Commission (CBRC), et al.	8/17/2016	Internet lending information intermediaries shall not engage in or accept entrusted equity crowdfunding and other businesses

continues on following page

Table 2. Continued

Title	Issuing Authority	Date Issued	Main Implications
The implementation plan for the special rectification of equity crowdfunding risks is hereinafter referred to as the implementation plan	China Securities Regulatory Commission, et al.	10/13/2016	The focus of rectification is as follows. The Internet equity financing platform (hereinafter referred to as the platform) engages in equity financing business in the name of "equity crowdfunding". The platform raises private equity investment funds in the name of "equity crowdfunding". The financiers on the platform publicly or publicly issue shares in a disguised form without approval. The platform conducts false publicity and misleads investors by fictitious or exaggerated methods such as platform strength, financing project information and return. Financiers on the platform fraudulently issue stocks and other financial products. The platform and its staff misappropriate or occupy investors' funds. Platforms, real estate development enterprises and real estate intermediaries engage in illegal fund-raising activities in the name of "equity crowdfunding". Eighth, licensed financial institutions such as securities companies, fund companies and futures companies cooperate with Internet enterprises to carry out business in violation of laws and regulations.
Provisions on the regulatory responsibilities of dispatched offices of CSRC	China Securities Regulatory Commission	5/15/2018	Article 9 the dispatched office shall be responsible for inspecting the following market entities: (9) equity crowdfunding financing service institutions, financiers and other entities related to equity crowdfunding financing activities
Regulations on prevention and disposal of illegal fund-raising	the State Council	2/11/2021	Unless otherwise provided by laws, administrative regulations and the state, the name and business scope of enterprises and individual industrial and commercial households shall not contain the words or contents of "equity crowdfunding"

Source: www.cbirc.gov.cn, www.sac.net.cn, www.csrc.gov.cn

CASE STUDIES

In this part we selected Chinese case studies that illuminate the different applications of crowdfunding for small enterprises or startups. The first successful case is Qingdao Raytheon Technology Company limited by shares. It is a very typical equity crowdfunding case for the method of lead and follow-up investment. Raytheon was founded in 2014. Raytheon had low market visibility and lacked competitiveness, so its later development could easily be blocked. At this time, a large number of customer groups needed to support the enterprise in order to solve the problem of its low market share. The current situation is applicable to the equity crowdfunding mode, the publicity and promotion of market consumers in the early stage and the feedback obtained, so as to stimulate the potential and make improvements in the subsequent product optimization. Consumers were informed about Raytheon's brand core and business characteristics by way of special publicity measures to build a good fan base.

The application of the crowdsourcing financing mode at this stage not only helps the company"s visibility, but also lays a solid foundation for the development funds of enterprises and creates a more stable base for the more mature stage of its products. However, although the public publicity method in the project operation is applicable to brand publicity and for attracting investors, there is also leakage of the project content and the risk of public publicity violating the provisions of National Security law. In this stage, the legitimacy of the project needs to be protected. The goal is to operate the project normally under the conditions permitted by law and have a legal pass in the financial market (Jiang, 2019).

Raytheon expanded the sales area coverage through product listing, continuously strengthened the product equipment functions, and took the lead in the gaming laptop market. Raytheon technology, with its rapid growth in recent years, has become a leader amongst China's small businesses or startups. Its success depends not only on its own business philosophy, but also on the external financial support obtained through financing. Raytheon has sufficient funds which not only depend on the return of funds through product pre-sale. Another reason for its success is that in the early stage of the company's establishment, the parent company gave corresponding financial support. In the later stage, through the financing methods of equity crowdfunding and product crowdfunding, it attracted the equity capital of Zihui venture capital. The available money increased and there was support in operations. Moreover, Raytheon conducted round B financing in 2016 and is the exclusive supplier of Tencent cfpl-s9. In 2017, round C raised 10.2 million dollar and was listed on the new third board in September of the same year, becoming the Number One Company in the Game Industry. In 2019, it issued additional shares on the new third board. Raytheon's development can be seen from the company's valuation. Compared with the beginning of its establishment, Raytheon's social status is stable and the company has better development prospects. Moreover, the selection of the equity crowdfunding mode of financing is just in line with the company's development mode, which ensures the company's operation and promotion in the medium term, and uses the advantages of financing to drive the development of the enterprise (Jiang, 2019).

In addition, there are few excellent financing projects like Raytheon, which have accumulated certain customers and fans in the early stage, and the total amount of crowd equity and reward-based crowdfunding is up to 50 million. Some mitigation practices require high upfront investments, which increases the total amount of crowdfunding required if the campaign were to cover the full investment costs (Kragt, Burton, Zahl-Thanem, Otte, 2021).

As shown in the following table of crowdfunding projects completed by Ant Angel platform from 2015 to 2017, the amount of equity crowdfunding is mainly visible in terms of the number of projects. Ant angel had 11, 24 and 15 projects in 2015, 2016 and 2017 respectively. The financing amount was 1,041,502 Dollar, 3,459,331 Dollar and 2,207,269 Dollar respectively, and the average financing amount of each project was 94,140 Dollar, 144,034 Dollar and 145,917 Dollar respectively. The average number of investors in each project is 15, 24 and 24, and the average financing amount of each person is 6,119 Dollar, 5,805 Dollar and 5,962 Dollar respectively. The financing amount and per capita investment amount of a single project are far lower than those of other entrepreneurial equity financing platforms in China (Table 3). It can be seen from the names of the projects that they are basically small and micro enterprises focused on scientific and technological innovation with need of early investments (Yuan, 2018).

Additionally, at present, many Chinese films have been successfully realized through crowdfunding equity. For instance, from the website www.360kuai.com, the crowdfunding of the return of the movie Monkey King: Hero is Back secured a total of 1.2 million Dollar of investment, and it obtained about 4.7 million Dollar of principal and interest at that time. Compared with the film and television crowdfunding of Alibaba, JD and Baidu, the crowdfunding of Monkey King: Hero is Back was not launched on a more formal crowdfunding platform for public investors. Its producer merely released the information of crowdfunding in the wechat circle of friends. This crowdfunding was mainly aimed at people who are familiar with the producer. In fact, the success of the crowdfunding of it was closely related to the industry influence of its producer, and his ability was greatly trusted by his friends.

Table 3. Crowdfunding projects completed by Ant Angel platform from 2014 to 2017

Project Name	Financing time	Financing amount (thousand dollar)	Number of investors	Operation status	Intellectual property right	New employment post	Project update Valuation (million dollar)	Latest equity value (thousand dollar)
Crazy Coffee (Internet + coffee)	2015	62.76	15	Terminated			0	0.00
Xiandu robot (industrial robot supplier)	2015	172.59	14	Continued	3	8		172.59
Rice music (music education 020)	2015	78.45	21	Terminated			0	0.00
Dimension C financial management (mobile internet financial management tool)	2015	149.055	12	Continued	10	100	23.5	1176.75
Smell the birds (K12 artificial online Q & a platform)	2015	65.898	14	Terminated				0.00
Le Bo culture (entertainment film and television)	2015	32.949	9	Continued	0	5		32.95
Men's clothing state (vertical men's clothing e-commerce application based on mobile terminal)	2015	78.45	7	Continued	2	12		78.45
Come and rent me (renter dating platform)	2015	180.435	29	Continued	1	10	9.41	721.74
And adjacent network (University supermarket 020 platform)	2015	78.45	1	Terminated				0.00
Here comes the wedding car (app for wedding business car rental)	2015	51.777	14	Continued	0	6		51.78

continues on following page

Table 3. Continued

Project Name	Financing time	Financing amount (thousand dollar)	Number of investors	Operation status	Intellectual property right	New employment post	Project update Valuation (million dollar)	Latest equity value (thousand dollar)
Shadow adult (financing platform for cultural industry)	2015	89.433	33	Continued	0	30		89.43
Jihuafang ("Internet +" high-end art potting)	2016	167.883	29	Continued	1	5		167.88
Poker expression (funny picture community)	2016	109.83	25	Terminated				0.00
Quxing people (robot education Internet platform)	2016	221.229	20	Continued	0	10		221.23
Mr. y (custom shirt)	2016	65.898	27	Continued	0	6		65.90
Smart smart (UAV core technology solution provider)	2016	329.49	49	Continued	31	10	3.138	658.98
Mi Qi (intelligent hardware)	2016	123.951	30	Continued	6	5	1.88	223.11
Kitano network (family interactive entertainment game)	2016	94.14	26	Terminated				0.00
Me (Internet light food)	2016	152.193	32	Terminated				0.00
Anada (gay Fetish social e-commerce)	2016	142.779	19	Terminated				0.00
Le Che paint (service provider for emergency repair and body painting)	2016	125.52	22	Continued				125.52

continues on following page

Table 3. Continued

Project Name	Financing time	Financing amount (thousand dollar)	Number of investors	Operation status	Intellectual property right	New employment post	Project update Valuation (million dollar)	Latest equity value (thousand dollar)
Senior staff (listen to bat's outstanding staff share practical experience)	2016	56.484	17	Terminated				0.00
Leyu (outdoor tourism Internet platform)	2016	78.45	28	Exit				0.00
Small wooden man (young people's Internet whole house solid wood customized brand)	2016	125.52	14	Continued	1	10		125.52
Qingxiang yuemian (sleep monitoring Pillow - Smart medical entrance)	2016	274.575	14	Continued	5	10		274.58
Abnormal theory (a mutual aid growth platform based on character)	2016	62.76	21	Continued	0	12		62.76
Human horse line (one-to-one fitness private education service platform for high-end users)	2016	174.159	24	Continued	26	35	9.414	451.87
Iparty (building a skilled online talent industry chain)	2016	98.847	16	Continued	4	7		98.85
Xianbaoduo (online supply chain linking thousands of community convenience stores)	2016	238.488	23	Terminated	0	0		0.00

continues on following page

Table 3. Continued

Project Name	Financing time	Financing amount (thousand dollar)	Number of investors	Operation status	Intellectual property right	New employment post	Project update Valuation (million dollar)	Latest equity value (thousand dollar)
Xingzhi (original content of theme travel IP + characteristic travel platform)	2016	127.089	29	Continued	0	9		127.09
Downhill (mobile app shopping platform)	2016	130.227	16	Terminated				0.00
Gaorenhui (a leading Zhouyi service platform)	2016	114.537	16	Continued	12	12		114.54
Street to Street (a service platform for street cultural groups)	2016	122.382	23	Continued	2	6	2.3535	158.47
Easy point fresh (domestic professional community fresh cloud platform)	2016	156.9	38	Terminated				0.00
Single account (third party financial risk control SaaS system of catering stores)	2016	164.745	30	Continued	1	10	5.9622	382.84
Fast sloth (the largest balance car operator and service provider in China)	2017	156.9	33	Continued	0	12		156.90
Source accelerator (accelerator for upgrading the Internet of traditional industries)	2017	122.382	22	Continued	0	3		122.38
Mowu (piano room solution and intelligent piano learning platform)	2017	78.45	29	Continued	0	5		78.45

continues on following page

Table 3. Continued

Project Name	Financing time	Financing amount (thousand dollar)	Number of investors	Operation status	Intellectual property right	New employment post	Project update Valuation (million dollar)	Latest equity value (thousand dollar)
Software factory (intelligent software generation platform)	2017	145.917	28	Continued	17	25		145.92
Big in laws (build the largest online dating platform in China)	2017	156.9	26	Continued	1	20	3.9225	359.30
Fast goods new retail (traditional retail + fast goods = new retail)	2017	120.813	15	Continued	1	6		120.81
Mukexin maker Education (maker education and steam Education)	2017	262.023	21	Continued	5	15		262.02
Fan Guo Education (the most professional steam maker event operator)	2017	238.488	49	Continued	0	25	10.983	1098.30
Fanggu.com (providing construction products and services for tourism and leisure)	2017	125.52	29	Continued	8	7		125.52
Accounting cat (insensible aggregate payment - shop opening collection artifact)	2017	103.554	22	Continued	3	10		103.55
Zhishan robot (solving the problem of high-frequency handling of factory logistics)	2017	144.348	21	Continued	15	18		144.35
Hello vending store (new retail)	2017	100.10	16	Continued	0	4		100.10

continues on following page

Table 3. Continued

Project Name	Financing time	Financing amount (thousand dollar)	Number of investors	Operation status	Intellectual property right	New employment post	Project update Valuation (million dollar)	Latest equity value (thousand dollar)
Shanghui information (rehabilitation medicine and nursing service platform)	2017	263.592	28	Continued	19	9		263.59
Hot question bank (practicing qualification examination content distribution platform)	2017	84.726	10	Continued	13	10		84.73
Yuntong AIS (agricultural big data)	2017	103.554	18	Continued	9	9		103.55
Total		6705.91	1124		196	496		8852.30

Source:mp.weixin.qq.com

CONCLUSION AND RECOMMENDATION

Under the national conditions of supporting emerging technologies and the background of the rapid development of financial technology, Startups should be optimistic about the current risks and challenges, seize opportunities and pay attention to the latest laws and regulations on FinTech (Yu et al., 2022) and blockchain technology (Yu, P., Gong, R., & Sampat, M., 2022 and Yu, P., Lu, S., Hanes, E., & Chen, Y., 2022). What"s more, among all financial technologies, crowdfunding is the easiest to use and there are several successful cases to learn. Additionally, at present, there are crowdfunding platforms with relatively perfect system and process, reducing the risk for investors and start-ups. The common point of the two examples mentioned above is that for some well-developed enterprises or mature projects, the projects involved in crowdfunding have first been screened by the platform, and lead investor have participated, so it will progress smoothly in the later stage of the enterprise or project.

In addition to the fact that donated crowdfunding has been well known and widely used by the public, solving many family difficulties and high medical expenses, other types of crowdfunding still have many limitations. An additional point which came out here was the lack of flexibility that crowdfunding provided as the funding provided is a fixed amount (Laffey, Durkin, Cummins, & Gandy, 2021). For example, the publicity of incentive crowdfunding is not strong enough, resulting in indirect inaccurate market survey data. In addition, the incentive crowdfunding group is highly targeted and often faces the situation that it cannot raise enough funds within the specified period, resulting in a full refund. For another example, the limitation of equity crowdfunding is that there is no targeted crowdfunding law, and there is no bill to support the public's small investment and equity crowdfunding. Thus, it can only take the form of non-public. Secondly, it also limits the group of investors, requiring that only individuals who meet the financial assets of no less than 0.47 million Dollar or the average annual income of no less than 78450 Dollar in the last three years can participate in equity crowdfunding. Perceived ease of use shows to be an essential feature influencing intention to buy on Kickstarter. The Internet is transformed from cable connectivity to wireless, and via mobile devices, people access technology frequently without limitation to time or place. This shows the perceived ease and usefulness of using an adopted innovation. Crowdfunding platforms must, therefore, increase their support to give stakeholders access easy (Djimesah et al., 2021).

In the recent progress of FinTech, a crowdfunding platform primarily based on blockchain technology, Ether Invest allows investors to take part in the ongoing projects offered by entrepreneurs on its platform (Nguyen et al., 2021). The combined application of crowdfunding platform and blockchain shows that the reputation of crowdfunding is increasing in the future, and more high-quality projects and enterprises will be encouraged to adopt financial technology to solve financing problems. Thus, in the future research discussion, we should pay attention to the changes of crowdfunding under the leadership of emerging technologies.

ACKNOWLEDGMENT

The authors extend sincere gratitude to:

- Our colleagues from Soochow University, the Australian Studies Centre of Shanghai University and Krirk University as well as the independent research colleagues who provided insight and expertise that greatly assisted the research, although they may not agree with all of the interpretations/conclusions of this chapter.
- China Knowledge for supporting our research.

REFERENCES

Benazzouz, S. M., Malkinson, J., & Toth, G. (2021). Current trends in and future potential of crowdfunding to finance R&D of treatments for neglected tropical diseases. *Drug Discovery Today*, *26*(7), 1563–1568. doi:10.1016/j.drudis.2021.02.021 PMID:33640317

Berné-Martínez, J. M., Ortigosa-Blanch, A., & Planells-Artigot, E. (2021). A semantic analysis of crowdfunding in the digital press. *Technological Forecasting and Social Change*, *173*, 121175. Advance online publication. doi:10.1016/j.techfore.2021.121175

Bollaert, H., Lopez-de-Silanes, F., & Schwienbacher, A. (2021). Fintech and access to finance. *Journal of Corporate Finance*, *68*, 101941. Advance online publication. doi:10.1016/j.jcorpfin.2021.101941

Chandler, J. A., Short, J. C., & Wolfe, M. T. (2021). Finding the crowd after exogenous shocks: Exploring the future of crowdfunding. *Journal of Business Venturing Insights*, *15*, e00245. Advance online publication. doi:10.1016/j.jbvi.2021.e00245

China Association of Small and Medium Enterprises. (2021). *In the first quarter of 2021, the development index of small and medium-sized enterprises continued to rise, reaching the highest level since the first quarter of last year, fully demonstrating the effectiveness of epidemic prevention and control and stabilizing the economy this year.* https://www.ca-sme.org/content/Content/index/id/31666

Djimesah, I. E., Zhao, H., Okine, A. N. D., Li, Y., Duah, E., & Kissi Mireku, K. (2021). Analyzing the technology of acceptance model of Ghanaian crowdfunding stakeholders. *Technological Forecasting and Social Change*. Advance online publication. doi:10.1016/j.techfore.2021.121323

Gloor, P. A., Fronzetti Colladon, A., Grippa, F., Hadley, B. M., & Woerner, S. (2020). The impact of social media presence and board member composition on new venture success: Evidences from VC-backed U.S. startups. *Technological Forecasting and Social Change*, *157*, 120098. Advance online publication. doi:10.1016/j.techfore.2020.120098

Hoang, H. V., Nguyen, C., & Nguyen, D. K. (2022). Corporate immunity, national culture and stock returns: Startups amid the COVID-19 pandemic. *International Review of Financial Analysis*, *79*, 101975. Advance online publication. doi:10.1016/j.irfa.2021.101975

Junge, L. B., Laursen, I. C., & Nielsen, K. R. (2021). Choosing crowdfunding: Why do entrepreneurs choose to engage in crowdfunding? *Technovation*. Advance online publication. doi:10.1016/j.technovation.2021.102385

Kragt, M. E., Burton, R., Zahl-Thanem, A., & Otte, P. P. (2021). Farmers'' interest in crowdfunding to finance climate change mitigation practices. *Journal of Cleaner Production*, *321*, 128967. Advance online publication. doi:10.1016/j.jclepro.2021.128967

Kshetri, N. (2015). Success of Crowd-based Online Technology in Fundraising: An Institutional Perspective. *Journal of International Management*, *21*(2), 100–116. doi:10.1016/j.intman.2015.03.004

Kuckertz, A., Brändle, L., Gaudig, A., Hinderer, S., Morales Reyes, C. A., Prochotta, A., Steinbrink, K. M., & Berger, E. S. C. (2020). Startups in times of crisis – A rapid response to the COVID-19 pandemic. *Journal of Business Venturing Insights*, *13*, e00169. Advance online publication. doi:10.1016/j.jbvi.2020.e00169

Laffey, D., Durkin, M., Cummins, D., & Gandy, A. (2021). A shift in power? Value co-creation through successful crowdfunding. *Technological Forecasting and Social Change*, *172*, 121035. Advance online publication. doi:10.1016/j.techfore.2021.121035

Leong, C., Tan, B., Xiao, X., Tan, F. T. C., & Sun, Y. (2017). Nurturing a FinTech ecosystem: The case of a youth microloan startup in China. *International Journal of Information Management*, *37*(2), 92–97. doi:10.1016/j.ijinfomgt.2016.11.006

Li, X., & Hasan, I. (2021). VC Participation and failure of startups: Evidence from P2P lending platforms in China. *Finance Research Letters*, *40*, 101726. Advance online publication. doi:10.1016/j.frl.2020.101726

Men, L. R. (2021). The impact of startup CEO communication on employee relational and behavioral outcomes: Responsiveness, assertiveness, and authenticity. *Public Relations Review*, *47*(4), 102078. Advance online publication. doi:10.1016/j.pubrev.2021.102078

Nguyen, L. T. Q., Hoang, T. G., Do, L. H., Ngo, X. T., Nguyen, P. H. T., Nguyen, G. D. L., & Nguyen, G. N. T.D. L. (2021). The role of blockchain technology-based social crowdfunding in advancing social value creation. *Technological Forecasting and Social Change*, *170*, 120898. Advance online publication. doi:10.1016/j.techfore.2021.120898

Oliva, F. L., Teberga, P. M. F., Testi, L. I. O., Kotabe, M., Giudice, M. D., Kelle, P., & Cunha, M. P. (2021). Risks and critical success factors in the internationalization of born global startups of industry 4.0: A social, environmental, economic, and institutional analysis. *Technological Forecasting and Social Change*. Advance online publication. doi:10.1016/j.techfore.2021.121346

Pabst, S., Wayand, M., & Mohnen, A. (2021). Coordinating contributions in crowdfunding for sustainable entrepreneurship. *Journal of Cleaner Production*, *319*, 128677. Advance online publication. doi:10.1016/j.jclepro.2021.128677

People's Bank of China. (2021). *Policy research | research on improving the financing system of small, medium and micro enterprises*. https://mp.weixin.qq.com/s/AJUZLDKVQe79aIiLE55fCg

Tiba, S., van Rijnsoever, F. J., & Hekkert, M. P. (2021). Sustainability startups and where to find them: Investigating the share of sustainability startups across entrepreneurial ecosystems and the causal drivers of differences. *Journal of Cleaner Production*, *306*. Advance online publication. doi:10.1016/j.jclepro.2021.127054

Xia Shuo Official Account. (2021). *Financing development report of China's small, medium and micro enterprises in 2021*. https://mp.weixin.qq.com/s/Pk0VDx5UTjMe9R77xuOVFw

Xu, B., Zhang, S., & Chen, X. (2021). Uncertainty in financing interest rates for startups. *Industrial Marketing Management*, *94*, 150–158. doi:10.1016/j.indmarman.2020.02.026

Yao, L., & Hu, Y. (2020). The impact of urban transit on nearby startup firms: Evidence from Hangzhou, China. *Habitat International, 99*. Advance online publication. doi:10.1016/j.habitatint.2020.102155

Yu, P., Gong, R., & Sampat, M. (2022). Blockchain Technology in China's Digital Economy: Balancing Regulation and Innovation. In P. Tehrani (Ed.), Regulatory Aspects of Artificial Intelligence on Blockchain (pp. 132–157). IGI Global. https://doi.org/10.4018/978-1-7998-7927-5.ch007.

Yu, P., Hu, Y., Waseem, M., & Rafay, A. (2021). Regulatory Developments in Peer-to-Peer (P2P) Lending to Combat Frauds: The Case of China. In A. Rafay (Ed.), *Handbook of Research on Theory and Practice of Financial Crimes* (pp. 172-194). IGI Global. doi:10.4018/978-1-7998-5567-5.ch010

Yu, P., Li, C., Sampat, M., & Chen, Z. (2022). How the Development of FinTech Can Bolster Financial Inclusion Under an Era of Disruptive Innovation?: Case Study on China. In M. Anshari, M. Almunawar, & M. Masri (Eds.), FinTech Development for Financial Inclusiveness (pp. 135–167). IGI Global. https://doi.org/10.4018/978-1-7998-8447-7.ch009.

Yu, P., Lu, S., Hanes, E., & Chen, Y. (2022). The Role of Blockchain Technology in Harnessing the Sustainability of Chinese Digital Finance. In Blockchain Technologies for Sustainable Development in Smart Cities (pp. 155–186). IGI Global. https://doi.org/ 10.4018/978-1-7998-9274-8.ch009.

Yu, S., & Fleming, L. (2021). Regional crowdfunding and high tech entrepreneurship. *Research Policy*. Advance online publication. doi:10.1016/j.respol.2021.104348

Yuan, Y. (2018). *Shanghai practice of equity crowdfunding to help innovative small and micro enterprises solve financing difficulties -- Taking Ant angel crowdfunding platform as an example*. https://mp.weixin.qq.com/s/wsOuzipBIVLL6RdE71HP1w

Zhang, J., & Gu, Q. (2021). Turning a curse into a blessing: Contingent effects of geographic distance on startup–VC partnership performance. *Journal of Business Venturing, 36*(4). Advance online publication. doi:10.1016/j.jbusvent.2021.106108

Compilation of References

Abbaraju, R. R., Dasgupta, N., & Virkar, A. V. (2008). Composite Nafion Membranes Containing Nanosize TiO[sub 2]/SnO[sub 2] for Proton Exchange Membrane Fuel Cells. *Journal of the Electrochemical Society, 155*(12), B1307. doi:10.1149/1.2994079

Abou-Shanab, R. A. I., Ji, M. K., Kim, H. C., Paeng, K. J., & Jeon, B. H. (2013). Microalgal species growing on piggery wastewater as a valuable candidate for nutrient removal and biodiesel production. *Journal of Environmental Management, 115*, 257–264. doi:10.1016/j.jenvman.2012.11.022 PMID:23270891

Adewuyi, A. O., & Awodumi, O. B. (2017). Biomass energy consumption, economic growth and carbon emissions: Fresh evidence from West Africa using a simultaneous equation model. *Energy, 119*, 453–471. doi:10.1016/j.energy.2016.12.059

Adib, R. (2015). *Renewables 2015 Global Status Report Distrbuted Renewable Energy For Energy Access*. Academic Press.

Akadiri, P. O., Chinyio, E. A., & Olomolaiye, P. O. (2012). Design of a sustainable building: A conceptual framework for implementing sustainability in the building sector. *Buildings, 2*(2), 126–152. doi:10.3390/buildings2020126

Akeke, G. A., Ephraim, M. E., Akobo, I. Z. S., & Ukpata, J. O. (2013). Structural properties of rice husk ash concrete. *International Journal of Engineering, 3*(3), 8269.

Ali, I., Peng, C., Khan, Z. M., Naz, I., Sultan, M., Ali, M., Abbasi, I. A., Islam, T., & Ye, T. (2019). Overview of microbes based fabricated biogenic nanoparticles for water and wastewater treatment. *Journal of Environmental Management, 230*(August), 128–150. doi:10.1016/j.jenvman.2018.09.073

Ali, N., Jaffar, A., Anwer, M., Alwi, S. K. K., Anjum, M. N., Ali, N., Raja, M. R., Hussain, A., & Ming, X. (2015). The Greenhouse Gas Emissions Produced by Cement Production and Its Impact on Environment: A Review of Global Cement Processing. *International Journal of Research, 2*(2), 488. http://internationaljournalofresearch.org

Alley, Berntsen, Bindoff, Chen, Chidthaisong, Friedlingstein, Gregory, Hegerl, Heimann, Hewitson, Hoskins, Joos, Jouzel, Kattsov, & Lohmann. (2007). Climate Change 2007 - The physical science basis: Contribution of working group I to the fourth assessment rep. *Bulletin Fuer Angewandte Geologie*, *18*(2), 5–19.

Amziane, S., & Sonebi, M. (2016). Overview on Biobased Building Material made with plant aggregate. *RILEM Technical Letters*, *1*, 31–38. doi:10.21809/rilemtechlett.2016.9

Ancaş, A. D., Aşchilean, I., Profire, M., Ţurcanu, F. E., & Felseghi, R. A. (2021). Experimental Study on the Behaviour of Seismic Actions on a Flexible Glass-Reinforced Plastic Structure Used in Water Transport Pipes. *Materials (Basel)*, *14*(11), 2878. doi:10.3390/ma14112878 PMID:34072081

Ancas, A. D., & Atanasiu, G. M. (2011). Seismic Risk Management considering the Urban Lifeline Existing System. *Business Excellence* vol. I, *Proceedings of the 6th International Conference on Business Excellence*, 20-23.

Anjum, M., Miandad, R., Waqas, M., Gehany, F., & Barakat, M. A. (2019). Remediation of wastewater using various nano-materials. *Arabian Journal of Chemistry*, *12*(8), 4897–4919. doi:10.1016/j.arabjc.2016.10.004

Anonymous. (2021a). *Carding the whole hydrogen energy industry chain and regional development hot spots map*. https://baijiahao.baidu.com/s?id=1702880835003475539&wfr=spider&for=pc

Anonymous. (2021b). *Blockbuster! Policy Summary and Interpretation of Hydrogen Energy Industry in China and 31 Provinces and Cities in 2021 (full)*. https://www.sohu.com/a/473410075_473133

Anonymous. (2021c). *The new round of investment in the new energy industry is expected to reach 12 trillion yuan, with consumption still a key issue*. https://baijiahao.baidu.com/s?id=1689778958293283188&wfr=spider&for=pc

Anonymous. (2021d). *Analysis on market status and competition pattern of China's hydrogen energy industry in 2021*. https://baijiahao.baidu.com/s?id=1706510978416834977&wfr=spider&for=pc

Anonymous. (2021e). *Summary and interpretation of hydrogen energy industry policies in China and 31 provinces and cities in 2021*. https://baijiahao.baidu.com/s?id=1702142255001709111&wfr=spider&for=pc

Arevalo, S., & Tomlinson, D. (2020). Experimental thermal bowing response of precast concrete insulated wall panels with stiff shear connectors and simple supports. *Journal of Building Engineering*, *30*, 101319. doi:10.1016/j.jobe.2020.101319

Arroyo, F., Luna-Galiano, Y., Leiva, C., Vilches, L. F., & Fernández-Pereira, C. (2020). Environmental risks and mechanical evaluation of recycling red mud in bricks. *Environmental Research*, *186*, 109537. PMID:32315825

Asgari, A., & Asgari, R. (2021). How circular economy transforms business models in a transition towards circular ecosystem: The barriers and incentives. *Sustainable Production and Consumption*, *28*, 566–579. doi:10.1016/j.spc.2021.06.020

Aslam, M. S., Huang, B., & Cui, L. (2020). Review of construction and demolition waste management in China and USA. *Journal of Environmental Management*, *264*, 110445. doi:10.1016/j.jenvman.2020.110445 PMID:32217323

Aste, N., Manfren, M., & Marenzi, G. (2017). Building Automation and Control Systems and performance optimization: A framework for analysis. *Renewable & Sustainable Energy Reviews*, *75*, 313–330. doi:10.1016/j.rser.2016.10.072

Aye, L., Ngo, T., Crawford, R. H., Gammampila, R., & Mendis, P. (2012). Life cycle greenhouse gas emissions and energy analysis of prefabricated reusable building modules. *Energy and Building*, *47*, 159–168. doi:10.1016/j.enbuild.2011.11.049

Azúa, G., González, M., Arroyo, P., & Kurama, Y. (2019). Recycled coarse aggregates from precast plant and building demolitions: Environmental and economic modeling through stochastic simulations. *Journal of Cleaner Production*, *210*, 1425–1434. doi:10.1016/j.jclepro.2018.11.049

Badur, S., & Chaudhary, R. (2008). Utilization of hazardous wastes and byproducts as a green concrete material through S/S process: A review. *Reviews on Advanced Materials Science*, *17*, 42–61.

Bagarić, M., Banjad Pečur, I., & Milovanović, B. (2021). Application of developed facade panel from recycled CDW: A case study. *Journal of Physics: Conference Series*, *2069*(1), 012187. doi:10.1088/1742-6596/2069/1/012187

Bahoria, B., Parbat, D. K., Nagarnaik, P. B., & Waghe, U. P. (2013, November). Effect of replacement of natural sand by quarry dust and waste plastic on compressive and split tensile strength of M20 concrete. *Proceedings of the International Conference on Engineering (NUiCONE 2013)*, 28-30).

Balat, M. (2007). Hydrogen in fueled systems and the significance of hydrogen in vehicular transportation. *Energy Sources. Part B, Economics, Planning, and Policy*, *2*(1), 49–61. doi:10.1080/15567240500400911

Balog, A. A., Cobirzan, N., Aciu, C., & Ilutiu-Varvara, D. A. (2014). Valorification of volcanic tuff in constructions and materials manufacturing industry. *Procedia Technology*, *12*, 323–328. doi:10.1016/j.protcy.2013.12.493

Basheer, A. A., & Ali, I. (2019). Water photo splitting for green hydrogen energy by green nanoparticles. *International Journal of Hydrogen Energy*, *44*(23), 11564–11573. doi:10.1016/j.ijhydene.2019.03.040

Bayer, P., Rybach, L., Blum, P., & Brauchler, R. (2013). Review on life cycle environmental effects of geothermal power generation. *Renewable & Sustainable Energy Reviews*, *26*, 446–463. doi:10.1016/j.rser.2013.05.039

Baykara, S. Z. (2004). Experimental solar water thermolysis. *International Journal of Hydrogen Energy*, *29*(14), 1459–1469. doi:10.1016/j.ijhydene.2004.02.011

Benazzouz, S. M., Malkinson, J., & Toth, G. (2021). Current trends in and future potential of crowdfunding to finance R&D of treatments for neglected tropical diseases. *Drug Discovery Today*, *26*(7), 1563–1568. doi:10.1016/j.drudis.2021.02.021 PMID:33640317

Bendaikha, O., & Larbi, S. (2013). Hydrogen production system analysis using direct photo-electrolysis process in Algeria. *Proceedings of 2013 International Conference on Renewable Energy Research and Applications, ICRERA 2013,* 1123–1128. 10.1109/ICRERA.2013.6749921

Berné-Martínez, J. M., Ortigosa-Blanch, A., & Planells-Artigot, E. (2021). A semantic analysis of crowdfunding in the digital press. *Technological Forecasting and Social Change*, *173*, 121175. Advance online publication. doi:10.1016/j.techfore.2021.121175

Bilgili, F., & Ozturk, I. (2015). Biomass energy and economic growth nexus in G7 countries: Evidence from dynamic panel data. *Renewable & Sustainable Energy Reviews*, *49*, 132–138. doi:10.1016/j.rser.2015.04.098

Błaszczyński, T., & Król, M. (2015). Usage of green concrete technology in civil engineering. *Procedia Engineering*, *122*, 296–301. doi:10.1016/j.proeng.2015.10.039

Blazejczak, J., Braun, F. G., Edler, D., & Schill, W. P. (2014). Economic effects of renewable energy expansion: A model-based analysis for Germany. *Renewable & Sustainable Energy Reviews*, *40*, 1070–1080. doi:10.1016/j.rser.2014.07.134

Bocken, N. M. P., & Short, S. W. (2021). Unsustainable business models - Recognising and resolving institutionalised social and environmental harm. *Journal of Cleaner Production*, *312*, 127828. doi:10.1016/j.jclepro.2021.127828

Bollaert, H., Lopez-de-Silanes, F., & Schwienbacher, A. (2021). Fintech and access to finance. *Journal of Corporate Finance*, *68*, 101941. Advance online publication. doi:10.1016/j.jcorpfin.2021.101941

Boopathi, Y., & Doraikkannan, J. (2016). Study on MSAND as a partial replacement of fine aggregate in concrete. *Int. J. Adv. Res. Trends Eng. Technol.*, *13*(1), 31–48.

Botici, A. A., Ungureanu, V., Ciutina, A., Botici, A., & Dubina, D. (2013). Sustainable retrofitting of large panel prefabricated concrete residential buildings. *Proceedings of Central Europe Towards Sustainable Building 2013 Conference - Sustainable Building and Refurbishment for Next Generations*, 39-42. http://odolnesidliska.sk/wp-content/uploads/2018/10/CESB13_SUSTAINABLE-RETROFITTING-OF-LARGE-PANEL.pdf

Brandes, M. R., & Kurama, Y. C. (2018). Behavior of shear-critical prestressed concrete beams with recycled concrete aggregates under ultimate loads. *Engineering Structures*, *165*, 237–246. doi:10.1016/j.engstruct.2018.03.029

Brennan, L., & Owende, P. (2010). Biofuels from microalgae-A review of technologies for production, processing, and extractions of biofuels and co-products. *Renewable & Sustainable Energy Reviews*, *14*(2), 557–577. doi:10.1016/j.rser.2009.10.009

Briefs, S., & Energy, I. N. (n.d.). *Solar Based H2 Prod Systems*. Academic Press.

Burgherr, P. (2011). Accidents and risks BT. *IPCC Special Report on Renewable Energy Sources and Climate Change Mitigation, 9.3.4.7.*

Burkhard, R., Deletic, A., & Craig, A. (2000). Techniques for water and wastewater management: A review of techniques and their integration in planning. *Urban Water*, *2*(3), 197–221. doi:10.1016/S1462-0758(00)00056-X

C155. (2013). *Normativ Privind Producerea Betoanelor Ușoare (Regulation on the Production of Lightweight Concrete); ASRO—Asociatia de standardizare din Romania*. Romanian Standardization Association.

Cal, F. D., Garrido-Marijuan, A., Eguiarte, O., Arregi, B., Romero-Amorrortu, A., Mezzasalma, G., ... Bernardi, A. (2021). Energy Performance Assessment of Innovative Building Solutions Coming from Construction and Demolition Waste Materials. *Materials (Basel)*, *14*(5), 1226. doi:10.3390/ma14051226 PMID:33807904

Cao, X., Li, X., Zhu, Y., & Zhang, Z. (2015). A comparative study of environmental performance between prefabricated and traditional residential buildings in China. *Journal of Cleaner Production*, *109*, 131–143. doi:10.1016/j.jclepro.2015.04.120

Catuneanu, M. V. (1990). *Optimizarea fiabilității sistemelor* Ed. Academiei.

CEN. (2010). *Sustainability of Construction Works - Sustainability Assessment of Buildings — Part 1: General Framework; EN 15643–1:2010–12*. European Committee for Standardization.

CEN. (2011). *EN 15978:2011–11 Sustainability of Construction Works - Assessment of Environmental Performance of Buildings — Calculation Method.* European Committee for Standardization.

CEN. (2012). *EN 15643–4:2012–04 Sustainability of Construction Works — Assessment of Buildings - Part 4: Framework for the Assessment of Economic Performance*. European Committee for Standardization.

CEN. (2013). EN 15804:2012+A1:2013 Sustainability of construction works—Environmental product declarations—Core rules for the product category of construction products. European Committee for Standardization.

Chabannes, M., Nozahic, V., & Amziane, S. (2015). Design and multi-physical properties of a new insulating concrete using sunflower stem aggregates and eco-friendly binders. *Mater. Struct. Constr.*, *48*(6), 1815–1829. doi:10.161711527-014-0276-9

Chandler, J. A., Short, J. C., & Wolfe, M. T. (2021). Finding the crowd after exogenous shocks: Exploring the future of crowdfunding. *Journal of Business Venturing Insights*, *15*, e00245. Advance online publication. doi:10.1016/j.jbvi.2021.e00245

Chauhan, J.S. (2019). Create Opportunity Victory In Disaster (COVID), SLOGAN-COVID-19, Samrat Ashok Technological Institute, Vidisha, M.P.

Chaurasia, A. K., & Mondal, P. (2021a). *Hydrogen Production From Waste and Renewable Resources*. doi:10.4018/978-1-7998-4945-2.ch002

Chaurasia, A. K., & Mondal, P. (2021b). *Simultaneous Removal of Organic Load and Hydrogen Gas Production Using Electrodeposits Cathodes in MEC*. doi:10.1007/978-981-15-6595-3_34

Chaurasia, A. K., & Mondal, P. (2022). Enhancing biohydrogen production from sugar industry wastewater using Ni, Ni–Co and Ni–Co–P electrodeposits as cathodes in microbial electrolysis cells. *Chemosphere, 286*(November). doi:10.1016/j.chemosphere.2021.131728

Chaurasia, A. K., Goyal, H., & Mondal, P. (2020). Hydrogen gas production with Ni, Ni–Co and Ni–Co–P electrodeposits as potential cathode catalyst by microbial electrolysis cells. *International Journal of Hydrogen Energy*, *45*(36), 18250–18265. doi:10.1016/j.ijhydene.2019.07.175

Chaurasia, A. K., Shankar, R., & Mondal, P. (2021). Effects of nickle, nickle-cobalt and nickle-cobalt-phosphorus nanocatalysts for enhancing biohydrogen production in microbial electrolysis cells using paper industry wastewater. *Journal of Environmental Management*, *298*(July), 113542. doi:10.1016/j.jenvman.2021.113542 PMID:34426219

Chaurasia, A. K., Siwach, P., Shankar, R., & Mondal, P. (2021). Shankar, · Ravi, & Mondal, · Prasenjit. (2021). Effect of pre-treatment on mesophilic anaerobic co-digestion of fruit, food and vegetable waste. *Clean Technologies and Environmental Policy*, *1*, 1–14. doi:10.100710098-021-02218-5

Chen, S. T. (2021). *SASAC: More than one third of central enterprises are formulating the layout of the whole hydrogen energy industry chain.* https://baijiahao.baidu.com/s?id=170543550010 1935720&wfr=spider&for=pc

Chen, C. C., Lin, C. L., & Chen, L. C. (2015). Functionalized carbon nanomaterial supported palladium nano-catalysts for electrocatalytic glucose oxidation reaction. *Electrochimica Acta*, *152*, 408–416. doi:10.1016/j.electacta.2014.11.116

Cheng, B., Cai, Y., & Looi, D. T. W. (2021). Experiment and numerical study of a new bolted steel plate horizontal joints for precast concrete shear wall structures. *Structures, 32,* 760-777. doi:10.1016/j.istruc.2021.03.043

Cheng, S., & Logan, B. E. (2007). Sustainable and efficient biohydrogen production via electrohydrogenesis. *Proceedings of the National Academy of Sciences of the United States of America*, *104*(47), 18871–18873. doi:10.1073/pnas.0706379104 PMID:18000052

Chen, Q. X., Liu, D. N., Lin, J., He, J. J., & Wang, Y. (2015). Business Models and Market Mechanisms of Energy Internet (1). *Power System Technology*, *39*(11), 3050–3056.

China Association of Small and Medium Enterprises. (2021). *In the first quarter of 2021, the development index of small and medium-sized enterprises continued to rise, reaching the highest level since the first quarter of last year, fully demonstrating the effectiveness of epidemic prevention and control and stabilizing the economy this year.* https://www.ca-sme.org/content/Content/index/id/31666

Chong, A. M. Y., Wong, Y. S., & Tam, N. F. Y. (2000). Performance of different microalgal species in removing nickel and zinc from industrial wastewater. *Chemosphere, 41*(1–2), 251–257. doi:10.1016/S0045-6535(99)00418-X PMID:10819208

Cioc, D., & Anton, A. (2001). *Reţele hidraulice: calcul, optimizare, siguranţă.* O. Universitare.

Cobîrzan, N., Balog, A.-A., & Moşonyi, E. (2015). Investigation of the natural pozzolans for usage in cement industry. *Procedia Technology, 19,* 506-511. doi:10.1016/j.protcy.2015.02.072

Cobîrzan, N., Muntean, R., Thalmaier, G., & Felseghi, R. A. (2022). Recycling of Mining Waste in the Production of Masonry Units. *Materials (Basel), 15*(2), 594. doi:10.3390/ma15020594 PMID:35057312

Cong, T. Y., Raj, A., Chanaphet, J., Mohammed, S., Ibrahim, S., & Al Shoaibi, A. (2016). A detailed reaction mechanism for hydrogen production via hydrogen sulphide (H2S) thermolysis and oxidation. *International Journal of Hydrogen Energy, 41*(16), 6662–6675. doi:10.1016/j.ijhydene.2016.03.053

CR0-2012. (2012). Design code. Basics of structure design. M. Of. 555bis/2013, in *B.C.*, 10/2012.

Creţu, M., Czumbil, L., Bargauan, B., Ceclan, A., Berciu, A., Polycarpou, A., Rizzo, R., & Micu, D. D. (2020). Modelling and Evaluation of the Baseline Energy Consumption and the KPIs in TUCN Buildings within a Demand Response Programme – A Case Study. *IET Renewable Power Generation Journal, 14*(15), 2864–2875. doi:10.1049/iet-rpg.2020.0096

Csoknyai, T., Hrabovszky-Horváth, S., Georgiev, Z., Jovanovic-Popovic, M., Stankovic, B., Villatoro, O., & Szendrő, G. (2016). Building stock characteristics and energy performance of residential buildings in Eastern-European countries. *Energy and Building, 132*, 39–52. doi:10.1016/j.enbuild.2016.06.062

Cuevas, F., Zhang, J. X., & Latroche, M. (2021). The Vision of France, Germany, and the European Union on Future Hydrogen Energy Research and Innovation. *Engineering, 7*(6), 715–718. doi:10.1016/j.eng.2021.04.010

Cui, Y., Wang, F., Iqbal, M. Z., Wang, Z., Li, Y., & Tu, J. (2015). Synthesis of novel 3D SnO flower-like hierarchical architectures self-assembled by nano-leaves and its photocatalysis. *Materials Research Bulletin, 70*, 784–788. doi:10.1016/j.materresbull.2015.06.021

CWEA. (2021). *The 6th global offshore wind power conference was held in Nanning, Guangxi.* https://news.bjx.com.cn/html/20210712/1163357.shtml

Dai, K., Lv, J., Lu, L., Liu, Q., Zhu, G., & Li, D. (2014). Synthesis of micro-nano heterostructure AgBr/ZnO composite for advanced visible light photocatalysis. *Materials Letters*, *130*, 5–8. doi:10.1016/j.matlet.2014.04.190

Daniel Ronald Joseph, J., Prabakar, J., & Alagusundaramoorthy, P. (2019). Experimental studies on through-thickness shear behavior of EPS based precast concrete sandwich panels with truss shear connectors. *Composites. Part B, Engineering*, *166*, 446–456. doi:10.1016/j.compositesb.2019.02.030

Danielsen, S. W., & Kuznetsova, E. (2016). *Resource management and a Best Available Concept for aggregate sustainability*. doi:10.1144/SP416.7

Davis, P., Aziz, F., Newaz, M. T., Sher, W., & Simon, L. (2021). The classification of construction waste material using a deep convolutional neural network. *Automation in Construction*, *122*, 103481. doi:10.1016/j.autcon.2020.103481

De Gioannis, G., Muntoni, A., Polettini, A., & Pomi, R. (2013). A review of dark fermentative hydrogen production from biodegradable municipal waste fractions. *Waste Management (New York, N.Y.)*, *33*(6), 1345–1361. doi:10.1016/j.wasman.2013.02.019 PMID:23558084

Di Lorenzo, M., Scott, K., Curtis, T. P., & Head, I. M. (2010). Effect of increasing anode surface area on the performance of a single chamber microbial fuel cell. *Chemical Engineering Journal*, *156*(1), 40–48. doi:10.1016/j.cej.2009.09.031

Dincer, I. (2012). Green methods for hydrogen production. *International Journal of Hydrogen Energy*, *37*(2), 1954–1971. doi:10.1016/j.ijhydene.2011.03.173

Directiva cadru 2000/60/CE. *Cadrul comunitar de acțiune în domeniul strategiei apelor.*

Directive, E. C. (2008). 98/EC of the European Parliament and of the Council, on waste and repealing certain Directives. Official Journal of the European Union, (312), 3-30.

Djimesah, I. E., Zhao, H., Okine, A. N. D., Li, Y., Duah, E., & Kissi Mireku, K. (2021). Analyzing the technology of acceptance model of Ghanaian crowdfunding stakeholders. *Technological Forecasting and Social Change*. Advance online publication. doi:10.1016/j.techfore.2021.121323

Dogan, E., & Inglesi-Lotz, R. (2017). Analyzing the effects of real income and biomass energy consumption on carbon dioxide (CO2) emissions: Empirical evidence from the panel of biomass-consuming countries. *Energy*, *138*, 721–727. doi:10.1016/j.energy.2017.07.136

Doloi, H. (2013). Cost overruns and failure in project management: Understanding the roles of key stakeholders in construction projects. *Journal of Construction Engineering and Management*, *139*(3), 267–279. doi:10.1061/(ASCE)CO.1943-7862.0000621

Drucker, P. F., & Joseph, A. M. (2006). The Daily Drucker 366 Days of Insight and Motivation for Getting the Right Things Done. Shanghai Translation Publishing House.

Drucker, P. F. (2012). *The practice of management*. Lixin Accounting Publishing House. doi:10.4324/9780080942360

Du, Q. M. (2022). *More than 1000 demonstration vehicles help the Winter Olympics: hydrogen fuel cell vehicles meet the business inflection point.*https://baijiahao.baidu.com/s?id=1724184264204144351&wfr=spider&for=pc

Edler, D., Ottm, M., & Lehr, U. (2009). *Bruttobeschäftigung durch erneuerbare Energien in Deutschland im Jahr 2008*. Academic Press.

Einea, A., Salmon, D. C., Fogarasi, G. J., Culp, T. D., & Tadros, M. K. (1991). State-of-the-art of Precast Concrete Sandwich Panels. *PCI Journal*, *36*(6), 78–92. doi:10.15554/pcij.11011991.78.98

Eleftheriadis, S., Mumovic, D., & Greening, P. (2017). Life cycle energy efficiency in building structures: A review of current developments and future outlooks based on BIM capabilities. *Renewable & Sustainable Energy Reviews*, *67*, 811–825. doi:10.1016/j.rser.2016.09.028

Elkington, J. (2020). *Green Swans: The coming boom in regenerative capitalism*. Fast Company Pr.

Ellen MacArthur Foundation. (2013). Towards the circular economy, economic and business rationale for an accelerated transition. Ellen MacArthur Foundation.

EN 12390-3:2019 Testing Hardened Concrete Part. 3: Compressive Strength of Test. Specimens; *European Committee for Standardization*: Brussels, Belgium, 2019.

EN 12390-5:2019 Testing Hardened Concrete Part. 3: Flexural Tensile Strength of Test. Specimens; *European Committee for Standardization*: Brussels, Belgium, 2019.

EN 12390-6:2010 Testing Hardened Concrete; Part 6: Split Tensile Strength of Test Specimens; *European Committee for Standardization*: Brussels, Belgium, 2010.

EN 12390-7:2019 Testing Hardened Concrete Part. 7: Density of Hardened Concrete; European Committee for Standardization: Brussels, Belgium, 2019.

EN 13412:2006 Products and Systems for the Protection and Repair of Concrete Structures. Test Methods. Determination of Modulus of Elasticity in Compression; *European Committee for Standardization*: Brussels, Belgium, 2006.

EN 197-1:2011. Cement. Composition, Specifications and Conformity Criteria for Common Cements; *European Committee for Standardization*: Brussels, Belgium, 2011.

Escapa, A., Gil-Carrera, L., García, V., & Morán, A. (2012). Performance of a continuous flow microbial electrolysis cell (MEC) fed with domestic wastewater. *Bioresource Technology*, *117*, 55–62. doi:10.1016/j.biortech.2012.04.060 PMID:22609714

Eskizeybek, V., Sari, F., Gülce, H., Gülce, A., & Avci, A. (2012). Preparation of the new polyaniline/ZnO nanocomposite and its photocatalytic activity for degradation of methylene blue and malachite green dyes under UV and natural sun lights irradiations. *Applied Catalysis B: Environmental*, *119–120*, 197–206. doi:10.1016/j.apcatb.2012.02.034

Estokova, A., Vilcekova, S., & Porhincak, M. (2017). Analyzing Embodied Energy, Global Warming and Acidification Potentials of Materials in Residential Buildings. *Procedia Engineering*, *180*, 1675–1683. doi:10.1016/j.proeng.2017.04.330

European Comission. (2014). Communication from the commission to the european parliament, the council, the european economic and social committee and the committee of the regions Towards a circular economy: A zero waste programme for Europe /COM/2014/398 final. In *Com* (Vol. 398, pp. 1–8). https://eur-lex.europa.eu/legal-content/EN/TXT/?uri=celex%3A52014DC0398

European Comission. (2015). *Communication from the commission to the european parliament, the council, the european economic and social committee and the committee of the regions Closing the loop - An EU action plan for the Circular Economy*. COM/2015/0614 final.

European Commission. (2018). A Clean Planet for all. A European long-term strategic vision for a prosperous, modern, competitive and climate neutral economy. In *Com(2018) 773* (p. 25). https://eur-lex.europa.eu/legal-content/EN/TXT/PDF/?uri=CELEX:52018DC0773&from=EN

European Commission. (2019). *The European Green Deal*. COM/2019/640 final. https://eur-lex.europa.eu/legal-content/EN/TXT/?uri=CELEX:52019DC0640%0Ahttps://ec.europa.eu/info/strategy/priorities-2019-2024/european-green-deal_en

European Commission. (2020). Communication from the commission to the european parliament, the council, the European economic and social committee and the committee of the regions - A new Circular Economy Action plan for a cleaner and more competitive Europe. In COM(2020) 98 final (pp. 233–249).

European Parliament and Council. (2008). Directive 2008/98/EC of the European Parliament and of the Council of 19 November 2008 on waste and repealing certain directives (Waste framework). In LexUriServ. do (pp. 3–30). .32008L0098 doi:2008/98/EC

European Parliament. (2018). Directive (EU) 2018/844 of the European Parliament and of the Council of 30 May 2018 Amending Directive 2010/31/EU on the Energy Performance of Buildings and Directive 2012/27/EU on Energy Efficiency (Text with EEA Relevance). In *Official Journal of the European Union* (Vol. 156, Issue 19.6.2018, pp. 75–91). https://data.europa.eu/eli/dir/2018/844/oj/eng

Fallah, A., Vafaei, M., & Alih, S. C. (2020). A Review of Seismic Response of Precast Structures. *International Journal of Engineering Research & Technology (Ahmedabad)*, *13*(10), 2694–2705. doi:10.37624/IJERT/13.10.2020.2694-2705

Fang, J., Fan, H., Ma, Y., Wang, Z., & Chang, Q. (2015). Surface defects control for ZnO nanorods synthesized by quenching and their anti-recombination in photocatalysis. *Applied Surface Science*, *332*, 47–54. doi:10.1016/j.apsusc.2015.01.139

Farias, R. D., García, C. M., Cotes Palomino, T., Andreola, F., Lancellotti, I., & Barbieri, L. (2017). Valorization of agro-industrial wastes in lightweight aggregates for agronomic use: Preliminary study. *Environmental Engineering and Management Journal*, *16*(8), 1691–1699. doi:10.30638/eemj.2017.184

Faruqi, M. H. Z., & Siddiqui, F. Z. (2020). A mini review of construction and demolition waste management in India. *Waste Management & Research*, *38*(7), 708–716. doi:10.1177/0734242X20916828 PMID:32347194

Fatimah, Y. A., Govindan, K., Murniningsih, R., & Setiawan, A. (2020). Industry 4.0 based sustainable circular economy approach for smart waste management system to achieve sustainable development goals: A case study of Indonesia. *Journal of Cleaner Production*, *269*, 122263. doi:10.1016/j.jclepro.2020.122263

Fearnside, P., Rosa, L. P., Saut, P., Guiana, F., Paris, T., & Nations, U. (2006). Methane Quashes Green Credentials of Hydropower Preprint Analysis Quantifies Scientific Plagiarism. *Nature*, *444*(November), 524–525.

Fehrer, J. A., & Wieland, H. (2021). A systemic logic for circular business models. *Journal of Business Research*, *125*, 609–620. doi:10.1016/j.jbusres.2020.02.010

Ferdous, W., Bai, Yu., Ngo, T. D., Manalo, A., & Mendis, P. (2019). New advancements, challenges and opportunities of multi-storey modular buildings – A state-of-the-art review. *Engineering Structures, 183,* 883-893. doi:10.1016/j.engstruct.2019.01.061

Ferroudj, N., Nzimoto, J., Davidson, A., Talbot, D., Briot, E., Dupuis, V., Bée, A., Medjram, M. S., & Abramson, S. (2013). Maghemite nanoparticles and maghemite/silica nanocomposite microspheres as magnetic Fenton catalysts for the removal of water pollutants. *Applied Catalysis B: Environmental*, *136–137*, 9–18. doi:10.1016/j.apcatb.2013.01.046

Fiala, J., Mikolas, M., Junior, J. F., & Krejsova, K. (2019). History and Evolution of Full Bricks of Other European Countries. *IOP Conference Series. Materials Science and Engineering*, *603*(3), 32097.

Fishbein, B. K. (1998). *Building for the future: strategies to reduce construction and demolition waste in municipal projects*. INFORM, Incorporated.

Formoso, C. T., Soibelman, L., De Cesare, C., & Isatto, E. L. (2002). Material waste in building industry: Main causes and prevention. *Journal of Construction Engineering and Management*, *128*(4), 316–325. doi:10.1061/(ASCE)0733-9364(2002)128:4(316)

Fořt, J., & Černý, R. (2020). Transition to circular economy in the construction industry: Environmental aspects of waste brick recycling scenarios. *Waste Management (New York, N.Y.)*, *118*, 510–520. doi:10.1016/j.wasman.2020.09.004 PMID:32980730

Gálvez-Martos, J. L., Styles, D., Schoenberger, H., & Zeschmar-Lahl, B. (2018). Construction and demolition waste best management practice in Europe. *Resources, Conservation and Recycling*, *136*, 166–178. doi:10.1016/j.resconrec.2018.04.016

Gao, J., & Zhang, L. (2021). Does biomass energy consumption mitigate CO2 emissions? The role of economic growth and urbanization: Evidence from developing Asia. *Journal of the Asia Pacific Economy*, *26*(1), 96–115. doi:10.1080/13547860.2020.1717902

Gasparatos, A., Doll, C. N. H., Esteban, M., Ahmed, A., & Olang, T. A. (2017). Renewable energy and biodiversity: Implications for transitioning to a Green Economy. *Renewable and Sustainable Energy Reviews, 70*(December), 161–184. doi:10.1016/j.rser.2016.08.030

GCCA. (2021). *Concrete Future-The GCCA 2050 Cement and Concrete Industry Roadmap for Net Zero Concrete*. https://gccassociation.org

Geissdoerfer, M., Morioka, S. N., de Carvalho, M. M., & Evans, S. (2018). Business models and supply chains for the circular economy. *Journal of Cleaner Production, 190*, 712–721. doi:10.1016/j.jclepro.2018.04.159

Geissdoerfer, M., Pieroni, M. P. P., Pigosso, D. C. A., & Soufani, K. (2020). Circular business models: A review. *Journal of Cleaner Production, 277*, 123741. doi:10.1016/j.jclepro.2020.123741

Ghangrekar, M. M., & Shinde, V. B. (2007). Performance of membrane-less microbial fuel cell treating wastewater and effect of electrode distance and area on electricity production. *Bioresource Technology, 98*(15), 2879–2885. doi:10.1016/j.biortech.2006.09.050 PMID:17107789

Ghasemi, M., Daud, W. R. W., Hassan, S. H. A., Oh, S. E., Ismail, M., Rahimnejad, M., & Jahim, J. M. (2013). Nano-structured carbon as electrode material in microbial fuel cells: A comprehensive review. *Journal of Alloys and Compounds, 580*, 245–255. doi:10.1016/j.jallcom.2013.05.094

Ghasemi, M., Ismail, M., Kamarudin, S. K., Saeedfar, K., Daud, W. R. W., Hassan, S. H. A., Heng, L. Y., Alam, J., & Oh, S. E. (2013). Carbon nanotube as an alternative cathode support and catalyst for microbial fuel cells. *Applied Energy, 102*, 1050–1056. doi:10.1016/j.apenergy.2012.06.003

Ghasemi, M., Shahgaldi, S., Ismail, M., Kim, B. H., Yaakob, Z., & Wan Daud, W. R. (2011). Activated carbon nanofibers as an alternative cathode catalyst to platinum in a two-chamber microbial fuel cell. *International Journal of Hydrogen Energy, 36*(21), 13746–13752. doi:10.1016/j.ijhydene.2011.07.118

Glazer, B., Graber, C., Roose, C., Syrett, P., & Youssef, C. (2011). *Fly ash in concrete*. The Perkins+Will Team Publisher.

Gleick, P. H. (2013). The World's Water Volume 7. *Berkeley Planning Journal, 26*(1), •••.

GlobalABC/IEA/UNEP (Global Alliance for Buildings and Construction, International Energy Agency, the United Nations Environment Programme). (2020). *GlobalABC Roadmap for Buildings and Construction: Towards a zero-emission, efficient and resilient buildings and construction sector*. IEA. https://www.iea.org/reports/globalabc-roadmap-for-buildings-and-construction-2020-2050

Gloor, P. A., Fronzetti Colladon, A., Grippa, F., Hadley, B. M., & Woerner, S. (2020). The impact of social media presence and board member composition on new venture success: Evidences from VC-backed U.S. startups. *Technological Forecasting and Social Change, 157*, 120098. Advance online publication. doi:10.1016/j.techfore.2020.120098

Goh, W. J., Makam, V. S., Hu, J., Kang, L., Zheng, M., Yoong, S. L., Udalagama, C. N. B., & Pastorin, G. (2012). Iron oxide filled magnetic carbon nanotube-enzyme conjugates for recycling of amyloglucosidase: Toward useful applications in biofuel production process. *Langmuir, 28*(49), 16864–16873. doi:10.1021/la303046m PMID:23148719

Guan, S., Hao, L., Lu, Y., Yoshida, H., Pan, F., & Asanuma, H. (2016). Fabrication of oxygen-deficient TiO2 coatings with nano-fiber morphology for visible-light photocatalysis. *Materials Science in Semiconductor Processing, 41*, 358–363. doi:10.1016/j.mssp.2015.09.018

Guerrero Delgado, M., Sánchez Ramos, J., Álvarez Domínguez, S., Tenorio Ríos, J. A., & Cabeza, L. F. (2020). Building thermal storage technology: Compensating renewable energy fluctuations. *Journal of Energy Storage, 27*, 101147. doi:10.1016/j.est.2019.101147

Guo, W., Zhai, Z., Cui, Y., Yu, Z., & Wu, X. (2019). Seismic performance assessment of low-rise precast wall panel structure with bolt connections. *Engineering Structures, 181*, 562-578. doi:10.1016/j.engstruct.2018.12.060

Gupta, U., & Rao, C. N. R. (2017). Hydrogen generation by water splitting using MoS2 and other transition metal dichalcogenides. *Nano Energy, 41*(July), 49–65. doi:10.1016/j.nanoen.2017.08.021

Habeeb, G. A., & Mahmud, H. B. (2010). Study on properties of rice husk ash and its use as cement replacement material. *Materials Research, 13*(2), 185–190. doi:10.1590/S1516-14392010000200011

Haghighi, A., Samani, S. M. V., & Samani, Z. M. V. (2011). GA-ILP Method for Optimization of Water Distribution Networks. *Water Resources Management, 25*(7), 1791–1808. doi:10.100711269-011-9775-4

Hamwi, M., Lizarralde, I., & Legardeur, J. (2021). Demand response business model canvas: A tool for flexibility creation in the electricity markets. *Journal of Cleaner Production, 282*, 124539. doi:10.1016/j.jclepro.2020.124539

Harbulakova, V. O., Estokova, A., Luptakova, A., & Korenova, S. (2014). Study of durability of fibrous cement based materials exposed to microorganisms. *The 9th International Conference "Environmental Engineering"*. http://enviro.vgtu.lt

He, Y. H., & Lu, S. T. (2019). *Integrated energy service provider: User centered, one-stop solution.* https://www.sohu.com/a/335720319_269757

He, J. T., & Ortiz, J. (2021). Sustainable business modeling: The need for innovative design thinking. *Journal of Cleaner Production, 298*, 126751. doi:10.1016/j.jclepro.2021.126751

He, Y., Sutton, N. B., Rijnaarts, H. H. H., & Langenhoff, A. A. M. (2016). Degradation of pharmaceuticals in wastewater using immobilized TiO2 photocatalysis under simulated solar irradiation. *Applied Catalysis B: Environmental, 182*, 132–141. doi:10.1016/j.apcatb.2015.09.015

Hoang, H. V., Nguyen, C., & Nguyen, D. K. (2022). Corporate immunity, national culture and stock returns: Startups amid the COVID-19 pandemic. *International Review of Financial Analysis, 79*, 101975. Advance online publication. doi:10.1016/j.irfa.2021.101975

Hosseini, S. E., & Wahid, M. A. (2016). Hydrogen production from renewable and sustainable energy resources: Promising green energy carrier for clean development. *Renewable & Sustainable Energy Reviews*, *57*, 850–866. Advance online publication. doi:10.1016/j.rser.2015.12.112

Hrabovszky-Horváth, S., & Szalay, Z. (2014). Environmental assessment of a precast concrete building stock in a time perspective Environ. *Engineering Management Journal*, *13*(11), 2997–2804. doi:10.30638/eemj.2014.312

Huang, R. L., Pu, T. J., Liu, K. W., Yang, Z. Y., & Chen, N. S. (2015). Functional system and application scheme design of urban energy Internet. *Automation of Electric Power Systems, 9*, 26-33+40.

Huarachi, D. A. R., Gonçalves, G., de Francisco, A. C., Canteri, M. H. G., & Piekarski, C. M. (2020). Life cycle assessment of traditional and alternative bricks: A review. *Environmental Impact Assessment Review*, *80*, 106335.

Huntzinger, D. N., Eatmon, T. D. (2009). A life-cycle assessment of Portland cement manufacturing: Comparing the traditional process with alternative technologies. *Journal of Cleaner Production, 17*, 668–675. ro.2008.04.007 doi:10.1016/j.jclep

Hussein, A. K. (2015). Applications of nanotechnology in renewable energies - A comprehensive overview and understanding. *Renewable & Sustainable Energy Reviews*, *42*, 460–476. doi:10.1016/j.rser.2014.10.027

Huuhka, S., Kaasalainen, T., Hakanen, J. H., & Lahdensivu, J. (2015). Reusing concrete panels from buildings for building: Potential in Finnish 1970s mass housing. *Resources, Conservation and Recycling*, *101*, 105–121. doi:10.1016/j.resconrec.2015.05.017

Hu, W., Yin, J., Deng, B., & Hu, Z. (2015). Application of nano TiO2 modified hollow fiber membranes in algal membrane bioreactors for high-density algae cultivation and wastewater polishing. *Bioresource Technology*, *193*, 135–141. doi:10.1016/j.biortech.2015.06.070 PMID:26125613

Idriss, H. (2021). Toward Large-Scale Hydrogen Production from Water: What Have We Learned and What Are the Main Research Hurdles to Cross for Commercialization? *Energy Technology (Weinheim)*, *9*(2), 1–9. doi:10.1002/ente.202000843

IEA. (2019). *Global Status Report for Buildings and Construction 2019*. UN Environment Programme. https://www.iea.org/reports/global-status-report-for-buildings-and-construction-2019

IEA. (2019). *The Future of Hydrogen – Analysis*. IEA.

IEA. (2020). *(2020). World Energy Outlook 2020*. IEA.

Ignjatović, I., Sas, Z., Dragaš, J., Somlai, J., & Kovács, T. (2016). Radiological and material characterization of high volume fly ash concrete. *Journal of Environmental Radioactivity*, *168*, 38–45. doi:10.1016/j.jenvrad.2016.06.021 PMID:27400654

Imran Jafri, R., Sujatha, N., Rajalakshmi, N., & Ramaprabhu, S. (2009). Au-MnO2/MWNT and Au-ZnO/MWNT as oxygen reduction reaction electrocatalyst for polymer electrolyte membrane fuel cell. *International Journal of Hydrogen Energy*, *34*(15), 6371–6376. doi:10.1016/j.ijhydene.2009.05.084

Indian Standard. (2005). *Common burnt clay building bricks - Specification (IS 1077 (1992), Fifth Revision.* Retrieved from https://law.resource.org/pub/in/bis/S03/is.1077.1992.pdf

ISO. (2006a). *ISO 14040 Environmental management — Life cycle assessment — Principles and framework.* ISO.

ISO. (2006b). *14044 (2006) Environmental management—life cycle assessment—requirements and guidelines.* ISO.

Jadhav, P. A., & Kulkarni, D. K. (2013). Effect of replacement of natural sand by manufactured sand on the properties of cement mortar. *International Journal of Civil and Structural Engineering*, *3*(3), 621.

Jaffar, N., Tharim, A. A., & Shuib, M. N. (2011). Factors of conflict in construction industry: A literature review. *Procedia Engineering*, *20*, 193–202. doi:10.1016/j.proeng.2011.11.156

Jain, S., Singhal, S., & Jain, N. K. (2021). Construction and demolition waste (C&DW) in India: Generation rate and implications of C&DW recycling. *International Journal of Construction Management*, *21*(3), 261–270. doi:10.1080/15623599.2018.1523300

Jaishankar, P. (2015). Experimental Investigation on Nano particles in High Performance Concrete. *International Journal of Chemtech Research*, *8*(4), 1666–1670.

Jang, J. S., Kim, H. G., Joshi, U. A., Jang, J. W., & Lee, J. S. (2008). Fabrication of CdS nanowires decorated with TiO2 nanoparticles for photocatalytic hydrogen production under visible light irradiation. *International Journal of Hydrogen Energy*, *33*(21), 5975–5980. doi:10.1016/j.ijhydene.2008.07.105

Jayabalan, T., Matheswaran, M., Preethi, V., & Naina Mohamed, S. (2020). Enhancing biohydrogen production from sugar industry wastewater using metal oxide/graphene nanocomposite catalysts in microbial electrolysis cell. *International Journal of Hydrogen Energy*, *45*(13), 7647–7655. doi:10.1016/j.ijhydene.2019.09.068

Jayabalan, T., Matheswaran, M., Radhakrishnan, T. K., & Naina Mohamed, S. (2021). Influence of Nickel molybdate nanocatalyst for enhancing biohydrogen production in microbial electrolysis cell utilizing sugar industrial effluent. *Bioresource Technology*, *320*, 124284. doi:10.1016/j.biortech.2020.124284 PMID:33137640

Jayaraman, A. (2014). Experimental Study on Partial Replacement of Natural Sand with M-Sand and Cement with Lime Stone Powder. *International Journal of Chemtech Research*, *6*, 948.

Jianqiang, W., Siwei, M., & D'Shawn, G. T. (2016). Correlation between hydration of cement and durability of natural fiber-reinforced cement composites. *Corrosion Science*, *106*, 1–15. doi:10.1016/j.corsci.2016.01.020

Jones, E. R., van Vliet, M. T. H., Qadir, M., & Bierkens, M. F. P. (2021). Country-level and gridded estimates of wastewater production, collection, treatment and reuse. *Earth System Science Data, 13*(2), 237–254. doi:10.5194/essd-13-237-2021

Joyce, A., & Paquin, R. (2016). The triple layered business model canvas: A tool to design more sustainable business models. *Journal of Cleaner Production, 135*, 1474–1486. doi:10.1016/j.jclepro.2016.06.067

Junge, L. B., Laursen, I. C., & Nielsen, K. R. (2021). Choosing crowdfunding: Why do entrepreneurs choose to engage in crowdfunding? *Technovation.* Advance online publication. doi:10.1016/j.technovation.2021.102385

K., S. V., M., P., & P., S. (2011). Use of Nanotechnology in Solar PV Cell. *International Journal of Chemical Engineering and Applications, 2*(2), 77–80.

Kadier, A., Chaurasia, A. K., Sapuan, S. M., Ilyas, R. A., Ma, P. C., Alabbosh, K. F. S., Rai, P. K., Logroño, W., Hamid, A. A., & Hasan, H. A. (2020). Essential Factors for Performance Improvement and the Implementation of Microbial Electrolysis Cells (MECs). In Bioelectrochemical Systems (pp. 139–168). Springer Singapore. doi:10.1007/978-981-15-6872-5_7

Kankam, C. K., Meisuh, B. K., Sossou, G., & Buabin, T. K. (2017). Stress-strain characteristics of concrete containing quarry rock dust as partial replacement of sand. *Case Studies in Construction Materials, 7*, 66-72.

Karim, M. M. (2000). Arsenic in groundwater and health problems in Bangladesh. *Water Research, 34*(1), 304–310. doi:10.1016/S0043-1354(99)00128-1

Katzner, T., Johnson, J. A., Evans, D. M., Garner, T. W. J., Gompper, M. E., Altwegg, R., Branch, T. A., Gordon, I. J., & Pettorelli, N. (2013). Challenges and opportunities for animal conservation from renewable energy development. *Animal Conservation, 16*(4), 367–369. doi:10.1111/acv.12067

Khaselev, O., Bansal, A., & Turner, J. A. (2001). High-efficiency integrated multijunction photovoltaic/electrolysis systems for hydrogen production. *International Journal of Hydrogen Energy, 26*(2), 127–132. doi:10.1016/S0360-3199(00)00039-2

Khatri, S., & Jain, R. (2021). Improvement of Index and Engineering Properties Of Black Cotton soil by Using RHA & WHA including Lime. *44th Indian Social Science Congress (ISSC) in Collaboration with SATI Vidisha.*

Khatri, D. S. (2014). Impact of admixture and rice husk ash in concrete mix design. *IOSR Journal of Mechanical and Civil Engineering, 11*(1), 13–17. doi:10.9790/1684-11141317

Kidd, I. M., Fischer, A., Chai, S., & Davis, J. A. (2015). A scenario-based approach to evaluating potential environmental impacts following a tidal barrage installation. *Ocean and Coastal Management, 116*, 9–19. doi:10.1016/j.ocecoaman.2015.06.016

Kim, E. S., Yu, Q., & Deng, B. (2011). Plasma surface modification of nanofiltration (NF) thin-film composite (TFC) membranes to improve anti organic fouling. *Applied Surface Science, 257*(23), 9863–9871. doi:10.1016/j.apsusc.2011.06.059

Kim, Y. J., & Allard, A. (2014). Thermal response of precast concrete sandwich walls with various steel connectors for architectural buildings in cold regions. *Energy and Building*, *80*, 137–148. doi:10.1016/j.enbuild.2014.05.022

Konieczny, A., Mondal, K., Wiltowski, T., & Dydo, P. (2008). Catalyst development for thermocatalytic decomposition of methane to hydrogen. *International Journal of Hydrogen Energy*, *33*(1), 264–272. doi:10.1016/j.ijhydene.2007.07.054

Kragt, M. E., Burton, R., Zahl-Thanem, A., & Otte, P. P. (2021). Farmers" interest in crowdfunding to finance climate change mitigation practices. *Journal of Cleaner Production*, *321*, 128967. Advance online publication. doi:10.1016/j.jclepro.2021.128967

Kshetri, N. (2015). Success of Crowd-based Online Technology in Fundraising: An Institutional Perspective. *Journal of International Management*, *21*(2), 100–116. doi:10.1016/j.intman.2015.03.004

Kuckertz, A., Brändle, L., Gaudig, A., Hinderer, S., Morales Reyes, C. A., Prochotta, A., Steinbrink, K. M., & Berger, E. S. C. (2020). Startups in times of crisis – A rapid response to the COVID-19 pandemic. *Journal of Business Venturing Insights*, *13*, e00169. Advance online publication. doi:10.1016/j.jbvi.2020.e00169

Kurniawan, W., Endarto, B., Prasetyo, I., Aliyyah, N., Helmi, R., Indriastuti, D. E., ... Kalbuana, N. (2021). The role of law in public private partnership: Indonesia. *Journal of Management Information and Decision Sciences*, *24*, 1–11.

Kyazze, G., Popov, A., Dinsdale, R., Esteves, S., Hawkes, F., Premier, G., & Guwy, A. (2010). Influence of catholyte pH and temperature on hydrogen production from acetate using a two chamber concentric tubular microbial electrolysis cell. *International Journal of Hydrogen Energy*, *35*(15), 7716–7722. doi:10.1016/j.ijhydene.2010.05.036

Kyropoulou, D., Heliades, E., Karalis, P., Diamantopoulos, G., Gougoura, S., & Dotsika, E. (2022). Technological Evolution of Historic Mortars: From Lime-Based Mortars to Roman Opus Caementicium. In I. Vayas & F. M. Mazzolani (Eds.), *Protection of Historical Constructions. PROHITECH 2021. Lecture Notes in Civil Engineering*. Springer., doi:10.1007/978-3-030-90788-4_8

Labib, S. M. (2017, November). Volunteer GIS (VGIS) based waste management: A conceptual design and use of web 2.0 for smart waste management in Dhaka City. In *2017 Third International Conference on Research in Computational Intelligence and Communication Networks (ICRCICN)* (pp. 137-141). IEEE. 10.1109/ICRCICN.2017.8234495

Laffey, D., Durkin, M., Cummins, D., & Gandy, A. (2021). A shift in power? Value co-creation through successful crowdfunding. *Technological Forecasting and Social Change*, *172*, 121035. Advance online publication. doi:10.1016/j.techfore.2021.121035

Langdon, D. (2009). *Designing out waste: a design team guide for buildings*. WRAP.

Langer, W. H., Drew, L. J., & Sachs, J. S. (2004). Aggregate and the environment. American Geological Institute Environmental Awareness Series. American Geological Institute.

Lauri, K (1992). *Application of the New Production Theory to Construction* (Vol. 72). Technical Report.

Leong, C., Tan, B., Xiao, X., Tan, F. T. C., & Sun, Y. (2017). Nurturing a FinTech ecosystem: The case of a youth microloan startup in China. *International Journal of Information Management, 37*(2), 92–97. doi:10.1016/j.ijinfomgt.2016.11.006

Letcher, T. M., & Vallero, D. (Eds.). (2019). *Waste: A handbook for management.* Academic Press.

Li, S. (2021). *The "hydrogen wind" of the energy transformation of the Guangdong-Hong Kong-Macao Greater Bay Area has arrived.* https://www.hbzhan.com/news/detail/143094.html

Li, Z. C., Xue, F., Li, X., & Hong, J. (2018). Geoffrey Qiping Shen, An Internet of Things-enabled BIM platform for on-site assembly services in prefabricated construction. *Automation in Construction, 89,* 146-161. doi:10.1016/j.autcon.2018.01.001

Li, Z., & Peng, M. J. (2021). *The Greater Bay Area (GBA) cooperates to solve problems in the whole hydrogen energy industry chain, and introduces business model innovation into the market.* https://baijiahao.baidu.com/s?id=1707844202191188126&wfr=spider&for=pc

Lieder, M., & Rashid, A. (2016). Towards circular economy implementation: A comprehensive review in context of manufacturing industry. *Journal of Cleaner Production, 115*, 36–51. doi:10.1016/j.jclepro.2015.12.042

Li, H. W., & Nishimiya, N. (2021). Insight from Japan's Hydrogen Strategy and Activities. *Engineering, 7*(6), 722–725. doi:10.1016/j.eng.2020.12.020

Li, J. (2019). Investigation of hydrogen evolution on electrodeposited Ni/P coated carbon paper electrode in microbial fuel cell. *International Journal of Electrochemical Science, 14*, 7582–7593. doi:10.20964/2019.08.04

Li, M., Ma, T., Liu, J., Li, H., Xu, Y., Gu, W., & Shen, L. (2019). Numerical and experimental investigation of precast concrete facade integrated with solar photovoltaic panels. *Applied Energy, 253*, 113509. doi:10.1016/j.apenergy.2019.113509

Lima, M. G. B., & Gupta, J. (2013). Studying Global Environmental Meetings. *Global Environmental Politics, 13*(August), 46–64. doi:10.1162/GLEP_a_00166

Lin, S. T., Thirumavalavan, M., Jiang, T. Y., & Lee, J. F. (2014). Synthesis of ZnO/Zn nano photocatalyst using modified polysaccharides for photodegradation of dyes. *Carbohydrate Polymers, 105*(1), 1–9. doi:10.1016/j.carbpol.2014.01.017 PMID:24708945

Li, Q. S., & Zhang, K. (2021). Research on the path of green energy development and utilization in China. *Engineering, 1*, 101–111.

Liu, D., Li, X., Chen, J., Jin, R. (2020). Real-Time Optimization of Precast Concrete Component Transportation and Storage. *Advances in Civil Engineering*. doi:10.1155/2020/5714910

Liu, Q., Zhang, H. L., Guan, S. F., & Li, Y. R. (2020). *Development status and trend of hydrogen energy industry in China.* https://www.163.com/dy/article/FS3FSOA005509YYU.html

Liu, Q., Zhang, H. L., Guan, S. F., & Li, Y. R. (2020). *Research on the development of hydrogen energy industry.* https://baijiahao.baidu.com/s?id=1683829736362232563&wfr=spider&for=pc

Liu, Z. M. (2019). *Build the 'Multi-Energy Integration' system.* https://www.zhonghongwang.com/show-278-151120-1.html

Liu, D. N., Zeng, M., Huang, R. L., Ji, L. H., Chen, Q. X., Duan, J. H., & Li, Y. F. (2015). Business Models and Market Mechanisms of Energy Internet (2). *Power System Technology*, *39*(11), 3057–3063.

Liu, H., Grot, S., & Logan, B. E. (2005). Electrochemically assisted microbial production of hydrogen from acetate. *Environmental Science & Technology*, *39*(11), 4317–4320. doi:10.1021/es050244p PMID:15984815

Liu, Y. D., Guo, H. X., & Ouyang, X. P. (2021). Development status and future prospect of hydrogen fuel cell technology. *Strategic Study of CAE*, *23*(04), 162–171. doi:10.15302/J-SSCAE-2021.04.019

Li, X., & Hasan, I. (2021). VC Participation and failure of startups: Evidence from P2P lending platforms in China. *Finance Research Letters*, *40*, 101726. Advance online publication. doi:10.1016/j.frl.2020.101726

Logan, B. E. (2009). Fuel Cells Fuel cells. *Fuel Cells: Technologies for Fuel Processing*, 1–32.

Logan, B. E., Call, D., Cheng, S., Hamelers, H. V. M., Sleutels, T. H. J. A., Jeremiasse, A. W., & Rozendal, R. A. (2008). Microbial electrolysis cells for high yield hydrogen gas production from organic matter. In Environmental Science and Technology (Vol. 42, Issue 23, pp. 8630–8640). doi:10.1021/es801553z

Luby, S. P., Biswas, D., Gurley, E. S., & Hossain, I. (2015). Why highly polluting methods are used to manufacture bricks in Bangladesh. *Energy for Sustainable Development*, *28*, 68–74.

Lüdeke-Freund, F. (2010). Towards a conceptual framework of business models for sustainability. *Environmental Management*, *49*, 1–28.

Lüdeke-Freund, F., Carroux, S., Joyce, A., Massa, L., & Breuer, H. (2018). The sustainable business model pattern taxonomy-45 patterns to support sustainability -oriented business model innovation. *Sustainable Production and Consumption*, *15*, 145–162. doi:10.1016/j.spc.2018.06.004

Lüdeke-Freund, F., Gold, S., & Bocken, N. (2019). A Review and Typology of Circular Economy Business Model Patterns. *Journal of Industrial Ecology*, *23*(1), 36–61. doi:10.1111/jiec.12763

Lu, J., & Chen, G. (2005). A time-varying complex dynamical network model and its controlled synchronization criteria. *IEEE Transactions on Automatic Control*, *50*(6), 841–846. doi:10.1109/TAC.2005.849233

Lu, L., Xing, D., Xie, T., Ren, N., & Logan, B. E. (2010). Hydrogen production from proteins via electrohydrogenesis in microbial electrolysis cells. *Biosensors & Bioelectronics*, *25*(12), 2690–2695. doi:10.1016/j.bios.2010.05.003 PMID:20537524

Lu, W., Yuan, L., & Xue, F. (2021). Investigating the bulk density of construction waste: A big data-driven approach. *Resources, Conservation and Recycling*, *169*, 105480. doi:10.1016/j.resconrec.2021.105480

Machado, N., & Morioka, S. N. (2021). Contributions of modularity to the circular economy: A systematic review of literature. *Journal of Building Engineering*, *44*, 103322. doi:10.1016/j.jobe.2021.103322

Madaeni, S. S., & Ghaemi, N. (2007). Characterization of self-cleaning RO membranes coated with TiO2 particles under UV irradiation. *Journal of Membrane Science*, *303*(1–2), 221–233. doi:10.1016/j.memsci.2007.07.017

Ma, L., & Zhang, W. X. (2008). Enhanced biological treatment of industrial wastewater with bimetallic zero-valent iron. *Environmental Science & Technology*, *42*(15), 5384–5389. doi:10.1021/es801743s PMID:18754450

Mália, M., de Brito, J., Pinheiro, M. D., & Bravo, M. (2013). Construction and demolition waste indicators. *Waste Management & Research*, *31*(3), 241–255. doi:10.1177/0734242X12471707 PMID:23315370

Manescu, A. (2010). *Aspecte legate de durabilitatea retelelor de canalizare realizate din mase plastice*. https://zdocs.ro/doc/durabilitatea-retelelor-de-canalizare-din-materiale-plastice

Manikandhan, K. U., Kumar, N. S., & Sakthivel, R. (2015). Effect of replacement of river sand by M-sand in high strength concrete. *International Journal of Modern Trends in Engineering and Research*, *2*(02).

Mao, C., Shen, Q., Shen, L., & Tang, L. (2013). Comparative study of greenhouse gas emissions between off-site prefabrication and conventional construction methods: Two case studies of residential projects. *Energy and Building*, *66*, 165–176. doi:10.1016/j.enbuild.2013.07.033

Martínez, F., López-Muñoz, M. J., Aguado, J., Melero, J. A., Arsuaga, J., Sotto, A., Molina, R., Segura, Y., Pariente, M. I., Revilla, A., Cerro, L., & Carenas, G. (2013). Coupling membrane separation and photocatalytic oxidation processes for the degradation of pharmaceutical pollutants. *Water Research*, *47*(15), 5647–5658. doi:10.1016/j.watres.2013.06.045 PMID:23863375

Mati-Baouche, N., De Baynast, H., Lebert, A., Sun, S., Lopez-Mingo, C. J. S., Leclaire, P., & Michaud, P. (2014). Mechanical, thermal and acoustical characterizations of an insulating bio-based composite made from sunflower stalks particles and chitosan. *Industrial Crops and Products*, *58*, 244–250. doi:10.1016/j.indcrop.2014.04.022

Maximous, N., Nakhla, G., & Wan, W. (2009). Comparative assessment of hydrophobic and hydrophilic membrane fouling in wastewater applications. *Journal of Membrane Science*, *339*(1–2), 93–99. doi:10.1016/j.memsci.2009.04.034

Meghashree, M., Gowda, A. K., & Roopa, K. (2016). Comparison of physical properties between natural sand and manufactured Sand. *International Journal for Innovative Research in Science and Technology*, *3*(7), 92–96.

Meher Kotay, S., & Das, D. (2008). Biohydrogen as a renewable energy resource-Prospects and potentials. *International Journal of Hydrogen Energy*, *33*(1), 258–263. doi:10.1016/j.ijhydene.2007.07.031

Meisuh, B. K., Kankam, C. K., & Buabin, T. K. (2018). Effect of quarry rock dust on the flexural strength of concrete. *Case Studies in Construction Materials, 8*, 16-22.

Men, L. R. (2021). The impact of startup CEO communication on employee relational and behavioral outcomes: Responsiveness, assertiveness, and authenticity. *Public Relations Review*, *47*(4), 102078. Advance online publication. doi:10.1016/j.pubrev.2021.102078

Mercader-Moyano, P., & Ramírez-de-Arellano-Agudo, A. (2013). Selective classification and quantification model of C&D Waste from material resources consumed in residential building construction. *Waste Management & Research*, *31*(5), 458–474. doi:10.1177/0734242X13477719 PMID:23446631

Miranville, A. (2020). Annual report 2019. *AIMS Mathematics*, *5*(1), i–v. doi:10.3934/math.2020i

Moghimifar, V., Raisi, A., & Aroujalian, A. (2014). Surface modification of polyethersulfone ultrafiltration membranes by corona plasma-assisted coating TiO2 nanoparticles. *Journal of Membrane Science*, *461*, 69–80. doi:10.1016/j.memsci.2014.02.012

Mondal, P., Kumari, P., Singh, J., Verma, S., Chaurasia, A. K., & Singh, R. P. (2017). Oil from algae. *Sustainable Utilization of Natural Resources*, 214–253. doi:10.1201/9781315153292

Monika, N. R., & Sarankokila, S. (2015). Partial Replacement of M-Sand on Concrete by Self Curing and Standard Curing Method. *International Journal of Innovative Research in Science, Engineering and Technology*, *4*(12).

Moore, R. (1969). An Early System of Large - Panel Construction. *Journal of the Royal Institute of British Architects*, 383.

More power from water. (2018). *Nature Nanotechnology, 13*(12), 1087. doi:10.1038/s41565-018-0340-7

Morseletto, P. (2020). Targets for a circular economy. *Resources, Conservation and Recycling*, *153*, 104553. doi:10.1016/j.resconrec.2019.104553

Mundra, S., Sindhi, P. R., Chandwani, V., Nagar, R., & Agrawal, V. (2016). Crushed rock sand–An economical and ecological alternative to natural sand to optimize concrete mix. *Perspectives in Science*, *8*, 345–347. doi:10.1016/j.pisc.2016.04.070

Munir, M. J., Kazmi, S. M. S., Gencel, O., Ahmad, M. R., & Chen, B. (2021). Synergistic effect of rice husk, glass and marble sludges on the engineering characteristics of eco-friendly bricks. *Journal of Building Engineering*, *42*, 102484. doi:10.1016/j.jobe.2021.102484

Na, S., Heo, S. J., & Han, S. (2021). Construction Waste Reduction through Application of Different Structural Systems for the Slab in a Commercial Building: A South Korean Case. *Applied Sciences (Basel, Switzerland)*, *11*(13), 5870. doi:10.3390/app11135870

National Development and Reform Commission. (2016). *Guidelines on promoting the development of "Internet Plus" Smart Energy*. http://www.nea.gov.cn/2016-02/29/c_135141026.htm

NE 012/1-2007. Normativ Pentru Producerea Betonului Şi Executarea Lucrarilor Din Beton, Beton Armat Şi Beton Precomprimat Partea 1: Producerea Betonului (Regulations for the Production of Concrete and the Execution of Concrete, Reinforced Concrete and Prestressed Concrete); ASRO—Asociatia de standardizare din Romania. Romanian Standardization Association, 2007.

Negoiţă, A., Focşa, V., Radu, A., Pop, I., Tutu, L., Dumitras, M., & Negoita, I. (1976). *Constructii Civile*. Editura Didactica si Pedagogica.

Nguyen, L. T. Q., Hoang, T. G., Do, L. H., Ngo, X. T., Nguyen, P. H. T., Nguyen, G. D. L., & Nguyen, G. N. T.D. L. (2021). The role of blockchain technology-based social crowdfunding in advancing social value creation. *Technological Forecasting and Social Change*, *170*, 120898. Advance online publication. doi:10.1016/j.techfore.2021.120898

Niall, D., Kinnane, O., West, R. P., & McCormack, S. (2017). Mechanical and thermal evaluation of different types of PCM–concrete composite panels. *Journal of Structural Integrity and Maintenance*, *2*(2), 100–108. doi:10.1080/24705314.2017.1318039

Nithyambigai, G., & Ash, F. (2015). Partial Replacement of Manufactured Sand and Fly Ash in Concrete. *International Journal of Emerging Technology and Advanced Engineering*, *5*(6), 166–170.

Nowotny, J., Atanacio, A. J., Bak, T., Belova, I. V., Fiechter, S., Ikuma, Y., Ionescu, M., Kennedy, B. J., Majewski, P., Murch, G. E., & Wachsman, E. D. (2014). Photosensitive oxide semiconductors for solar hydrogen fuel and water disinfection. *International Materials Reviews*, *59*(8), 449–478. doi:10.1179/1743280414Y.0000000039

Nowotny, J., Dodson, J., Fiechter, S., Gür, T. M., Kennedy, B., Macyk, W., Bak, T., Sigmund, W., Yamawaki, M., & Rahman, K. A. (2017, February). Towards global sustainability: Education on environmentally clean energy technologies. *Renewable & Sustainable Energy Reviews*, *81*, 2541–2551. doi:10.1016/j.rser.2017.06.060

Nowotny, J., Hoshino, T., Dodson, J., Atanacio, A. J., Ionescu, M., Peterson, V., Prince, K. E., Yamawaki, M., Bak, T., Sigmund, W., Veziroglu, T. N., & Alim, M. A. (2016). Towards sustainable energy. Generation of hydrogen fuel using nuclear energy. *International Journal of Hydrogen Energy*, *41*(30), 12812–12825. doi:10.1016/j.ijhydene.2016.05.054

Nozahic, V., Amziane, S., Torrent, G., Saïdi, K., & De Baynast, H. (2012). Design of green concrete made of plant-derived aggregates and a pumice–lime binder. *Cement and Concrete Composites*, *34*(2), 231–241. doi:10.1016/j.cemconcomp.2011.09.002

O'Brien, J. E., Stoots, C. M., Herring, J. S., McKellar, M. G., Sohal, M. S., & Condie, K. G. (2010). *High Temperature Electrolysis for Hydrogen Production from Nuclear Energy - Technology Summary*. Academic Press.

O'Hegarty, R., & Kinnane, O. (2020). Review of precast concrete sandwich panels and their innovations. *Construction & Building Materials*, *233*, 117145. Advance online publication. doi:10.1016/j.conbuildmat.2019.117145

O'Hegarty, R., Kinnane, O., Grimes, M., Newell, J., Clifford, M., & West, R. (2021). Development of thin precast concrete sandwich panels: Challenges and outcomes. *Construction & Building Materials*, *267*, 120981. Advance online publication. doi:10.1016/j.conbuildmat.2020.120981

Oh, B. T., Just, C. L., & Alvarez, P. J. J. (2001). Hexahydro-1,3,5-trinitro-1,3,5-triazine mineralization by zerovalent iron and mixed anaerobic cultures. *Environmental Science & Technology*, *35*(21), 4341–4346. doi:10.1021/es010852e PMID:11718353

Oliva, F. L., Teberga, P. M. F., Testi, L. I. O., Kotabe, M., Giudice, M. D., Kelle, P., & Cunha, M. P. (2021). Risks and critical success factors in the internationalization of born global startups of industry 4.0: A social, environmental, economic, and institutional analysis. *Technological Forecasting and Social Change*. Advance online publication. doi:10.1016/j.techfore.2021.121346

Omajuwa, C. E., & Ngwu, C. (2021). Assessment of Construction Managers' Competency Performance on Project Delivery in Delta State,(Case Study of Isoko Community). *International Journal of Progressive Research in Science and Engineering*, *2*(11), 76–90.

Ortega-Liebana, M. C., Hueso, J. L., Ferdousi, S., Yeung, K. L., & Santamaria, J. (2016). Nitrogen-doped luminescent carbon nanodots for optimal photo-generation of hydroxyl radicals and visible-light expanded photo-catalysis. *Diamond and Related Materials*, *65*, 176–182. doi:10.1016/j.diamond.2016.03.021

Osmani, M., & Villoria-Sáez, P. (2019, January). Current and emerging construction waste management status, trends and approaches. In Waste (pp. 365-380). Academic Press. doi:10.1016/B978-0-12-815060-3.00019-0

Osterwalder, A., & Pigneur, Y. (2010). *Business Model Generation: A handbook for visionaries, game changers, and challengers*. John Wiley and Sons.

Otanicar, T. P., Carlson, J. D., Golden, J. S., Kaloush, K. E., & Phelan, P. E. (2010). Impact of the Urban Heat Island on light duty vehicle emissions for the phoenix, AZ Area. *International Journal of Sustainable Transportation*, *4*(1), 1–13. doi:10.1080/15568310802428552

Otanicar, T. P., & Golden, J. S. (2009). Comparative environmental and economic analysis of conventional and nanofluid solar hot water technologies. *Environmental Science & Technology*, *43*(15), 6082–6087. doi:10.1021/es900031j PMID:19731722

P100-1/2013. (2013). Seismic design code - Part I - Design provisions for buildings (in romanian), 1–607, UTCB, Monitorul Oficial, Part I., No 558bis (2013)

P101-78. (1979). Instructiuni tehnice pentru proiectarea cladirilor de locuit cu structura de rezistenta din panouri mari [Technical instructions for the design of residential buildings with a structure made of large panels]. In *B.C.* (Vol. 5).

Pabst, S., Wayand, M., & Mohnen, A. (2021). Coordinating contributions in crowdfunding for sustainable entrepreneurship. *Journal of Cleaner Production, 319*, 128677. Advance online publication. doi:10.1016/j.jclepro.2021.128677

Padmini, A. K., Ramamurthy, K., & Mathews, M. S. (2009). Influence of parent concrete on the properties of recycled aggregate concrete, *Construction and Building Materials, 23*(2), 829-836. doi:10.1016/j.conbuildmat.2008.03.006

Panwar, N. L., Kaushik, S. C., & Kothari, S. (2011). Role of renewable energy sources in environmental protection: A review. *Renewable & Sustainable Energy Reviews, 15*(3), 1513–1524. doi:10.1016/j.rser.2010.11.037

People's Bank of China. (2021). *Policy research | research on improving the financing system of small, medium and micro enterprises*. https://mp.weixin.qq.com/s/AJUZLDKVQe79aIiLE55fCg

Pescari, S., Stoian, V., Tudor, D., & Măduța, C. (2015). Energy demands of the existing collective buildings with bearing structure of large precast concrete panels from Timisoara. *Journal of Applied Engineering Sciences, 5*(18), 75–81. doi:10.1515/jaes-2015-0010

Peste, F., Paula, A., da Silva, L. P., Bernardino, J., Pereira, P., Mascarenhas, M., Costa, H., Vieira, J., Bastos, C., Fonseca, C., & Pereira, M. J. R. (2015). How to mitigate impacts of wind farms on bats? A review of potential conservation measures in the European context. *Environmental Impact Assessment Review, 51*, 10–22. doi:10.1016/j.eiar.2014.11.001

Philip, S., Ajin, A., Shahul, F., Babu, L., & Tina, J. (2019). Fabrication and performance evaluation of lightweight eco-friendly construction bricks made with fly ash and bentonite. *International Journal of Engineering and Advanced Technology, 8*(6), 2090–2096.

Ponnada, M. R., & Kameswari, P. (2015). Construction and demolition waste management–A review. *Safety, 84*, 19-46.

Qu, X., Brame, J., Li, Q., & Alvarez, P. J. J. (2013). Nanotechnology for a safe and sustainable water supply: Enabling integrated water treatment and reuse. *Accounts of Chemical Research, 46*(3), 834–843. doi:10.1021/ar300029v PMID:22738389

Ram, V. G., & Kalidindi, S. N. (2017). Estimation of construction and demolition waste using waste generation rates in Chennai, India. *Waste Management & Research, 35*(6), 610–617. doi:10.1177/0734242X17693297 PMID:28566032

Rani, M., Shanker, U., & Chaurasia, A. K. (2017). Catalytic potential of laccase immobilized on transition metal oxides nanomaterials: Degradation of alizarin red S dye. *Journal of Environmental Chemical Engineering*, *5*(3), 2730–2739. Advance online publication. doi:10.1016/j.jece.2017.05.026

Ratnakar, R. R., Gupta, N., Zhang, K., Doorne, C. V., Fesmire, J., Dindoruk, B., & Balakotaiah, V. (2021). Hydrogen supply chain and challenges in largescale LH2 storage and transportation. *International Journal of Hydrogen Energy*, *46*(47), 24149–24168. doi:10.1016/j.ijhydene.2021.05.025

Rautray, P., Roy, A., Mathew, D. J., & Eisenbart, B. (2019). Bio-Brick-Development of sustainable and cost effective building material. *Proceedings of the Design Society: International Conference on Engineering Design*, *1*(1), 3171–3180.

Reike, D., Vermeulen, W. J., & Witjes, S. (2018). The circular economy: New or refurbished as CE 3.0?—exploring controversies in the conceptualization of the circular economy through a focus on history and resource value retention options. *Resources, Conservation and Recycling*, *135*, 246–264. doi:10.1016/j.resconrec.2017.08.027

Reis, I. F. G., Gonçalves, I., Lopes, M. A. R., & Autunes, C. H. (2021). Business models for energy communities: A review of key issues and trends. *Renewable & Sustainable Energy Reviews*, *144*, 111013. doi:10.1016/j.rser.2021.111013

Reyes-Veras, P. F., Renukappa, S., & Suresh, S. (2021). Challenges faced by the adoption of big data in the Dominican Republic construction industry: An empirical study. *Journal of Information Technology in Construction*, *26*, 812–831. doi:10.36680/j.itcon.2021.044

Ríos, S. D., Salvadó, J., Farriol, X., & Torras, C. (2012). Antifouling microfiltration strategies to harvest microalgae for biofuel. *Bioresource Technology*, *119*, 406–418. doi:10.1016/j.biortech.2012.05.044 PMID:22739578

Ritala, P., Huotari, P., Bocken, N., Albareda, L., & Puumalainen, K. (2018). Sustainable business model adoption among S&P 500 firms: A longitudinal content analysis study. *Journal of Cleaner Production*, *170*, 216–226. doi:10.1016/j.jclepro.2017.09.159

Roberts, J. J., Cassula, A. M., Osvaldo Prado, P., Dias, R. A., & Balestieri, J. A. P. (2015). Assessment of dry residual biomass potential for use as alternative energy source in the party of General Pueyrredón, Argentina. *Renewable & Sustainable Energy Reviews*, *41*, 568–583. doi:10.1016/j.rser.2014.08.066

Rozell, D. J., & Reaven, S. J. (2012). Water Pollution Risk Associated with Natural Gas Extraction from the Marcellus Shale. *Risk Analysis*, *32*(8), 1382–1393. doi:10.1111/j.1539-6924.2011.01757.x PMID:22211399

Ryan, J. F. (1929). The story of portland cement. Part I. *Journal of Chemical Education*, *6*(11), 1855–1868. doi:10.1021/ed006p1854

Sahu, J. N., Zabed, H., Karri, R. R., Shams, S., & Qi, X. (2019). Applications of nano-biotechnology for sustainable water purification. In *Industrial Applications of Nanomaterials*. Elsevier Inc. doi:10.1016/B978-0-12-815749-7.00011-6

Salman, G. A. S., & Salih, S. A. (n.d.). *Effect of Firing Temperatures on Physico-Mechanical Properties of Clay Bricks Containing Reeds*. Academic Press.

Santosh, P., & Shubha, K. (2018). Enhancement of index and engineering properties of black cotton soil using rice husk ash and wheat husk ash including lime. *International Journal for Research in Applied Science and Engineering Technology*, *6*(2), 1448–1455.

Sarkodie, S. A., Strezov, V., Weldekidan, H., Asamoah, E. F., Owusu, P. A., & Doyi, I. N. Y. (2019). Environmental sustainability assessment using dynamic Autoregressive-Distributed Lag simulations—Nexus between greenhouse gas emissions, biomass energy, food and economic growth. *The Science of the Total Environment*, *668*, 318–332. doi:10.1016/j.scitotenv.2019.02.432 PMID:30852209

Sathishkumar, P., Sweena, R., Wu, J. J., & Anandan, S. (2011). Synthesis of CuO-ZnO nanophotocatalyst for visible light assisted degradation of a textile dye in aqueous solution. *Chemical Engineering Journal*, *171*(1), 136–140. doi:10.1016/j.cej.2011.03.074

Saxe, J. P., Lubenow, B. L., Chiu, P. C., Huang, C.-P., & Cha, D. K. (2006). Enhanced Biodegradation of Azo Dyes Using an Integrated Elemental Iron-Activated Sludge System: II. Effects of Physical-Chemical Parameters. *Water Environment Research*, *78*(1), 26–30. doi:10.2175/106143005X84486 PMID:16553163

Schaltegger, S., Freund, F. L., & Hansen, E. G. (2012). Business cases for sustainability: The role of business model innovation for corporate sustainability. *International Journal of Innovation and Sustainable Development*, *6*(2), 95–119. doi:10.1504/IJISD.2012.046944

Selembo, P. A., Merrill, M. D., & Logan, B. E. (2009). The use of stainless steel and nickel alloys as low-cost cathodes in microbial electrolysis cells. *Journal of Power Sources*, *190*(2), 271–278. doi:10.1016/j.jpowsour.2008.12.144

Shams, A., Horstmann, M., & Hegger, J. (2014). Experimental investigations on textile-reinforced concrete (TRC) sandwich sections. *Composite Structures*, *118*(1), 643–653. doi:10.1016/j.compstruct.2014.07.056

Shanmugapriya, T., & Uma, R. N. (2012). Optimization of partial replacement of M-sand by natural sand in high performance concrete with silica fume. *International Journal of Engineering Sciences and Emerging Technologies*, *2*(2), 73–80.

Shariati, M., Armaghani, D. J., Khandelwal, M., Zhou, J., & Khorami, M. (2021). Assessment of longstanding effects of fly ash and silica fume on the compressive strength of concrete using extreme learning machine and artificial neural network. *Journal of Advanced Engineering and Computation*, *5*(1), 50–74. doi:10.25073/jaec.202151.308

Shrair, J. S. A. (2014). *Advances in Nanotechnology can Provide Clean Energy Resources and Sustainable Development*. Academic Press.

Shrivastava, S., & Chini, A. (2009). Construction materials and C&D Waste in India. *Lifecycle design of buildings, systems and materials, 72*.

Shushkewich, K. W. (2012). Eugène Freyssinet—Invention of Prestressed concrete and precast segmental construction. *Structural Engineering International, 22*(3), 415–420. https://doi.org/10.2749/101686612X13363869853338

Sijakova-Ivanova, T., Panov, Z., Blazev, K., & Zajkova-Paneva, V. (2011). Investigation of fly ash heavy metals content and physico chemical properties from thermal power plant, Republic of Macedonia. *International Journal of Engineering Science and Technology, 3*(12).

Smith, D. K., & Tardif, M. (2009). *Building information modeling: a strategic implementation guide for architects, engineers, constructors, and real estate asset managers*. John Wiley & Sons. doi:10.1002/9780470432846

Soharu, A., Naveen, B. P., & Sil, A. (2022). An Approach Towards Zero-Waste Building Construction. In *Advances in Construction Materials and Sustainable Environment* (pp. 239–257). Springer. doi:10.1007/978-981-16-6557-8_19

Song, H. F., Wan, J. B., & Ren Z. B. (2010). Research on the connotation characteristics, industrial selection and development policy of China's strategic emerging industries. *Science & Technology for Development, 9*, 7-14.

Spišáková, M., Mésároš, P., & Mandičák, T. (2021). Construction Waste Audit in the Framework of Sustainable Waste Management in Construction Projects—Case Study. *Buildings, 11*(2), 61. doi:10.3390/buildings11020061

SR 3518:2009. Încercări Pe Betoane. Determinarea Rezistenței La Îngheț-Dezgheț Prin Masurarea Variației Rezistenței La Compresiune¸ Si/Sau Modulului de Elasticitate Dinamic Relativ (Tests Hardened Concrete. Determination of Freeze-Thaw Resistance by Measuring the Variation of the ComPressive Strength and/or the Relatively Dynamic Modulus of Elasticity); *ASRO - Asociatia de standardizare din Romania (Romanian Standardization Association)*: Bucharest, Romania, 2009.

Stoica, S. (2007). *Managementul retelelor de alimentare cu apă și canalizare urbane*. Technical University of Civil Engineering Bucharest.

Stubbs, W., & Cocklin, C. (2008). Conceptualizing a "sustainability business model". *Organization & Environment, 21*(2), 103–107. doi:10.1177/1086026608318042

Sua-iam, G., & Makul, N. (2015). Rheological and mechanical properties of cement–fly ash self-consolidating concrete incorporating high volumes of alumina-based material as fine aggregate. *Construction & Building Materials, 95*, 736–747. doi:10.1016/j.conbuildmat.2015.07.180

Subramani, A. K., Ramani, S. E., & Selvasembian, R. (2021). Understanding the microstructure, mineralogical and adsorption characteristics of guar gum blended soil as a liner material. *Environmental Monitoring and Assessment, 193*(12), 1–17. doi:10.100710661-021-09644-4 PMID:34853926

Sucher, N. J., Carles, M. C., Nowotny, J., & Bak, T. (2012). Photocatalytic water disinfection on oxide semiconductors: Part 2 - Structure, functional properties and reactivity of microbial agents. *Advances in Applied Ceramics, 111*(1–2), 16–33. doi:10.1179/1743676111Y.0000000039

Sulaiman, C., Abdul-Rahim, A. S., & Ofozor, C. A. (2020). Does wood biomass energy use reduce CO2 emissions in European Union member countries? Evidence from 27 members. *Journal of Cleaner Production, 253*, 119996. doi:10.1016/j.jclepro.2020.119996

Sun, C. L. (2019). *Building a hydrogen energy industry economic belt in the Yangtze River Delta.* http://qwfb.chinareports.org.cn/qcfc/52963.html

Suresh, S., & Revathi, J. (2016). Effect of M-sand on setting time of high performance concrete. *Asian Journal of Research in Social Sciences and Humanities, 6*(10), 1648–1654. doi:10.5958/2249-7315.2016.01118.7

Su, Y., Liang, Y., Chai, L., Han, Z., Ma, S., Lyu, J., Li, Z., & Yang, L. (2019). Water degradation by China's fossil fuels production: A life cycle assessment based on an input-output model. *Sustainability (Switzerland), 11*(15), 4130. Advance online publication. doi:10.3390u11154130

Tabassum, A., Premalatha, M., Abbasi, T., & Abbasi, S. A. (2014). Wind energy: Increasing deployment, rising environmental concerns. *Renewable & Sustainable Energy Reviews, 31*, 270–288. doi:10.1016/j.rser.2013.11.019

Tan, X. G., & Yu, Z. P. (2020). Development status and prospect of fuel cell commercial vehicle industry. *Strategic Study of CAE, 22*(5), 152–158. doi:10.15302/J-SSCAE-2020.05.019

Tao, K., Arano, H., Zhang, P., Ai, P., Han, L., & Tsubaki, N. (2017). Enhanced Hydrogen Production from Steam Reforming of Vegetable Oil over Bimodal ZrO2-SiO2 Supported Ni Catalyst. *ChemistrySelect, 2*(1), 527–532. doi:10.1002lct.201601395

Tarrad, M. (2020). A Vision to Revive Mud Architecture, a Community Heritage Architecture in Jordan, for Low Income. *Journal, 15*(2), 269–275. Http://Iieta. Org/Journals/Ijdne

Taylor, R. A., Phelan, P. E., Otanicar, T. P., Walker, C. A., Nguyen, M., Trimble, S., & Prasher, R. (2011). Applicability of nanofluids in high flux solar collectors. *Journal of Renewable and Sustainable Energy, 3*(2), 023104. Advance online publication. doi:10.1063/1.3571565

Tenca, A., Cusick, R. D., Schievano, A., Oberti, R., & Logan, B. E. (2013). Evaluation of low cost cathode materials for treatment of industrial and food processing wastewater using microbial electrolysis cells. *International Journal of Hydrogen Energy, 38*(4), 1859–1865. doi:10.1016/j.ijhydene.2012.11.103

Tiba, S., van Rijnsoever, F. J., & Hekkert, M. P. (2021). Sustainability startups and where to find them: Investigating the share of sustainability startups across entrepreneurial ecosystems and the causal drivers of differences. *Journal of Cleaner Production*, *306*. Advance online publication. doi:10.1016/j.jclepro.2021.127054

Tingem, M., & Rivington, M. (2009). Adaptation for crop agriculture to climate change in Cameroon: Turning on the heat. *Mitigation and Adaptation Strategies for Global Change*, *14*(2), 153–168. doi:10.100711027-008-9156-3

Tolga Balta, M., Dincer, I., & Hepbasli, A. (2009). Thermodynamic assessment of geothermal energy use in hydrogen production. *International Journal of Hydrogen Energy*, *34*(7), 2925–2939. doi:10.1016/j.ijhydene.2009.01.087

Tremblay, A., Lambert, M., & Gagnon, L. (2004). Do hydroelectric reservoirs emit greenhouse gases? *Environmental Management*, *33*(S1), 509–517. doi:10.100700267-003-9158-6

Tyagi, V. K., Angériz Campoy, R., Álvarez-Gallego, C. J., & Romero García, L. I. (2014). Enhancement in hydrogen production by thermophilic anaerobic co-digestion of organic fraction of municipal solid waste and sewage sludge - Optimization of treatment conditions. *Bioresource Technology*, *164*, 408–415. doi:10.1016/j.biortech.2014.05.013 PMID:24880931

Udawatta, N., Zuo, J., Chiveralls, K., & Zillante, G. (2015). Improving waste management in construction projects: An Australian study. *Resources, Conservation and Recycling*, *101*, 73–83. doi:10.1016/j.resconrec.2015.05.003

UN. (2015). *Transforming Our World: The 2030 Agenda for Sustainable Development*. A/RES/70/1. https://sdgs.un.org/sites/default/files/publications/21252030%20Agenda%20for%20Sustainable%20Development%20web.pdf

UNEP. (2014). *A Guidance Manual for*. UNEP.

Ungureanu, V., & Muntean, D. M. (2019). Life Cycle Assessment of Retrofitting Large Prefabricated Panels Low-Rise Collective Dwellings. *IOP Conf. Ser. Materials Science and Engineering*, *471*, 112070. doi:10.1088/1757-899X/471/11/112070

USDA (United States Department of Agriculture). (2021). *World Agricultural Production*. Circular Series WAP 10-21 2021, 38. Available at https://apps.fas.usda.gov/psdonline/circulars/production.pdf

Usković, V. (2007). Nanotechnologies: What we do not know. *Technology in Society*, *29*(1), 43–61. doi:10.1016/j.techsoc.2006.10.005

Vaddiraju, S., Tomazos, I., Burgess, D. J., Jain, F. C., & Papadimitrakopoulos, F. (2010). Emerging synergy between nanotechnology and implantable biosensors: A review. *Biosensors & Bioelectronics*, *25*(7), 1553–1565. doi:10.1016/j.bios.2009.12.001 PMID:20042326

Venkrbec, V., & Klanšek, U. (2020). Suitability of recycled concrete aggregates from precast panel buildings deconstructed at expired lifespan for structural use. *Journal of Cleaner Production*, *247*, 119593. https://doi.org/10.1016/j.jclepro.2019.119593

Walczak, P., Szymański, P., & Różycka, A. (2015). Autoclaved Aerated Concrete based on fly ash in density 350 kg/m[3] as an environmentally friendly material for energy-efficient constructions. *Procedia Engineering*, *122*, 39–46. doi:10.1016/j.proeng.2015.10.005

Wan, Y. M. (2019). *Analysis of the coordinated path of Beijing-Tianjin-Heibei Urban Agglomeration in hydrogen energy development*. http://www.china-nengyuan.com/news/144437.html

Wang, R. F. (2016). *Background Material for the Action Plan of Internet Plus Smart Energy*. http://fangtan.china.com.cn/2016-06/24/content_38738456_2.htm

Wang, H. N., Wang, L. H., Zhou, Z. C., Wang, K. S., & Cui, J. (2020). Study on The Development Strategy of Emerging Industries (2035). *Engineering*, *22*(2), 1–8.

Wang, X., Li, W., Chen, Z., Waje, M., & Yan, Y. (2006). Durability investigation of carbon nanotube as catalyst support for proton exchange membrane fuel cell. *Journal of Power Sources*, *158*(1), 154–159. doi:10.1016/j.jpowsour.2005.09.039

Wankhade, R. L., Landage, A. B., & Konnur, B. A. (2016). Experimental Investigation on Combined Effect of SBR and Steel Fiber on Properties of Concrete. *International Journal of Engine Research*, *5*(1), 257–261.

Wasif, M., Sinha, A., Ahmed, Z., & Qin, Q. (2021). Effects of biomass energy consumption on environmental quality : The role of education and technology in Asia-Pacific Economic Cooperation countries. *Renewable & Sustainable Energy Reviews*, *142*(March), 110868. doi:10.1016/j.rser.2021.110868

Wei, M., Patadia, S., & Kammen, D. M. (2010). Putting renewables and energy efficiency to work: How many jobs can the clean energy industry generate in the US? *Energy Policy*, *38*(2), 919–931. doi:10.1016/j.enpol.2009.10.044

Weng, S. L., Huang, Z., Yu, L. J., Xie, X. M., You, T., & Zhang, T. T. (2021). Research on the construction strategy of modern energy system in the Yangtze River Delta. *Engineering*, *23*(01), 42–51.

WHO. (2005). The World Health Report 2005: Make every mother and child count. *World Health Report*.

Wright, G., Renewable, S., Law, E., Review, P., & Wright, G. (2018). *The International Renewable Energy Agency : A Global Voice for the Renewable Energy The International Renewable Energ A Global Voice for the Renewable*. Academic Press.

Wuebbles, D. J., & Jain, A. K. (2001). Concerns about climate change and the role of fossil fuel use. *Fuel Processing Technology*, *71*(1–3), 99–119. doi:10.1016/S0378-3820(01)00139-4

Xia Shuo Official Account. (2021). *Financing development report of China's small, medium and micro enterprises in 2021*. https://mp.weixin.qq.com/s/Pk0VDx5UTjMe9R77xuOVFw

Xu, W. W., & Shen, H. (2021). *The "Guangzhou Consensus" on hydrogen energy industry was released!* https://xw.qq.com/cmsid/20210928A0FPIE00

Xu, B., Zhang, S., & Chen, X. (2021). Uncertainty in financing interest rates for startups. *Industrial Marketing Management*, *94*, 150–158. doi:10.1016/j.indmarman.2020.02.026

Yang, M., Evans, S., Vladimirova, D., & Rana, P. (2017). Value uncaptured perspective for sustainable business model innovation. *Journal of Cleaner Production*, *140*, 1794–1804. doi:10.1016/j.jclepro.2016.07.102

Yao, L., & Hu, Y. (2020). The impact of urban transit on nearby startup firms: Evidence from Hangzhou, China. *Habitat International*, *99*. Advance online publication. doi:10.1016/j.habitatint.2020.102155

Yin, J., Zhu, G., & Deng, B. (2013). Multi-walled carbon nanotubes (MWNTs)/polysulfone (PSU) mixed matrix hollow fiber membranes for enhanced water treatment. *Journal of Membrane Science*, *437*, 237–248. doi:10.1016/j.memsci.2013.03.021

You, S. J., Zhang, Z., Zhou, Y., Du, G. M., Wan, H., Xu, P., & Bao, L. Q. (2019). Industrial policy of hydrogen energy pioneer countries and its enlightenment. *Oil Forum, 5*, 57-66.

Yu, P., Gong, R., & Sampat, M. (2022). Blockchain Technology in China's Digital Economy: Balancing Regulation and Innovation. In P. Tehrani (Ed.), Regulatory Aspects of Artificial Intelligence on Blockchain (pp. 132–157). IGI Global. https://doi.org/10.4018/978-1-7998-7927-5.ch007.

Yu, P., Hu, Y., Waseem, M., & Rafay, A. (2021). Regulatory Developments in Peer-to-Peer (P2P) Lending to Combat Frauds: The Case of China. In A. Rafay (Ed.), *Handbook of Research on Theory and Practice of Financial Crimes* (pp. 172-194). IGI Global. doi:10.4018/978-1-7998-5567-5.ch010

Yu, P., Li, C., Sampat, M., & Chen, Z. (2022). How the Development of FinTech Can Bolster Financial Inclusion Under an Era of Disruptive Innovation?: Case Study on China. In M. Anshari, M. Almunawar, & M. Masri (Eds.), FinTech Development for Financial Inclusiveness (pp. 135–167). IGI Global. https://doi.org/10.4018/978-1-7998-8447-7.ch009.

Yu, P., Lu, S., Hanes, E., & Chen, Y. (2022). The Role of Blockchain Technology in Harnessing the Sustainability of Chinese Digital Finance. In Blockchain Technologies for Sustainable Development in Smart Cities (pp. 155–186). IGI Global. https://doi.org/ 10.4018/978-1-7998-9274-8.ch009.

Yu, S., Liu, Y., Wang, D., Bahaj, A. B. S., Wu, Y., & Liu, J. (2021). Review of thermal and environmental performance of prefabricated buildings: Implications to emission reductions in China. *Renewable and Sustainable Energy Reviews, 137*. doi:10.1016/j.rser.2020.110472

Yuan, Y. (2018). *Shanghai practice of equity crowdfunding to help innovative small and micro enterprises solve financing difficulties -- Taking Ant angel crowdfunding platform as an example*. https://mp.weixin.qq.com/s/wsOuzipBIVLL6RdE71HP1w

Yuan, Y., Zhou, S., & Zhuang, L. (2010). Polypyrrole/carbon black composite as a novel oxygen reduction catalyst for microbial fuel cells. *Journal of Power Sources*, *195*(11), 3490–3493. doi:10.1016/j.jpowsour.2009.12.026

Yu, P., Gu, H., Zhao, Y., & Ahuja, A. (2022). Digital Transformation Driven by Internet Data Center: Case Studies on China. In D. Piaggesi, H. Landazuri, & B. Jia (Eds.), *Cases on Applying Knowledge Economy Principles for Economic Growth in Developing Nations* (pp. 203–230). IGI Global. doi:10.4018/978-1-7998-8417-0.ch011

Yu, P., Liu, Z., Hanes, E., & Mumtaz, J. (2022). Integration of IoT and Blockchain for Smart and Secured Supply Chain Management: Case Studies of China. In S. Goyal, N. Pradeep, P. Shukla, M. Ghonge, & R. Ravi (Eds.), *Utilizing Blockchain Technologies in Manufacturing and Logistics Management* (pp. 179–207). IGI Global. doi:10.4018/978-1-7998-8697-6.ch010

Yu, S., & Fleming, L. (2021). Regional crowdfunding and high tech entrepreneurship. *Research Policy*. Advance online publication. doi:10.1016/j.respol.2021.104348

Zare, Khanipour, Hojjat, Sarverstani, Iman, Kakavandi, Ehsan, Shokroo, Farniaei, Baghbani, & Shokroo. (2019). Hydrogen and carbon dioxide recovery from the petrochemical flare gas to methanol production using adsorption and absorption combined high-efficient method. *Applied Petrochemical Research, 9*(2), 127–145. doi:10.1007/s13203-019-0232-2

Zelmanov, G., & Semiat, R. (2008). Phenol oxidation kinetics in water solution using iron(3)-oxide-based nano-catalysts. *Water Research, 42*(14), 3848–3856. doi:10.1016/j.watres.2008.05.009 PMID:18657285

Zeng, K., & Zhang, D. (2010). Recent progress in alkaline water electrolysis for hydrogen production and applications. In Progress in Energy and Combustion Science (Vol. 36, Issue 3, pp. 307–326). doi:10.1016/j.pecs.2009.11.002

Zhang, R., Zhang, Y., Li, A., & Yang, T. Y. (2022). Experimental study on a new type of precast beam-column joint. *Journal of Building Engineering, 51*, 104252. doi:10.1016/j.jobe.2022.104252

Zhang, J., & Gu, Q. (2021). Turning a curse into a blessing: Contingent effects of geographic distance on startup–VC partnership performance. *Journal of Business Venturing, 36*(4). Advance online publication. doi:10.1016/j.jbusvent.2021.106108

Zhang, L. W., Sojobi, A. O., Kodur, V. K. R., & Liew, K. M. (2019). Effective utilization and recycling of mixed recycled aggregates for a greener environment. *Journal of Cleaner Production, 236*, 117600. doi:10.1016/j.jclepro.2019.07.075

Zhang, N., Zhang, H., Schiller, G., Feng, H., Gao, X., Li, E., & Li, X. (2021). Unraveling the Global Warming Mitigation Potential from Recycling Subway-Related Excavated Soil and Rock in China Via Life Cycle Assessment. *Integrated Environmental Assessment and Management, 17*(3), 639–650. doi:10.1002/ieam.4376 PMID:33241889

Zhang, X. Q. (2021). The development trend of and suggestions for China's hydrogen energy industry. *Engineering, 7*(6), 719–721. doi:10.1016/j.eng.2021.04.012

Zhang, Y., Wang, B. Y., Li, R., & Mi, X. R. (2019). Research on business model and development path of multi-station integration. *Distribution & Utilization, 6*, 62–66.

Zheng, L., Wu, H., Zhang, H., Duan, H., Wang, J., Jiang, W., Dong, B., Liu, G., Zuo, J., & Song, Q. (2017). Characterizing the generation and flows of construction and demolition waste in China. *Construction & Building Materials*, *136*, 405–413. doi:10.1016/j.conbuildmat.2017.01.055

Zhuang, L., Zhou, S., Wang, Y., Liu, C., & Geng, S. (2009). Membrane-less cloth cathode assembly (CCA) for scalable microbial fuel cells. *Biosensors & Bioelectronics*, *24*(12), 3652–3656. doi:10.1016/j.bios.2009.05.032 PMID:19556120

Zhu, H. X., Geng, X. Y., Xiao, Y., Cai, R., & Liu, X. M. (2021). Research on the strategic path of promoting energy revolution in Northwest China in the new era. *Engineer*, *23*(01), 92–100.

Zona, D., Janssens, I. A., Aubinet, M., Gioli, B., Vicca, S., Fichot, R., & Ceulemans, R. (2013). Fluxes of the greenhouse gases (CO2, CH4 and N2O) above a short-rotation poplar plantation after conversion from agricultural land. *Agricultural and Forest Meteorology*, *169*, 100–110. doi:10.1016/j.agrformet.2012.10.008

Zou, Y., Xiang, C., Yang, L., Sun, L. X., Xu, F., & Cao, Z. (2008). A mediatorless microbial fuel cell using polypyrrole coated carbon nanotubes composite as anode material. *International Journal of Hydrogen Energy*, *33*(18), 4856–4862. doi:10.1016/j.ijhydene.2008.06.061

Zsembik, B. A. (2006). Health issues in latino families and households. Handbook of Families and Health: Interdisciplinary Perspectives, 6(20), 40–61. doi:10.4135/9781452231631.n3

About the Contributors

Raluca - Andreea Felseghi is a Civil Engineer and Independent Scientific Researcher, she attended the postgraduate work (Ph.D.) at the Technical University of Cluj-Napoca, Romania, and obtained the Doctoral Diploma in an Engineering Sciences fundamental field - Civil Engineering. Her main areas of interest address clean technologies and sustainable development in the Civil Engineering domain, in particular the practical applicability of the building's energy efficiency, renewable energies integration in stationary applications, hydrogen fuel cell technology, hybrid power generation systems, the alternative energies impact on the environment and climate change. She is president of the Scientific, Innovation and Technical Education Commission and CEO of the Romanian Association of Building Services Engineering - Transylvania Branch, member of the Romanian Association for Hydrogen Energy, and member of the Women in Green Hydrogen community.

Nicoleta Cobîrzan is an Associate Professor at Faculty of Civil Engineering, Technical University of Cluj-Napoca, Romania. She has a master's degree in Energy, Energy Resources and Environment and a Ph.D. in Civil Engineering. The main fields of interest and research expertise areas are in structural analysis of masonry and concrete buildings, design of low energy buildings, structural and thermal rehabilitation of existing buildings, seismic vulnerability assessment, microscopical and macroscopical investigations of building materials, investigation of recovered materials, ecological and low intensive materials, circular economy issues, sustainability.

Maria Simona Raboaca is working as a Researcher at to National Research and Development Institute for Cryogenics and Isotopic Technologies ICSI Rm. Valcea, Hydrogen and Fuel Cell Department. Her Ph.D. is "Theoretical and practical Contribution regarding to sustain with hybrid energy a Passive House" in Faculty of Building Services Engineering in Technical University of Cluj-Napoca, Romania. Now, she is a project manager at ICSI to project "Smart conductive charging station, fixed and mobile, for electric propulsion transportation (SMiLE-EV)" proposes the deployment of fixed and mobile EV & PHEV charging stations to meet the mobility needs of tomorrow's society and to prepare active/potential industrial partners for knowledge/technology transfer at the component or system level in prepare launching new products. She has been contributing to the field of Renewable Energy, Green Buildings, Passive House Concept, Hydrogen Energy and Stationary and Mobile applications. She is the author and co-author of more technical papers in scientific conference proceedings and ISI journals.

* * *

Ana Diana Ancas, Ph.D Eng., "Gheorghe Asachi" Technical University of Iasi, Romania, with expertise both in installations and in seismic proven by published work and research contracts, as project manager will coordinate the entire activity of the experimental model to demonstration and dissemination and development of technology. Competence: critical infrastructure facilities, vital networks, seismic risk management, pipeline rehabilitation technologies for water transport, Coordonator of Multidisciplinary Center of Structural Engineering and Risk Management https://erris.gov.ro/Centrul-multidisciplinar-Ing, Faculty of Civil Engineering and Building Services, Department of Buildings Services.

Amit Kumar Chaurasia has 3+ years industrial experience along with an educational degree of B Tech, M Tech. and PhD from IIT Roorkee (2021). Previously, he has worked with VASCSC Ahmedabad and Thyrocare Technologies Navi Mumbai. His area of research interest on wastewater characterization/treatment and gained wide experience on bio-electrochemical systems (microbial electrolysis cells), fermentation, microbial cultures, isolation, and identification, and production of enzymes. He has published more than 14+ peer-reviewed international publications in high impact factors journals (IF: 5 to7.1, h-index 60-248) and presented his research work in 3 countries Paris, Algeria and Thailand. He has presented his research work in 6 international conferences (3 in abroad) and attained 11 national conferences. He is the session chair and scientific and organizing committee member in several international conferences. He is also a member of several professional bodies and a reviewer in various reputed journals.

Nelu Cristian Chereches, associate professor "Gheorghe Asachi" Technical University of Iasi, Romania, received PhD degree in 2006 from the University of Reims, France, in the field of Fluid Mechanics and Energetics, where he completed 2 years research mobility stage and from "Gheorghe Asachi" Technical University of Iasi in the field of Civil Engineering. The research member has 18 years of experience in the Building Services field. His main research areas are heat transfer and fluid flow applied to the cooling of electrical equipment, BIPV and double skin facade. His studies involved both experimental and numerical simulations, regarding energy efficiency and energy optimization through dimensional and configuration changes.

Rathishchandra Gatti is currently a researcher in the field of self-powered systems, energy harvesting and wave energy and also working as a Professor at Department of Mechanical Engineering, Sahyadri College of Engineering and Management. Prior to his role, he has worked as Research Academic and IP Analyst at Curtin University, Australia, NPI program manager at General Electric, Design Engineering roles at Meritor Automotive, SVI and Rapsri. He pursued his BE in Mechanical engineering from NITK Suratkal, MBA from IGNOU, PG Diploma in Global Strategic Management from ICFAI and PhD from Curtin University, Perth, Australia. His PhD work "Spatially varying multi-degree-of-freedom Electro-magnetic energy harvesting" was funded by Australian Post-Graduate Award fellowship by Australian Federal Government for PG students of exceptional research potential. During his PhD, he also pursued Graduate Certificate in Research Commercialization funded by GCRC fellowship from Commonwealth Australia. His awards during his tenure at GE IIC include - Management Award from GE India Innovation Centre for efficient execution of NPI projects, GE APPLAUSE award for patented Retrofit LV Switchgear, GE Centre of excellence contribution Award, Micro innovation award from GE India innovation Centre. His current projects include point absorber wave energy generator is funded by VGST, Government of India and Automated Robotic welding TIG and MIG funded by KCTU, Government of Karnataka state. His previous funded projects include "Design, Fabrication and characterization of cantilever based in-vitro body energy harvester using piezoelectric composites" funded by INUP, GoI,"Affordable Hybrid Solar-Mechanical Powered Agricultural Dryer" funded by TRIP-VGST and about 3 KSCST student projects. During his tenure at GE, he has worked on industrial R&D projects that include Zero Energy Home, Lexus Panel relevant to early stage smart grids (2004-2008). He has 4 international Patents 1 US Patent, 1 Mexican Patent,1 Canadian Patent commercialized at GE and 1 Indian Patent(filed) at SCEM. He has around 13 publications to his credit. He is a reviewer of IEEE transactions on Magnetics, Journal of Mechanical Systems and Signal Processing and Walailak Journal of Science and Technology.

Cătălina Mihaela Grădinaru is an civil engineer with PhD in "Contributions for the hygrothermal optimization of the ecological concrete used in the building envelope". The main research directions were related to the development of different ecological concrete recipes in which mineral aggregates were partially/totally replaced with renewable aggregates, of a vegetal nature (corn cobs and sunflower stalks) and the cement was partially replaced with materials with cementitious characteristics from the industrial waste category. The researches have shown a special impact on the environmental environment of the concrete recipes realized by envisaging the large-scale use of these concretes of a reduction of the load of the pollutant mass represented by greenhouse gases resulting from the mining extraction of the mineral aggregates, fact that it essentially translates into a better quality of life on earth. The research results were published in more than 40 scientific articles, a book subchapter with national and international ISBN, two books (https://www.researchgate.net/profile/Catalina_Gradinaru, https://scholar.google.com/citations?user=bjSFyUMAAAAJ&hl=ro).

Emanuela Hanes is an independent researcher. Her cooperations include the Vienna University FH BFI Campus Wien, University of Graz and University of Salzburg. Her research interests include China-EU business strategies, RegTech, FinTech, Cryptocurrencies, Geopolitics, Chinese Strategic Planning and Development Policies.

Rajmunni Hombal is currently a researcher in Construction materials working at Sahyaware technologies Pvt limited, Mangaluru.

Valeriu Sebastian Hudisteanu, assistant professor "Gheorghe Asachi" Technical University of Iasi, Romania, gained experience in the field of research and on the subject proposed in the project both during the doctoral and postdoctoral periods. The doctoral thesis entitled "Efficiency of the photovoltaic panels integrated in buildings with double skin facades", completed in 2016, represents "a reference work in terms of PV integration in buildings, identifying technologies to improve PV performance through cooling and numerical and experimental approaches on the evaluation of the functioning in different climatic conditions" (according to the doctoral committee report). He has an experience of 8 years in Energy field, proven by the published papers and the research contracts in which he participated as member. His main research areas concern the cooling of photovoltaic panels, heat transfer, fluid flow, solar radiation and numerical modeling. His studies involved both experimental and numerical simulations regarding energy efficiency and energy optimization of PV systems. He will coordinate the entire activity, from the numerical simulations and the design of the experimental stand to the testing and dissemination of the results.

Manoj Kumar Karnena received the B.S. degree in Environmental Science and the M.S. degree in Environmental science from GITAM (Deemed to be) University, India. Later, he pursued his Ph.D. in Water and Waste Water treatment at the same institution. He was the university topper and gold medalist in bachelors and post-graduation. He holds an Environmental Law degree from National Law University, Delhi, and Industrial Safety Degree from Annamalai University, Tamilnadu. Since 2019, he has been working as a teaching assistant with the Department of Environmental Science, GITAM Institute of Science, GITAM (Deemed to be) University, India. His greatest strengths are his research and communication skills. He started teaching and working in technical work in 2019 in various backgrounds. Manoj Kumar Karnena has written more than 50 papers since he started writing in 2017. He has written several book chapters and books on the topic within his research interest and holds two patents. His research interest is in Water and Wastewater Treatment, Natural Coagulants, Nanocatalysis, Storage Technologies, and membrane technology. Manoj Kumar Karnena had won several engineering designs and environmental protection awards in water and wastewater treatment.

Madhavi Konni holds a doctorate in Nanoscience and Nanotechnology in the Department of Chemistry from GITAM (Deemed to be University), Visakhapatnam, India. Currently, she is working as Associate Professor at Dadi Institute of Engineering and Technology, Anakapalle. Her research interests include synthesizing Metal-Organic Frameworks, Material Chemistry, Carbon Nano Materials, Hydrogen Storage Technologies, and Nano Catalysis. She published many research articles and book chapters in reputed international journals and holds two patents. Dr. Madhavi is a fellow and received HEAM Scholar Award (2018), the Young Scientist Award (December 2017 and March 2017), and few other awards from different societies.

Xinyi Liu is an independent researcher. Her research interests include FinTech, Chinese economy and business analytics.

Shengyuan Lu is an independent researcher. His research interests include RegTech, FinTech and Cryptocurrencies.

Ramya Mahendran is a managed innovation consultant. She is certified in design thinking, business modeling and jobs to be done. She has over 10 years of experience in the fields of managed innovation, startup incubation and acceleration, crowd sources idea management systems, design thinking and sustainability. She works with student entrepreneurs to build their business ideas into a successful business model. She works with some of India's leading Technology and Business Incubators, Institution Innovation Councils and Entrepreneurship Cells. She specializes in setting up innovation strategy, opportunity identification, large scale ideation campaigns and facilitating rapid prototyping events like design service jams and hackathons, organizing large-scale Innovation summits and global idea crowd-sourcing events.

Swati Mohapatra has been working in the field of biology and biotechnology research for more than 10 years and have more than 20 publications in peer reviewed journals.

Radu Muntean has been carrying out teaching and research activities since 2003 at the Faculty of Civil Engineering, Transilvania University of Brasov. He has a Ph.D. in Civil Engineering and a Master's degree in Rehabilitation and consolidation of buildings. He is the Project Manager of the international projects: Recover-IND - Ecological and innovative technologies for recovering industrial areas from LCA and Energy Efficiency point of view; UrbanBIM - Innovative Educational Integration of Urban Plannings based on BIM-GIS Technologies and Focused on Circular Economy Challenges and Local project coordinator of other international project like OERCO2, CircularBIM, BIMHealty, BKStone. He is Coordinator of the Civil Engineering Design Centre in R&D Institute of Transilvania University of Brașov (https://icdt.unitbv.ro/), and author of books, articles and conference papers published in national and international publishing houses and journals. He is a member of the organizing committee, editor and reviewer for international conferences and journals. He has expertise in: buildings' design and optimization; modelling, assessment and optimization of structural elements; management and economic assessment of buildings; research in civil engineering, new materials for sustainable constructions, environmental protection and energy saving.

Adrian Alexandru Șerbănoiu, PhD civil engineer, has teaching and research activities in Civil Engineering and Building Services since 2003. He has experience in coordinating research projects on the study of concrete with industrial and agricultural by-products or waste, "Eco friendly concrete by using technological intermediary products from recycled materials" (http://www.ecostprem.tuiasi.ro/), and „Sustainable solutions for eco-concrete obtaining from renewable raw materials" (http://www.ecorenew.tuiasi.ro/). He is the author of 10 books and a consistent number of scientific articles, many of them being related to the subject of ecological building materials, published in journals indexed in international databases, including Clarivate Analytics (https://scholar.google.ro/citations?user=PYb2sHcAAAAJ&hl=ro, https://www.researchgate.net/profile/Adrian_Serbanoiu/research).

Ravi Shankar has published more than 35 research articles in international peer-reviewed journals and professor now in engineering college.

Lokendra Singh Thakur is a professor in government engineering college and published more than 20 international articles in peer-reviewed journals.

Gyorgy Thalmaier is a lecturer, doctor engineer in materials engineering at the Faculty of Materials and Environmental Engineering, Technical University of Cluj-Napoca, head of The Porous Materials and Composites Research Group His research interest is manufacturing porous materials by powder metallurgy namely membranes for microfiltration and metallic foams manufactured by powder metallurgy from industrial grade and lately from recycled powders. The manufactured porosity was in a wide range, as needed by the specific application and adapted the pores size starting from submicron pores to pores in the millimeter range made by hollow particles or by space holders. Lately the focus shifted to the investigation of recovered materials' effect on the building materials. The obtained results were published in 33 WoS indexed articles and numerous conference papers.

Florin Emilian Turcanu, Ph.D. Eng, "Gheorghe Asachi" Technical Univeristy of Iasi with expertise in numerical simulation in buildings and facilities will be involved in the verification and validation of methods of numerical simulation using ANSYS software, interpretation of results, develop technology to improve the resilience seismic collection network Engineering and Risk Management https://erris.gov.ro/Centrul-multidisciplinar-Ing, Faculty of Civil Engineering and Building Services Competence in the project: numerical simulation in buildings and facilities.

Marina Verdes, associate professor "Gheorghe Asachi" Technical University of Iasi, Romania, is an experienced researcher with expertise in urban networks and risk management will be involved in designing and implementing experimental design, interpretation of results, develop technology to improve the resilience seismic collection network. Professor Verdes Marina Ph.D.Eng., member Multidisciplinary Center of Structural Engineering and Risk Management https://erris.gov.ro/Centrul-multidisciplinar-Ing, Faculty of Civil Engineering and Building Services, Department of Buildings Services - experienced researcher.

Aiqi Xu is an independent researcher. Her research interests include Blockchain, Fintech and Crowdfunding.

Poshan (Sam) Yu is a Lecturer in Accounting and Finance in the International Cooperative Education Program of Soochow University (China). He is also an External Professor of FinTech and Finance at SKEMA Business School (China), a Visiting Professor at Krirk University (Thailand) and a Visiting Researcher at the Australian Studies Centre of Shanghai University (China). Sam leads FasterCapital (Dubai, UAE) as a Regional Partner (China) and serves as a Startup Mentor for AIC RAISE (Coimbatore, India). His research interests include financial technology, regulatory technology, public-private partnerships, mergers and acquisitions, private equity, venture capital, start-ups, intellectual property, art finance, and China's "One Belt One Road" policy.

Index

H

I

L

M

N

P

R

S

T

V

W

www.ingramcontent.com/pod-product-compliance
Lightning Source LLC
LaVergne TN
LVHW061540290926
849162LV00004B/860
* 9 7 8 1 7 9 9 8 9 8 1 0 8 *